Mechanical Engineering Series

Frederick F. Ling
Series Editor

T0234462

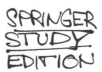

Mechanical Engineering Series

(continued after index)

Daniel Post   Bongtae Han
Peter Ifju

High Sensitivity Moiré

Experimental Analysis for
Mechanics and Materials

With 245 illustrations, 161 Fringe Patterns/Contour Maps

 Springer

Daniel Post
Professor Emeritus
Department of Engineering Science
  and Mechanics
College of Engineering
Virginia Polytechnic Institute
  and State University
Blacksburg, VA 24061 USA

Series Editor
Frederick F. Ling
Ernest F. Gloyna Regents Chair in
  Engineering
Department of Mechanical Engineering
The University of Texas at Austin
Austin, TX 78712-1063 USA
and
Distinguished William Howard Hart
  Professor Emeritus
Department of Mechanical Engineering,
  Aeronautical Engineering, and Mechanics
Rensselaer Polytechnic Institute
Troy, NY 12180-3590 USA

Bongtae Han
Clemson University
Mechanical Engineering Department
Clemson, SC 29634 USA

Peter Ifju
University of Florida
Aerospace Engineering, Mechanics, and
  Engineering Science
Gainesville, FL 32611-6250 USA

Library of Congress Cataloging-in-Publication Data
Post, Daniel
    High sensitivity moiré : experimental analysis for mechanics and
materials / Daniel Post, Bongtae Han, Peter Ifju.
       p. cm. -- (Mechanical engineering series)
    Includes bibliographical references and index.
    ISBN 0-387-94149-5 (New York). -- ISBN 3-540-94149-5 (Berlin)
    1. Non-destructive testing. 2. Interferometry. 3. Moiré methods.
4. Materials--Mechanical properties. I. Han, Bongtae. II. Ifju,
Peter. III. Title. IV. Series: Mechnical engineering series
(Berlin, Germany)
TA417.2.P67   1994
620.1'127--dc20   93-29790

Printed on acid-free paper.
This volume was originally published in hardcover. ISBN: 0-387-94149-5

Production managed by Karen Phillips, manufacturing supervised by Jacqui Ashri.
Camera-ready copy prepared from the authors' Microsoft Word files.
Printed and bound by Edwards Brothers Inc., Ann Arbor, MI.
Printed in the United States of America.

9 8 7 6 5 4 3 2 1

ISBN 0-387-98220-5 Springer-Verlag New York Berlin Heidelberg SPIN 10572601

Dedicated to

Nathan
Ian
Brian
Timothy
Randi
Will
may their accomplishments give them joy

Dongsoo
Dongjoo
Aeng Sun
for their patience and sacrifice

my parents
Mara
and
Geza

Series Preface

Mechanical engineering, an engineering discipline born of the needs of the industrial revolution, is once again asked to do its substantial share in the call for industrial renewal. The general call is urgent as we face profound issues of productivity and competitiveness that require engineering solutions, among others. The Mechanical Engineering Series features graduate texts and research monographs, intended to address the need for information in contemporary areas of mechanical engineering.

The series is conceived as a comprehensive one the will cover a broad range of concentrations important to mechanical engineering graduate education and research. We are fortunate to have a distinguished roster of consulting editors, each an expert in one of the areas of concentration. The names of the consulting editors are listed on the first page of the volume. The areas of concentration are applied mechanics, biomechanics, computational mechanics, dynamic systems and control, energetics, mechanics of materials, processing, thermal science, and tribology.

Professor Winer, the consulting editor for applied mechanics and tribology, and I are pleased to present this volume of the series: *High Sensitivity Moiré: Experimental Analysis for Mechanics and Materials* by Professor Post, Dr. Han and Dr. Ifju. The selection of this volume underscores again the interest of the Mechanical Engineering Series to provide our readers with topical monographs as well as graduate texts.

Austin, Texas   Frederick F. Ling

Preface/Acknowledgments

This book evolved largely from a course on Photomechanics given at Virginia Polytechnic Institute and State University (VPI&SU). We extend thanks to the Department of Engineering Science and Mechanics and the University for supporting the course and the Photomechanics Laboratory, and to Professor C. W. Smith (VPI&SU) who gave invaluable administrative and technical advice throughout the development and operation of the Photomechanics Laboratory.

Many of the techniques presented here and many of the fringe patterns and graphs stem from research in the Photomechanics Laboratory at VPI&SU. Credit for these accomplishments belongs to numerous former students and colleagues who participated in the laboratory. They include D. E. Bowles, W. A. Baracat, M. L. Basehore, E. M. Weissman, R. B. Watson, E. W. Brooks, G. Nicoletto, J. S. Epstein, S. J. Lubowinski, A. L. Highsmith, R. Czarnek, D. Joh, J. D. Wood, D. H. Mollenhauer, D. O. Adams, Y. Guo, R. G. Boeman, B. Han and P. G. Ifju; Visiting Scholars Peitai Ning, Mierong Tu and Yinyan Wang; Post-doctoral Fellow A. Asundi; Visiting Professors C. Ruiz (Oxford Univ., England), A. Livnat (Beer-Sheva Univ., Israel), K. Patorski (Warsaw Univ. of Technology, Poland), J. Wasowski (Warsaw Univ. of Technology, Poland), F.-L. Dai (Tsinghua Univ., P. R. China), J. McKelvie (Univ. of Strathclyde, Scotland), Y. Morimoto (Osaka Univ., Japan) and H. E. Gascoigne (California Polytechnic Univ., USA).

The research sponsors include the National Science Foundation, NASA Langley Research Center, Office of Naval Research, U. S. Air Force Rocket Propulsion Laboratory, Sandia National Laboratory, DARPA, Virginia Center for Innovative Technology, Micro-Measurements Div. of Measurements Group, Inc., and IBM Corporation.

Special assistance on various chapters was extended by our colleagues: Chap.2—critical review by G. Indebetouw (Physics Dept.); Chap. 3—critical review by V. J. Parks (Catholic Univ. of America) and advice by F.-P. Chiang (State Univ. of N.Y. at Stonybrook); Chap. 6—critical review, analysis and computations by J. McKelvie (Univ. of Strathclyde), G. Indebetouw (Physics Dept.) and G. A. Hagedorn (Mathematics Dept.), and assistance with SEM for Fig. 6.2 by S.

Willard (NASA Langley Research Center); Chap. 9—critical review by C. E. Harris (NASA Langley Research Center), and assistance with FEM analysis for Fig. 9.7 by D. S. Dawicke (NASA Langley Research Center); Chap. 10—critical review by Y. Guo (IBM) and advice by C. K. Lim (IBM); Chap. 11—specimen for Fig. 11.1 courtesy of G. M. Holloway, GE Aircraft Engines; Chap. 12—critical review by C. W. Smith (Engineering Science and Mechanics Dept.), specimen for Fig. 12.4 courtesy of W. G. Reuter of the Idaho National Engineering Laboratory (INEL), and specimen for Fig. 12.10 courtesy of M. S. Dadkhah (Rockwell Science Center); Chap. 13—information on the absolute frequency of Bausch & Lomb gratings by E. G. Loewen (Milton Roy Co.), advice on statistics by A. Heller, statistician, R. A. Heller (Engineering Science and Mechanics Dept.), and R. G. Krutchkoff (Statistics Dept.) and assistance with the experiment by R. B. Watson (Measurements Group, Inc.) and A. L. Wang (Engineering Science and Mechanics Dept.). The departments listed above are at VPI&SU.

Figures that originated outside the authors' affiliations (VPI&SU, NASA and IBM) were provided by colleagues; they are acknowledged in the figure captions. Figures 12.5 and 12.6 are republished by courtesy of the Society for Experimental Mechanics. Other figures provided by colleagues are in the public domain by virtue of sponsorship by the U. S. Government.

Permission was extended by the Society for Experimental Mechanics (K. A. Galione, Managing Director and Publisher) to use published material that we authored. Portions of Ref. 1 in Chap. 7 were used with permission of the J. Composite Materials (H. T. Hahn, editor). Portions of Ref. 2 in Chap. 8 were used with permission of the American Society of Mechanical Engineers. Research in Chap. 9 was conducted at the NASA Langley Research Center and included with permission. Research in Chap. 10 was conducted at the IBM Corporation and included with permission.

Professor J. W. Dally (Univ. of Maryland) reviewed the entire book and gave us numerous important suggestions. His experience, wisdom and fresh point of view were shared most graciously.

The seemingly endless efforts of Frieda Post, who typed, proofread and critiqued the manuscript, is deeply appreciated. Her support and encouragement has been wonderful.

The assistance of all who contributed to this book is gratefully acknowledged.

Contents

List of Symbols

A	electric field strength of light wave; absorptance
a	amplitude of field strength
C	velocity of light in free space
C'	velocity of light in a transparent material
CF	calibration factor
CTE	coefficient of thermal expansion
c	superscript denoting carrier fringes
D	distance; superscript denoting mechanical differentiation; deviation
d	distance
E	modulus of elasticity
e	subscript denoting error; combined error
F	fringe frequency, fringe gradient, fringes per mm
F_{xx}	$\partial N_x / \partial x$
F_{xy}	$\partial N_x / \partial y$
f	frequency of a grating, lines per mm; frequency of virtual reference grating
f_m	frequency of a virtual reference grating formed in a refractive material
f_s	specimen grating frequency
FL	focal length
G	pitch of fringes, $1/F$; shear modulus
GF	gage factor
g	pitch of a grating, $1/f$
h	height; fringe half-width
I	intensity of light
J	joule, unit of energy
K	constant; function of S
k	constant
L	index number of curve in a parametric family of curves

M	magnification of image; index number of curve in a parametric family
m	diffraction order; an integer; subscript denoting a refractive medium
N	fringe order, index number
N_x	fringe order in the x displacement field
N^+	fringe order of additive moiré fringes
N^D	fringe order in pattern of mechanical differentiation
N^*	fringe contour number for O/DFM
NA	numerical aperture of lens
n	index of refraction; a number; an integer
OPL	optical path length
P	point on specimen; external pressure; load or force
ppm	parts per million
psi	pounds per square inch
R	reflectance; ratio of bar and space width; electrical resistance
r	polar coordinate
S	distance between wave fronts of equal phase
SD	standard deviation
s	diameter of Airy's disk
T	period of light wave; transmittance; temperature
t	time; time since the head of the wave train passed by; time since an arbitrary datum time; superscript denoting *total*
U	in-plane displacement, x component of displacement
V	in-plane displacement, y component of displacement
W	out-of-plane displacement, z component of displacement
w	wave front
x,y,z	Cartesian coordinates
α	angle of incidence; coefficient of thermal expansion
α'	angle of reflection
α''	angle of refraction
β	angle of diffracted ray; angle of observation; fringe multiplication factor
γ	shear strain
Δ	prefix meaning *change of*; displacement of a fringe

δ optical path length; displacement; rigid-body displacement; misalignment error

ε strain

η strain gage reading

Θ angle of wedge prism; angle of incidence for 3-beam moiré interferometer

θ half-angle between two beams, rays, or wave fronts; angle between two coplanar gratings; polar coordinate; angular misalignment

λ wavelength of light in vacuum

λ_m wavelength of light in refractive material

ν Poisson's ratio

σ stress; superscript denoting *stress induced*

τ shear stress

ϕ phase, or fractional part of a periodic cycle; angle of the fringe vector

Ψ angle of out-of-plane rotation

ψ angle of in-plane rotation of grating with respect to the plane of incidence (Fig. 2.37)

ω frequency of light wave, cycles/sec

1

Introduction

1.1 Our Subject

We live in an amazing period of technological and scientific expansion. The rapid advance of computer modeling and computer simulation is largely responsible, together with advancing techniques of physical measurement and efficient data analysis. The book is devoted to one element in this spectrum, the physical measurement.

Our purpose is to teach high-sensitivity moiré—principally moiré interferometry—including its *theory and practice.* Our focus is on the mechanics and micromechanics of materials and structural elements. The applications introduced here are in that category, but the reader can look beyond these to investigate phenomena in other disciplines of engineering and science.

The moiré methods taught and illustrated here utilize visible light. The data are received as contour maps of displacement fields. Moiré interferometry raises the sensitivity of traditional geometrical moiré to the level of optical interferometry. For most of the illustrations in the book, the sensitivity corresponds to moiré with 2400 lines/mm (60,960 lines/in.). The corresponding contour interval is 0.417 µm (16.4 µin.) per fringe order. For micromechanics, 4800 lines/mm (~122,000 lines/in.) is used and displacement maps of 17.4 nm (0.684 µin.) per contour are produced by computer analysis of moiré fringe patterns.

The contour maps of moiré interferometry represent *in-plane* displacement fields, i.e., the displacements parallel to the surface of the specimen. They are distinct from the contour maps produced by classical interferometry and holographic interferometry, which are most effective for determining the out-of-plane displacements. The distinction is important for strain and stress analysis, since engineering strains are determined by in-plane displacements.

Moiré is often called a method of strain analysis. We prefer to call it a method of experimental mechanics, since it is not always necessary to extract the strains. Frequently the displacement field itself is the desired result, e.g., when experimental and numerical solutions are compared.

Computer modeling is an instrument of design. Design is the application of theory, and theory must rest upon a solid *physical foundation*. High-sensitivity moiré has emerged as a powerful tool to strengthen and extend the foundation.

1.2 Scope and Style

The book is divided into two parts, Fundamentals and Applications. The division is somewhat artificial in that some aspects of each are found in both parts. Both parts are intended to teach concepts and practice. The intent, too, is to produce a self-contained treatment, one that minimizes the need to refer to outside sources.

Part I begins with a review of optics. It covers those aspects that are pertinent to moiré, and it does so at a level usually found in a first course on optics at colleges and universities. Physical concepts (physical models) are emphasized. They are reinforced by mathematical derivations, and they are reinforced by numerical examples of the physical variables.

A review of geometric moiré comes next. An extensive body of literature exists on this subject, and only a small portion is addressed. It is the portion that we consider most directly related to subsequent coverage of high-sensitivity moiré and engineering practice.

Chapters 4-13 treat moiré interferometry and microscopic moiré interferometry. The subjects are developed and much attention is given to their actual practice. Alternative configurations are described to assure the reader that basic concepts dominate—rather than push-button apparatus—and to inspire initiative and creative design of experiments.

Part II addresses diverse applications. Some chapters treat specific experimental analyses in great detail, discussing special techniques, procedures, data and results. Others address a broad array of experiments. Throughout Part II, however, emphasis continues on the growth of knowledge beneficial to the experimentalist. It is designed to stimulate inquiry and initiative. The book should serve engineers and scientists who are concerned with measurements of real phenomena; and it should provide a vehicle that stimulates students to understand experimental analyses and their practical results.

A special objective is to produce a reader-friendly volume. The language is natural, uncomplicated. We endeavored to couple words and equations with abundant and genuinely helpful illustrations. Conceptual models and analogies are used to aid understanding. Repetition is admitted where it might aid the connection and assimilation of ideas.

We learn, too, by imitation. Almost all of the fringe patterns are excellent examples, with a very high signal-to-noise ratio. They show the quality of experimental data that can be produced. They represent a standard that can be and should be achieved by every experimentalist. Fringe pattern excellence is not merely for aesthetic pride, but it is a style that yields the maximum extraction of data, and it is a style that inspires confidence in the validity of the data.

1.3 History

Sources of technical information will be cited by references, but we will not dwell on history. Tracing contributions to original sources is outside the scope of the book. History is addressed broadly in other literature. A recent volume is by Patorski[1] (1993) who authored a handbook on moiré that treats a great variety of techniques. It describes them in mathematical and physical terms. Its coverage of strain analysis and moiré interferometry is lean, but it states in the Preface that the subject of strain analysis is found elsewhere. Patorski's knowledge of the moiré literature emanating from numerous parts of the world is extensive. The volume is a historical reference as well as a valuable technical reference.

An article on the history of moiré interferometry was written by Walker[2] (1994). It discusses moiré fringe multiplication as a predecessor of present-day practice. It cites important contributions in the Japanese literature that parallels and often predates developments of high-sensitivity moiré in Western countries.

A current volume by Indebetouw and Czarnek[3] (1992) is one in a series that collects and reprints important articles on the evolution of various branches of optics. Like the handbook by Patorski, it addresses diverse disciplines that utilize moiré phenomena.

The *Physics of Moiré Metrology* by Kafri and Glatt[4] appeared in 1990. It is largely a mathematical treatment and highlights the application of Talbot fringes (Sec. 2.5.3) in a moiré method called *deflectometry*.

The definitive monograph by Guild[5] (1956) is a classic for its insight and detail. It treats, among other things, the effects of multiple-beam interference when contributions from numerous diffraction orders comprise the moiré pattern. The style is largely mathematical. It demonstrates moiré with grating frequencies up to 567 lines/mm (14,400 lines/in.).

Many experimentalists concentrated on moiré methods for strain and stress analysis. Publications that are well known to the American community include those by: Theocaris[6] (1969) and Durelli and Parks[7] (1970), who published books addressed to strain analysis almost simultaneously; Sciammarella[8] (1982), who produced an extensive literature review; Dally and Riley,[9] who addressed moiré methods in their textbook on experimental stress analysis. The Society for Experimental Mechanics published two manuals and a handbook that teaches the techniques of experimental mechanics. In these, Chiang[10] and Parks[11] addressed geometrical moiré for in-plane and out-of-plane measurements, and Post[12,13] addressed high-sensitivity moiré interferometry. All of these individuals (among many others), made extensive contributions to the theory, techniques and application of moiré for strain analysis over many years.

Recent articles by Dally and Read[14-16] (1993) highlight the potential of electron beam radiation. Using a scanning electron microscope (SEM), they produced 10,000 lines/mm gratings on their specimens. The specimens were loaded by a miniature machine inside the SEM. The deformed specimen grating was viewed in the SEM, whereby the scanning roster acted as a pseudo reference grating to produce fringes of geometrical moiré.

Extremely high-sensitivity moiré is used in a completely different discipline, namely the study of matter on the atomic level. Interferometers that are slight variations of those shown later in Figs. 4.20 and 4.21 are employed. Instead of visible light, however, X-ray or neutron beams are used. Instead of the fabricated gratings of Chapters 2 and 4, the lattice structure of single crystals is used to diffract the beams. The field is called *X-ray and neutron interferometry*, and a broad review is given in Ref. 17. Using wavelengths near 1 Angstrom Unit (10^{-10}m), the sensitivity is 3 to 4 orders of magnitude higher than that achieved by the moiré interferometry of Chapter 4. Yet, the principles are basically the same. Applications include measurements of lattice parameters, studies of dislocations and lattice defects, nuclear scattering and the influence of magnetic and gravity fields. The techniques were reported in 1965, which predates most of the optical techniques of moiré interferometry described in this book.

1.4 Comment on Exercises

Numerous exercises are given. Some are routine, but many are intriguing and challenging. They are posed to develop insights beyond those extracted from the text. Several exercises are open-ended, i.e, they have no fixed solution. Instead, they are intended to stimulate original thought and invention, and the realistic analysis of options. Experience has shown that diverse correct solutions are possible, many with special merit.

1.5 References

1. K. Patorski, *Handbook of the Moiré Fringe Technique*, Elsevier, New York (1993).

2. C. A. Walker, "A Historical Review of Moiré Interferometry," *Experimental Mechanics*, (to be published).

3. G. Indebetouw and R. Czarnek, editors, *Selected Papers on Optical Moiré and Applications*, SPIE Vol. MS 64, SPIE Optical Engineering Press, Bellingham, Washington (1992).

4. O. Kafri and I. Glatt, *The Physics of Moiré Metrology*, John Wiley & Sons, New York (1990).

5. J. Guild, *The Interference Systems of Crossed Diffraction Gratings, Theory of Moiré Fringes*, Oxford University Press, New York (1956).

6. P. S. Theocaris, *Moiré Fringes in Strain Analysis*, Pergamon Press, Elmsford, New York (1969).

7. A. J. Durelli and V. J. Parks, *Moiré Analysis of Strain*, Prentice-Hall, Englewood Cliffs, New Jersey (1970).

8. C. A. Sciammarella, "The Moiré Method, A Review," *Experimental Mechanics*, Vol. 22, No. 11, pp. 418-433 (1982).

9. J. W. Dally and W. F. Riley, *Experimental Stress Analysis*, 3rd edition, McGraw-Hill, New York (1991).

10. F.-P. Chiang, "Moiré Methods of Strain Analysis," Chap. 7, *Manual on Experimental Stress Analysis*, 5th edition, J. F. Doyle and J. W. Phillips, editors, Society for Experimental Mechanics, Bethel, Connecticut (1989).

11. V. J. Parks, "Geometric Moiré," Chap. 6, *Handbook on Experimental Mechanics*, 2nd Edition, A. S. Kobayashi, editor, VCH Publishers, New York (1993).

12.   D. Post, "Moiré Interferometry," Chap. 7, *Handbook on Experimental Mechanics*, 2nd Edition, A. S. Kobayashi, editor, VCH Publishers, New York (1993).

13.   D. Post, "Moiré Interferometry for Composites," *Manual on Experimental Methods for Mechanical Testing of Composites*, R. L. Pendleton and M. E. Tuttle, editors, Society for Experimental Mechanics, Bethel, Connecticut, pp. 67-80 (1989).

14.   J. W. Dally and D. T. Read, "Electron Beam Moiré," *Proc. 1993 SEM Spring Conference on Experimental Mechanics*, pp. 627-635, Society for Experimental Mechanics, Bethel, Connecticut (1993); also *Experimental Mechanics* (to be published).

15.   D. T. Read and J. W. Dally, "Theory of Electron Beam Moiré," *ibid.*, pp. 636-645.

16.   D. T. Read and J. W. Dally, "Electron Beam Moiré Study of Fracture of a GFRP Composite," *ibid.*, pp. 320-329; also ASME *J. Applied Mechanics* (to be published).

17.   U. Bonse and W. Graeff, "X-Ray and Neutron Interferometry," Chap. 4, *Topics in Applied Physics, Vol. 22: X-Ray Optics*, H.-J. Queisser, editor, Springer-Verlag, New York (1977).

PART I

FUNDAMENTALS

Elements of Optics
Geometric Moiré
Moiré Interferometry
Microscopic Moiré Interferometry
On the Limits of Moiré Interferometry

2
Elements of Optics

2.1 The Nature of Light

This introduction is a brief treatment of the concepts and tools of optics encountered in the following chapters. Standard textbooks on optics can be consulted for more information. The wave theory of light is sufficient to explain all the characteristics of moiré. In what follows, we will describe a model of light consistent with the wave theory. A parallel beam of light propagating in the z direction is depicted at a given instant as a train of regularly spaced disturbances that vary with z as

$$A = a \cos 2\pi \frac{z}{\lambda} \qquad (2.1)$$

For mathematical convenience, this expression is often replaced by the real part of the complex equation

$$A = a e^{i2\pi \frac{z}{\lambda}} \qquad (2.1a)$$

The symbol A describes the strength of the disturbance, which is usually viewed as the strength of an electromagnetic field at a point in space, particularly, the electric field strength; A will be called *field strength*. The coefficient a is a constant called the amplitude of the field strength. The field strength varies harmonically along z, where the distance between neighboring maxima is λ, called the *wavelength*.

Length z is not endless—that would represent a light that shines forever—but it is very long compared to λ. In the case of laser light, length z of the wave train may be millions of wavelengths. For classical light sources (not lasers), many short wave trains coexist simultaneously in the beam. They overlap along the length of the beam, such that any cross section of the beam cuts through numerous wave trains. Consequently, light from real sources is not portrayed accurately by Eq. 2.1, but the equation provides a model that represents light sufficiently well for our purposes.

However, the wave train is not stationary. It travels or propagates through space with a very high constant velocity C. At any fixed point along the path of the wave train, the disturbance is a periodic variation of field strength. There, the field strength varies through one full cycle in the brief time period T (seconds/cycle), where

$$T = \frac{\lambda}{C} \tag{2.2}$$

The frequency of this variation is ω (cycles/second), given by

$$\omega = \frac{C}{\lambda} \tag{2.3}$$

During the passage of the wave train through any fixed point $z = z_0$, the field strength varies with time t as

$$A = a \cos 2\pi \frac{C}{\lambda} t = a \cos 2\pi \omega t \tag{2.4}$$

or as the real part of

$$A = a\, e^{i2\pi\omega t} \tag{2.4a}$$

Figure 2.1 offers a graphical interpretation of the equations. At a fixed time $t = t_0$, the field strength A varies along the length of the light ray as shown by the harmonic curve. The curve represents a *wave train*, with the head of the wave train on the right side. At point z_0 the field strength is less than a at this instant. However, the wave train moves along z with the velocity of light, so the field strength at z_0 varies with time between $\pm a$ with the very high frequency ω.

The argument of the cosine function is called the *phase* of the disturbance. Here, the phase is $2\pi\omega t$. The phase at a point z_0 is 2π times the number of wave cycles that passed the point, where the first cycle is at the head of the wave train. An implication is that t is the time since the head of the wave train passed z_0. In our work we will be interested in the relative phase of two wave trains of the same frequency; the relative phase is independent of t. Accordingly, we can interpret t as the time since an arbitrary starting time, t_0, during the passage of the wave train.

We will define a *wave front* as any continuous surface in space along which the phase is constant, that is, along which

$$\omega t = k \tag{2.5}$$

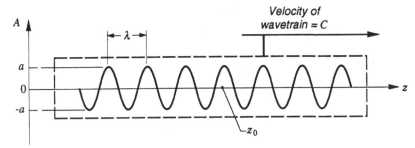

Fig. 2.1   At any instant t_0, the electric field strength of the light wave varies cyclically along z. At any fixed point z_0, the strength varies cyclically with time as the wave train propagates with velocity C.

where k may be any numerical value. In a parallel beam (also called a *collimated* beam), the wave fronts are always plane cross sections of the beam, as illustrated in Fig. 2.2a. These move along the length of the beam with the speed of light.

In a conical beam (Fig. 2.2b), the wave fronts are spherical and their radii of curvature increase with distance from the origin. The shape of the wave front changes constantly along the length of the beam, but all wave fronts have a fixed shape as they intercept any fixed point z_0 in their path. In an increment of time, the wave front through z_0 moves to a new location in space, while a wave front further back in the wave train moves to z_0.

Beams may acquire irregular shapes (Fig. 2.2c), for example by reflection from an irregular mirror. Again the wave fronts constantly change shape along the beam, and they may even fold and become double-valued, as shown. This is a consequence of the rule that light propagates in the direction of the *wave normal*, that is, in the direction perpendicular to the wave front; the rule is always true for light propagation in homogeneous media. The word *ray* is often used synonymously with wave normal to describe the path of an infinitesimally narrow part of the light beam. As before, every wave front in the beam has the same unique shape as it propagates through a fixed point z_0. Wave fronts of smooth but irregular shapes are called *warped* wave fronts.

2.1.1 More Details

Light trains propagate with an astounding velocity of about 3×10^8 meters per second or 186,000 miles per second in free space. The visible spectrum encompasses the wavelength range from about 400 nm (i.e., nanometers or 10^{-9} m) for blue light to about 700 nm for red light.

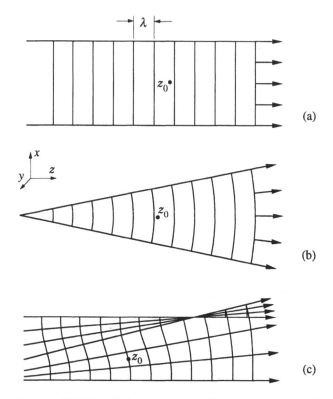

Fig. 2.2   Beams of light and their corresponding wave fronts: (a) plane; (b) spherical; (c) irregular, or warped wave fronts.

The visible spectrum is only a tiny portion of the electromagnetic spectrum in which wavelengths extend at least from 10^{-14} m to 10^3 m, a span of 17 decades.   All the phenomena discussed here— interference, diffraction, and so on—apply for the entire electro- magnetic spectrum.

When a wave train travels through a fixed point in space, its frequency of oscillation, $\omega = C/\lambda$, is about 6×10^{14} Hz (cycles per second) for visible light. No instruments can detect individual cycles in this frequency range. Instead, receivers like the eye, photographic film, and photoelectric cells respond to time-averaged energy.

When a wave train, or a beam of light, is intercepted by a receiver of unit area, the *energy* available to be dissipated into the receiver is the *intensity* of the light multiplied by the exposure time. During the passage of the wave train in a volume of space, the light has an intensity everywhere in that space. The energy associated with it is a potential energy. This potential is realized only when the light is intercepted by a receiver and its energy is dissipated into the receiver.

It is shown in electromagnetic theory that the intensity, I, of light of field strength $A = a \cos 2\pi\omega t$ is proportional to a^2. We will assume that the units chosen for field strength make the factor of proportionality equal to one, such that

$$I = a^2 \tag{2.6}$$

The intensity of light is the rate of energy flow per unit cross-sectional area of the beam; it is energy/second/m² (J/s·m²). Power is the energy output per unit time, measured in watts or milliwatts for lasers. The energy absorbed by a unit area of a receiver is intensity multiplied by exposure time and multiplied by the absorption coefficient of the receiver; photographic films are designed to maximize the absorption coefficient. The word *irradiance* is used synonymously with intensity.

TABLE 2.1 Wavelength and Color

Wavelength range in nanometers (10^{-9}m)	Color
400 - 450	violet
450 - 480	blue
480 - 510	blue-green
510 - 550	green
550 - 570	yellow-green
570 - 590	yellow
590 - 630	orange
630 - 700	red

Different wavelengths in the visible range each stimulate different responses in the human eye, and are interpreted as different colors and shades of color. For the normal human eye, colors are interpreted approximately as indicated in Table 2.1. Photographic films are formulated to transmit the same color stimulation to the eye as the original scene. Light from a source that emits a continuous spectrum with approximately equal energy for every wavelength is seen as white light. A monochromatic source is one that emits light of a single wavelength; a truly monochromatic source cannot exist, but sources that emit wavelengths within a narrow range—e.g., a range of a few nanometers and smaller—are usually called monochromatic sources.

The velocity C' of light propagation in a transparent material is less than its velocity in free space. The relationship is

$$C' = \frac{C}{n} \qquad (2.7)$$

where n is a property of the material called the *refractive index* or *index of refraction*. As a rough generalization, n increases with the density of the material, but there are many exceptions. For visible light, n is only slightly greater than unity for gasses; for liquids, values between 1.3 and 1.5 are common; for solids, values between 1.5 and 1.7 are common.

While n is a material constant at any wavelength, its numerical value varies slightly with wavelengths in the visible range. This characteristic is termed *dispersion of refractive index with wavelength*, or more simply, *dispersion*. The change of refractive index of most transparent solids is of the order of 2% in the visible range, with the higher values of refractive index occurring at shorter wavelengths.

Combining Eqs. 2.3 and 2.7,

$$C' = \frac{\omega \lambda}{n}$$

For any wave train in the electromagnetic spectrum, its frequency ω is invariant, i.e., its frequency is the same in all materials. Therefore, the wavelength must decrease by the same ratio as the velocity, such that

$$\lambda_m = \frac{\lambda}{n} \qquad (2.8)$$

where λ_m is the wavelength of light in the refractive material. Light propagates with shorter wavelengths in transparent materials than in vacuum.

Optical path length (*OPL*) is defined as the mechanical path length of a ray multiplied by the index of refraction of the medium traversed by the ray. In Fig. 2.3, the optical path length between a and b is

$$OPL = D_1 + n_2 D_2 + D_3 + n_4 D_4 + D_5$$

where D is the length, or mechanical distance, along the ray. Generalizing, OPL is the summation of nD for each element in the path, i.e.,

$$OPL = \sum_{1}^{i=q} n_i D_i \qquad i = 1, 2, 3, \cdots, q \qquad (2.9)$$

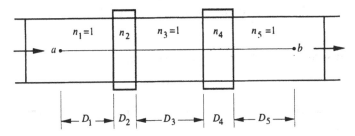

Fig. 2.3   The optical path length between a and b is the sum of $n_i D_i$ for the 5 sections.

In refractive materials, the *OPL* is longer than the mechanical path length (which is merely the summation of D_i).

Path lengths can be measured in units of wavelengths. The mechanical path length between two points in any material is the number of wave cycles between the points (as the wave train travels through the material) times the wavelength in that material. The *OPL* is the same number of wave cycles times the wavelength in vacuum. The concept of *OPL* is useful because it defines the relative phase of two wave trains and determines their state of constructive or destructive interference (Sec. 2.3.8).

Since the number of wave cycles between two wave fronts is a fixed number, an important corollary is this: *all rays between two wave fronts travel the same optical path length.*

Polarization

Sometimes we are interested in the *polarization* of the light. Light propagates as a transverse wave, with its wave-like character confined to a plane called the *plane of polarization*. More specifically, the electric field is a two-dimensional phenomenon. The oscillating electric field acts in the plane of polarization, for example in the y direction, while it propagates in the z direction.

Most lasers emit wave trains of relatively pure polarized light. Most other light sources emit each train with a different plane of polarization; their disturbances, taken together, constitute unpolarized light. For some applications, Eqs. 2.1 and 2.4 can be assumed to represent unpolarized light. They show A as functions of z and t, but independent of x and y. With plane-polarized light, the field strength is unidirectional, say in the y direction, and the disturbance is written as

$$A_y = a \cos 2\pi\omega t = \text{Re}\,[ae^{i2\pi\omega t}] \qquad (2.10)$$

Then field strength in the x direction is zero, or

$$A_x = 0$$

which means that there is no component of the electrical field parallel to x.

Circular and *elliptical* polarizations are special cases. They occur when plane polarized light is divided into two orthogonally polarized components (for example, by the action of a photoelastic material or a birefringent crystal) and the phase of one of the components is shifted by a fraction of a wave cycle. Then, the combined electric field of the two components can be represented by a vector. The direction of the vector varies through 360° for each wave cycle and its tip generates an ellipse. When the phase shift is 1/4 of a cycle, the vector tip generates a circle.

Coherence

Ideally, a beam of light is *coherent* if it emerges from a single point in space. In addition, it must be comprised to only one wave train, which means all the waves must have the same frequency and identical polarization and the wave train must be infinitely long. True coherence cannot be achieved, but it is approximated to various degrees by real light sources (Sec. 2.2.1).

Two beams of light are said to be *mutually* coherent if they have the same frequency and polarization. In practice, this requirement is satisfied only when the two beams originate from the same source, e.g., when the two beams are divided from a single beam by means of mirrors or other optical elements. Optical interference (Sec. 2.3) requires beams of mutually coherent light.

2.1.2 Metallic Reflection

When light strikes a metallic surface, a portion is absorbed and a portion is reflected. The *reflectance* of a surface is the ratio of the reflected intensity to the transmitted intensity. It depends upon the material and the wavelength of the light. Aluminum and silver have high reflectance ($R > 0.9$) in the visible range of wavelengths, although silver tarnishes quickly and loses its high reflectance. The reflectance of chromium is about 0.6 at normal incidence, but it is very hard and much more resistant to scratches and wear. Reflection is illustrated in Fig. 2.4. The incident and reflected rays lie

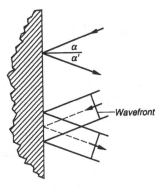

Fig. 2.4   For reflection, $\alpha' = \alpha$. Rays are drawn to represent beams.

in the *plane of incidence*, defined as the plane containing the incident ray and the normal to the surface. The angle of reflection is equal to the angle of incidence, or

$$\alpha' = \alpha \tag{2.11}$$

The reflectance depends upon the angle of incidence and the polarization of the incident light. In addition, *phase changes* occur upon reflection and the phase changes are different for polarizations parallel and perpendicular to the plane of incidence. As a consequence, an incident beam of plane polarized light can be changed to elliptical polarization by reflection from metals. When the incident beam is polarized either parallel or perpendicular to the plane of incidence, however, the reflected beam remains plane polarized.

Mirrors are frequently made by applying metallic coatings to the surface of plates. The coatings can be so thin that the mirror reflects a portion of the incident light and transmits another portion. Partially reflecting mirrors can be made this way, with reflectance R and transmittance T related by

$$R + T + A = 1 \tag{2.12}$$

where A is the absorptance of the metal film. Curiously, the absorptance can be larger for a partially transparent coating than for an opaque coating of the same metal.

2.1.3 Dielectric Reflection

When light is incident upon an interface between two *dielectric materials* (electrically non-conducting materials, e.g., air, water, glass), some of the light is reflected at the boundary surface. At normal incidence, the intensity of reflected light is given by

$$I_r = I_0 R = I_0 \left(\frac{n_2 - n_1}{n_2 + n_1} \right)^2 \tag{2.13}$$

where I_0 and I_r are the intensities of incident and reflected light, respectively, and R is the reflectance. At an air-glass interface, where the refractive index of the glass is 1.5, the reflected intensity is 4% of the incident intensity; reflectance for a glass window is about 8%, considering reflections from both surfaces.

At oblique incidence, intensities of reflected light vary with the angle of incidence and the polarization of incident light. Figure 2.5a shows this variation at an interface where $n_2/n_1 = 1.5$, including the case of external reflection in air ($n_1 = 1$) when $n_2 = 1.5$. *External* reflection occurs on the low index side of the boundary, as illustrated in the figure. Figure 2.5b shows internal reflection. When the plane

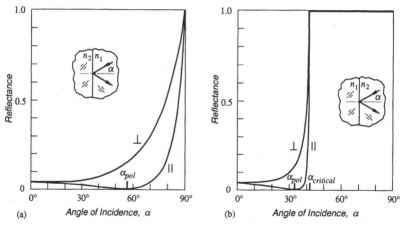

Fig. 2.5   Reflectance (reflected intensity) at the interface between two dielectric bodies, when $n_2/n_1 = 1.5$;   (a) external reflection and (b) internal reflection; α_{pol} is the polarizing angle and $\alpha_{critical}$ is the angle of total internal reflection.

of polarization lies in the plane of incidence, the curve marked ‖ applies. When the plane of polarization is perpendicular to the plane of incidence, the curve marked ⊥ applies. Since any state of polarization can be resolved into its vector components in these two directions, all polarizations are treated. Total internal reflection is indicated in (b) when α exceeds $\alpha_{critical}$.

Another important phenomenon is the *phase change upon reflection*. For external reflection, the light experiences a phase change of π for every angle α between 0 and α_{pol}. For internal reflection at angles smaller than the polarizing angle, the phase change is zero. As a physical model, a phase change of π can be equated to an increase of OPL by $\lambda/2$; thus the reflection can be imagined to occur beneath the actual interface. (Certain textbooks on optics present a mathematical analysis that yields zero phase change upon reflection for light polarized parallel to the plane of incidence. This discrepancy results from a particular sign convention. The physical interpretation of the mathematical analysis, however, prescribes the phase change of π for both the parallel and perpendicular polarizations.)

2.1.4 Refraction

When light *crosses* a boundary between materials of different refractive indices, the direction of propagation changes in accord with Snell's law of refraction

$$\frac{\sin \alpha''}{\sin \alpha} = \frac{n_1}{n_2} \tag{2.14}$$

where the variables are specified in Fig. 2.6a. Equation 2.14 applies for the case illustrated ($n_2 > n_1$) and also for the case of $n_1 > n_2$, for which $\alpha'' > \alpha$. The directions of propagation and the normal to the boundary surface lie in a common plane. If one medium is air, its refractive index can be taken as 1.000

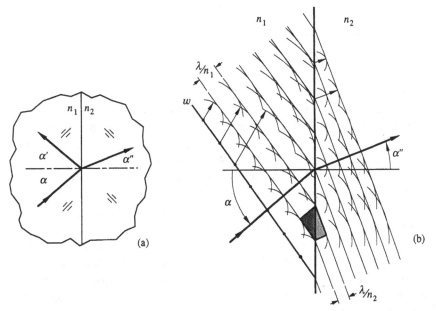

Fig. 2.6   (a) Refraction and reflection at a dielectric interface. (b) Huygen's construction to derive Snell's law.

Some portion of the incident light is reflected at the interface, while the remainder is refracted and transmitted into the second material. The reflected light is discussed in the previous section. Total internal reflection is governed by Snell's law, too; the *critical angle* occurs when $n_1 > n_2$ and $\alpha'' = 90°$, whereby

$$\sin \alpha_{critical} = \frac{n_2}{n_1} \qquad (2.15)$$

For angles of incidence larger than $\alpha_{critical}$, no light can be refracted and total internal reflection prevails.

It is instructive to view refraction in terms of *Huygen's wavelet principle*. It states that each point in a wave front can be assumed to act as a new source, with light radiating from each point. In Fig. 2.6b, let w represent a plane wave front approaching the boundary surface obliquely, at angle α. Wavelets originating along w have a radius of one wavelength λ_1, or by Eq. 2.8, λ/n_1; the surface tangent to these wavelets defines the new wave front. In succeeding time intervals of $1/\omega$, the wavelets and corresponding wave fronts advance by increments of one wavelength to new positions. In the medium of refractive index n_2, however, the wavelet radius becomes λ/n_2 and this leads to a discontinuity of slope of the wave surface at the boundary. From the geometry of the two shaded triangles, the ratio of the sines of α and α'' gives Snell's law, Eq. 2.14.

2.2 Optical Elements

2.2.1 Light Sources

The choice of a light source for a given application involves three main variables: its spectrum, or band of wavelengths present in the light; its delivered power, or energy per unit time; and its physical size, or the area of the light emission region. Representative spectra are illustrated in Fig. 2.7. A thermal source, such as an incandescent tungsten filament lamp, radiates light of about equal intensity for all the visible wavelengths. The spectrum can be narrowed by *color* filters, which transmit a band of wavelengths and absorbs the energy of the remaining wavelengths. *Interference filters* can isolate very narrow spectral bands; in this case, the rejected light is not absorbed, but instead it is reflected back while the narrow spectral band is transmitted. Of course, a great amount of light energy is lost when filters are used to isolate a narrow part of a broad spectrum.

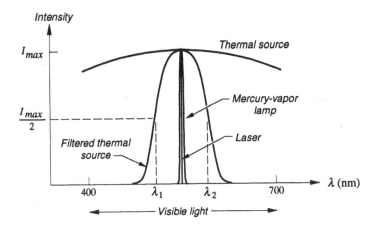

Fig. 2.7   Schematic graphs illustrating the spectrum of various light sources. The spectral bandwidth is $\lambda_2 - \lambda_1$.

The mercury-vapor lamp is a common laboratory light source. Light of the mercury spectrum is emitted when an electric current passes through a mercury gas. Strong emission bands are centered at 435.8 nm (violet) and 546.1 nm (green), and at the doublet 577.0 and 579.1 nm (yellow). The spectral bands are sufficiently far apart that they can be isolated by color or interference filters. The spectral width of the individual bands depends upon the pressure and temperature of the mercury gas. The half-width $\lambda_2 - \lambda_1$ is about 5 nm for high-pressure mercury lamps, 0.1 to 1 nm for medium-pressure lamps, and about 10^{-3} nm for low-pressure lamps. The high-pressure lamps are brightest, whereas the light output of low-pressure lamps is relatively weak.

Before the advent of the laser, the cadmium-vapor lamp was noted for its very narrow spectral bandwidth. It was used with an interferometric technique to determine the number of wavelengths in the length of the standard meter bar, and its wavelength became a standard unit of measurement. Today, laser wavelengths have become standard units of measurement.

The bandwidths of *continuous-wave gas lasers*, such as the helium-neon laser and the argon-ion laser, can be of the order of 10^{-6} nm (or 10^{-15} m). This is so small that the output can be considered to be a single wavelength for most practical purposes. Popular wavelengths are 632.8 nm (red) for the helium-neon laser, and 514.5 nm (green) and 488.0 nm (blue-green) for the argon-ion laser. Other choices are becoming increasingly available as laser technology advances.

The issue of delivered power puts the classical thermal and metal-vapor lamps at a disadvantage for two reasons. Light is emitted in all directions, while only a small part of the total can be channeled into most optical systems. Secondly, many applications require a small source size, such that light from only a small part of the emission region can be used. On the other hand, the stimulated emission that takes place in a laser cavity gives the light strict directional properties and nearly all the output can be channeled into the optical system.

For many applications, a point source or a reasonable approximation to a point source of light is needed. The classical thermal and metal-vapor sources are considered to be extended sources, since light emerges from a broad emission region. Even the small tungsten filament of an incandescent lamp and the short arc length of a mercury-vapor arc lamp are in the millimeter size scale. They are broad sources compared to the micrometer size scale that is often required. Again the laser offers superiority. The output beam is so well directed that the light can be focused virtually to a point; the laser can be treated as a point source.

The remarkable qualities of laser light stem from self-stimulated emissions within the laser cavity. Successive emissions are stimulated by light already propagating within the cavity. It is stimulated in such a way that the additional light propagates in the same direction and with the same phase as the original light beam. This action creates the very long wave trains of laser light, with extremely narrow spectral bandwidth and extremely smooth wave front surfaces.

Laser light is said to be coherent light. *Temporal* coherence relates to time. It refers to the extremely narrow spread of period or frequency of laser light waves (i.e., the high monochromatic purity) and the concomitant narrow wavelength spread and long wave trains. *Spatial* coherence relates to space. It refers to the systematic directional properties whereby laser light can be considered to emerge from a point in space. Laser light has high temporal and spatial coherence.

Is high coherence always desirable? While it is a great asset for most techniques of optical interferometry, it creates certain problems too. Ghost patterns, or faint unintentional interference patterns superimposed upon the intended patterns, are frequently encountered when coherent light is used. They must be eliminated by careful laboratory practices or accepted as optical noise in the field of view.

2.2.2 Wedges, Prisms, Mirrors

Wedges, prisms, and mirrors are used to change the direction of an incident beam by refraction, reflection, or both. All have plane surfaces, as illustrated in Fig. 2.8. The direction of emergence is determined by applying the simple laws of reflection and refraction, Eqs. 2.11 and 2.14 to the rays encountered at each surface.

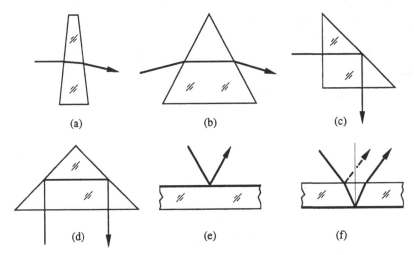

Fig. 2.8   Wedges, prisms, and mirrors.

Refracting elements that deviate the beam through relatively small angles are called wedges, as in (a). Larger deviations are accomplished with prisms by refraction (b) or internal reflection (c and d); reflecting surfaces of prisms can be coated with thin metal films, or they can be bare and depend upon total internal reflection. A front-surface mirror is illustrated in (e), where the reflecting surface can be the polished surface of a metal plate, but more usually it is a polished glass plate coated with a metallic film. The common household mirror, or rear-surface mirror (f), is a glass plate with a metal film applied to its back surface; these often suffer from double imagery as a result of the light reflected from the front surface.

2.2.3 Partial Mirrors

A partial mirror is a mirror that reflects a fraction, R, of the light intensity incident upon it and transmits a fraction, T, as illustrated by rays in Fig. 2.9. Such mirrors are also called beam splitters or semi-transparent mirrors. Metallic coatings of controlled thickness can be applied to polished glass by vacuum deposition. The coating can be so thin that it is semi-transparent; the transmittance decreases with coating thickness.

Vacuum deposition, or *evaporation*, is depicted in Fig. 2.10. The operation takes place in a chamber evacuated to a low absolute pressure. The plate is supported with its surface to be coated facing a reservoir of coating material. The coating material is heated electrically and it evaporates rapidly. In the absence of air resistance, the vapor travels long distances in the chamber and deposits on the plate, where it sublimes (or else condenses and freezes) to solid particles. With metallic coatings, coating thickness is usually a small fraction of a wavelength of visible light. Uniformity of thickness is nearly perfect, and surface smoothness duplicates the substrate smoothness.

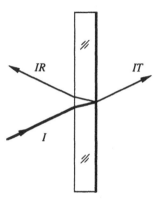

Fig. 2.9   Partial mirrors split the incident light into transmitted and reflected parts.

Nonmetallic partial mirrors are widely used. Dielectric coatings produce excellent partial mirrors, since absorption is practically zero. A single layer of a high refractive index coating is illustrated in Fig. 2.11, where beams a and b are reflected at the air/coating interface and the coating/glass interface, respectively. If the coating thickness is controlled such that the phase difference between a and b is 2π or an integral multiple of 2π, then a and b will combine in constructive interference (Sec. 2.3) and enhance the reflected intensity. Here, S includes both the change of *OPL* and the phase changes upon reflection, where a phase change of π is equal to a path length of $\lambda/2$ (Sec. 2.1.3). As an example, if a coating of zinc sulphide ($n_H = 2.32$) is applied to an optical glass ($n_G = 1.52$), constructive interference between beams a and b in Fig. 2.11 provides a reflectance of about 35%.

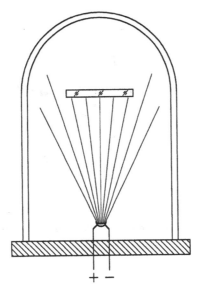

Fig. 2.10   Metallic and dielectric coatings can be produced by evaporation. The coating material is heated in a high vacuum environment; it evaporates and deposits onto the substrate.

Multilayer dielectric coatings can produce virtually any reflectance. The coatings are illustrated schematically in Fig. 2.12, where alternating layers of high and low refractive index coatings are applied. Magnesium fluoride ($n_L = 1.38$) is an example of a suitable low index material. A portion of the advancing incident beam is reflected at each interface. When these reflected beams combine in constructive interference the total reflected intensity can be very large, approaching 100% reflectance. Multilayer dielectric coatings are used extensively for laser cavity mirrors and many other diverse applications.

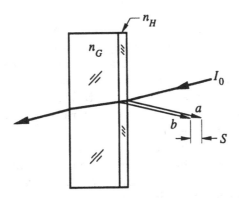

Fig. 2.11   Single layer dielectric mirror coating.

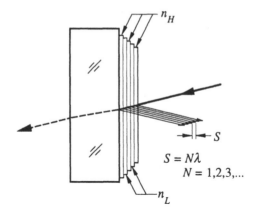

Fig. 2.12   Multilayer dielectric mirror with alternating layers of high and low refractive index.

2.2.4 Antireflection Coatings

The dielectric reflection at air/glass interfaces is frequently detrimental. Examples are reflections from air/glass interfaces of lenses and reflection from the second (nonmirror) surface of beam splitters. Such reflections can be suppressed by dielectric coatings similar to those used for partial mirrors, but the coating thickness and the distance S in Figs. 2.11 and 2.12 must be controlled to produce destructive interference (Sec. 2.3).

An alternative approach for beam splitters is to apply the partial mirror coating to one surface of a wedge. Then, the unwanted reflection from the back surface of the wedge becomes inclined to the main reflected beam and the unwanted light becomes separated from the primary beam.

2.2.5 Lenses

Lenses change the direction of light by refraction. Snell's law, Eq. 2.14, determines the path of every ray as it crosses a lens. Figure 2.13 depicts a lens that changes a divergent beam to a collimated beam. The spherical wave fronts of the divergent beam are converted to plane wave fronts. Since the optical path lengths of all rays between two wave fronts are identical (Sec. 2.1.1), rays oa and oa' have equal OPL. Obviously their mechanical path lengths are different, but the optical thickness of the lens varies in just the right way to compensate for the changes of mechanical path length. An important property is the constant OPL of all rays through a lens between an object point and its image.

The required variation of lens thickness can be accommodated by spherical surfaces (almost always). This allows the manufacture of lenses with relatively simple generating and lapping machines. For special applications, lenses with nonspherical surfaces may be required, and these *aspheric* lenses are made by more sophisticated machinery or manual methods.

Lens action should be explained differently for *matte* objects and *specular* objects. However, the explanations are supplementary, not conflicting. Matte objects have rough surfaces when viewed on the wavelength scale. When illuminated, each local region of the surface reflects or (for transparent objects) refracts the light in all directions, just as each region of a classical light source radiates in all directions. The law of reflection (Eq. 2.11) and Snell's law (Eq. 2.14)

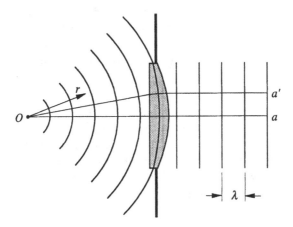

Fig. 2.13   A lens introduces systematic changes in wave fronts.

remain effective on a microscopic scale, but the surface is so rough that the local angle α changes randomly within any small region. The reflected or refracted light is nearly the same as light from a self-luminous object, where light diverges from every point with spherical wave fronts, each with random phase relative to its neighbors.

A specular object has smooth, mirror-like surfaces. It reflects or refracts light in a systematic way, again governed by Eqs. 2.11 and 2.14. When a beam with a smooth wave front is incident upon a specular object, the reflected or refracted beam also exhibits a smooth wave front.

The two cases are pertinent to our studies. The techniques of geometric moiré utilize objects with matte surfaces, while moiré interferometry deals mostly with specular objects.

Fig. 2.14 illustrates the focusing properties of a lens. In Fig. 2.14a, the illuminated matte object reflects light from each point. The portion of the light emerging from a that is collected by the lens is redirected to converge to point a'. Point a' is the image of a; they are called *conjugate* points, a word that implies they are equivalent or interchangeable. Light from any off-axis point b in the object is collected and converged (or focused) to point b'. Thus, all points on the object are focused to conjugate points in the image plane, where an image or visual replica can be seen. In the case of a camera, a photographic film is located in the image plane to record the image.

When the object is at infinity, the lens converges the light to a point a'. Its distance from the lens is defined as the *focal length, FL,* of the lens. This condition is illustrated in (b) and (c) of Fig. 2.14, but in (c) a *virtual* image is formed. The light is not concentrated at a', but rays that emerge on the right side of the negative lens appear to come from a'. Of course, the objects we will consider do not lie at infinity, but in the common optical system illustrated in (d) the light received by the second lens appears to come from infinity. The system depicts a point light source that is collimated by a first lens and decollimated or converged by the second lens.

Note that the lenses in Fig. 2.14 are drawn with their larger curvature facing the side of smallest obliquity of rays. To reduce aberrations, simple lenses should be oriented such that the two surfaces of the lens refract a ray by approximately equal angles.

An extremely valuable relationship between object and image distances is the thin lens formula

$$\frac{1}{D_1} + \frac{1}{D_1} = \frac{1}{FL} \tag{2.16}$$

where D_1 and D_2 are distances from the lens to the object plane and image plane, respectively. D_1 and D_2 are positive when they lie on opposite sides of the lens. The relationship is not exact since real lenses have thickness, but where the distances are large compared to the lens thickness the approximation is valid. This relationship applies for all the cases illustrated in Fig. 2.14. In (b) for example, D_1 is infinity, its reciprocal is zero, and $D_2 = FL$. In (c), the focal length of a negative lens is a negative distance, which is equal to D_2. Magnification M of the image, or the ratio of length $a'b'$ to ab, is given by

$$M = \frac{D_2}{D_1} \tag{2.17}$$

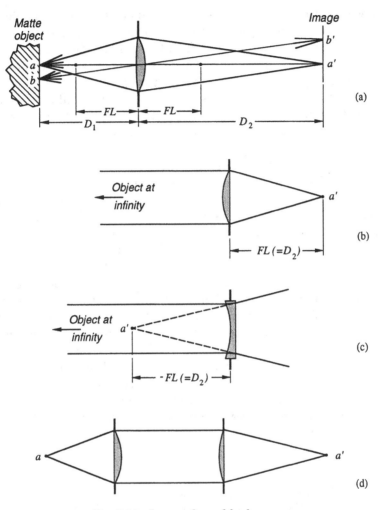

Fig. 2.14   Images formed by lenses.

Equations 2.16 and 2.17 can be applied, too, for *compound* lens systems, or a series of thin lenses. The equations are applied individually for each thin lens element, with the image produced by the first element becoming the object for the second element, etc. For example, in Fig. 2.14d, the image distance D_2 from the first lens is determined by Eq. 2.16 to be infinity. This becomes the object plane for the second lens; for the second lens D_1 is minus infinity, its reciprocal is zero, and by Eq. 2.16 the image distance equals FL, the focal length of the second lens. (See Exercise 2.3.)

These relationships apply equally to images of specular objects
and images of matte objects. Additional lens properties that bear
mentioning involve the quality or fidelity of the image. One pertains
to the finite aperture of the lens. The lens in Fig. 2.14a intercepts only
a portion of the wave front radiating from point a. The light that
strikes the lens aperture (or lens boundary) is diffracted and some of
it falls outside the geometric projection of a. As a result, the image of
point a is not simply a mathematical point a', but instead, it is a
distribution of light across a finite area surrounding point a' in the
image plane; most of the energy is concentrated within a zone called
Airy's disk. This limits the resolution of the lens, or its ability to
resolve fine details. The size of Airy's disk becomes smaller as the
lens size is increased, i.e., as the solid angle of the cone of light
accepted by the lens is increased. Thus, lenses with large apertures
provide higher resolution than smaller lenses, when all other
variables are kept constant.

Image degradation also results from inherent imperfections in
the lens, called *aberrations*. Light from a matte object point is spread
in the image plane across an area called the *circle of confusion*. For
high quality focusing lenses, designers minimize the circle of
confusion until it is smaller than Airy's disk. Such lenses are called
diffraction limited lenses. Their design and fabrication is increasingly difficult as the aperture inceases. The resolution of high quality
focusing lenses is usually specified by the manufacturer in terms of
the contrast of closely spaced line pairs. The resolution limitations of
a lens apply to images of matte objects, *not* specular objects.

Specular Objects

With specular objects, the performance is different. Consider the
arrangement in Fig. 2.15, where the object is a plane mirror. Let a
beam of light with plane wave fronts approach the object at normal
incidence. To accomplish this, a beam splitter is used to direct the
light toward the object. The object reflects the light as indicated by
arrows attached to it, without distorting the wave fronts. However, at
the boundaries of the object and also at scratches or marks on the
object, light is reflected as it would from a matte object.

Light from the boundaries of the object radiates out and fills the
entire lens. It is then focused by the lens to display an image of the
boundaries in the image plane. The image of the boundaries suffers
the resolution limitations discussed above.

Light from all non-boundary parts of the specular object advances
toward the lens as a collimated beam with plane wave fronts. The

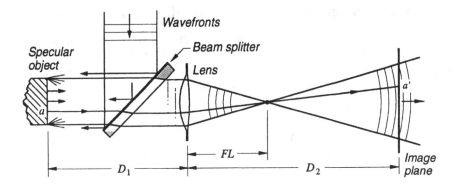

Fig. 2.15 Light from a point on a specular (mirror-like) object traverses the lens along a single line. Boundary points on the object are exceptions; light emerges from each boundary point in all directions, it fills the lens, and some of it is diffracted by the lens boundaries.

lens converges the beam through an apex at the focal plane of the lens, whereupon the beam diverges with spherical wave fronts that cross the image plane. Within the image of the boundary of the object, the image plane is uniformly illuminated; no detail is seen.

Why are we interested in these images, which now seem blank? The answer lies in Sec. 2.3.6, which treats image formation when two specular beams emerge from the object, each with differently warped wave fronts. The two beams interfere constructively and destructively with each other and form an interference fringe pattern in the image plane. The light comprising such patterns can travel through the lens without intercepting the lens aperture or boundary, so it is not influenced by boundary effects of the lens. Consequently, the interference fringe pattern projected on the image plane exhibits full fringe contrast. The only requirement is that the two beams with warped wave fronts pass through the lens.

The task of an imaging lens is very different for specular objects, compared to the task for matte objects. With matte objects, light from each object point fills the lens, and the lens must bend all these rays to reunite them at the image point. With specular objects (illuminated by light having smooth wave fronts), the light from each object point emerges as a single ray; it enters the lens at a single point and it traverses the lens along a single path as it travels to the image point. In Fig. 2.15, all the light that emerges from the object point a travels along the singular path aa' to reach the image point. The task is much less complex.

2.2.6 Laser-beam Expander, Spatial Filter, Optical Fibers

The output of a typical gas laser is a narrow, intense beam of coherent light. Being narrow, the beam is often called a *pencil* of light. A beam-expander is used to enlarge it into a divergent beam. As illustrated in Fig. 2.16a, a beam expander is a small, short-focal-length lens. Microscope lenses are frequently used and the power of the lens determines the angular divergence of the beam.

A spatial filter (Fig. 2.16b) is a beam expander with a pinhole aperture at the node (or waist) between the convergent and divergent light. The diameter of the aperture is small, usually 10 - 50 times the wavelength. The pinhole passes light from the main beam of the laser, but it blocks light from extraneous sources such as light scattered from dust particles on the laser window and beam expander lens, and light that experiences multiple reflections at the lens surfaces. By filtering out the extraneous light, the spatial filter provides a *clean* beam, i.e., without random variations of intensity across its field. Instead the intensity distribution within the beam approaches the theoretical Gaussian distribution, which prescribes a

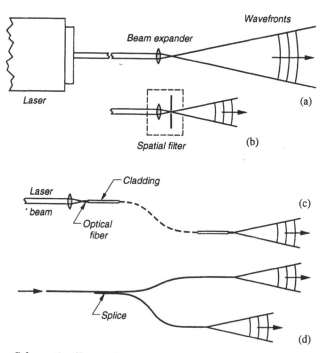

Fig. 2.16   Schematic illustrations of (a) a beam expander, which converts a narrow laser beam into a diverging beam; (b) a spatial filter, which is a beam expander plus a pinhole aperture; (c) an optical fiber, which conducts light to a new location where it emerges as a divergent beam; (d) spliced fibers, which provide two mutually coherent beams.

smooth variation of intensity near the center of the beam. Usually, only the central part of the divergent beam is admitted into an optical system, providing a nearly uniform beam intensity; the remainder of the beam is wasted.

The optimum pinhole aperture size depends upon the power of the beam expander. The lens and pinhole arrangement is available in sets, with mounting fixtures and fine adjusting screws to position the aperture at the node.

A single-mode optical fiber (Fig. 2.16c) acts very much like a pinhole filter, while at the same time it conducts the light to a new location. Special adjustment fixtures are used to align the fiber with the concentrated laser beam. Optical fibers are often used to separate a bulky laser from the rest of the optical system. In addition to convenience, this separation can isolate the optical system from troublesome vibrations that are sometimes produced by water cooled or forced-air cooled lasers.

Single-mode fibers preserve the temporal and spatial coherence of the laser light. Single-mode polarization-preserving fibers are also available, which preserves the state of plane polarization when polarized light is input.

Optical fibers can be spliced together, as illustrated in Fig. 2.16d, to provide two mutually coherent output beams. These spliced fibers are very useful for the assembly of various two-beam interferometer systems.

2.2.7 Parabolic Mirrors

A parabolic mirror provides an efficient way to form a collimated or parallel beam of light. Parabolic mirrors are usually fabricated from glass or pyrex (for its low thermal expansion). The mirror surface is polished to a smooth contour that matches a paraboloid within a fraction of a wavelength, e.g. $\lambda/8$, and the surface is coated with aluminum for high reflectance.

As illustrated in Fig. 2.17, a diverging beam from a light source located in the focal plane of the mirror is reflected as a parallel beam. The source could be one of the configurations of Fig. 2.16. In Fig. 2.17a, the mirror is slightly inclined so that the source does not obstruct the collimated beam. In (b), the mirror is used on-axis, which means the collimated beam is parallel to the axis of the paraboloidal surface of the mirror; large parabolic mirrors are manufactured as indicated by the dashed lines, but then they are cut into smaller mirrors for use as shown in the diagram. The wave front distortion of the on-axis design is smaller, but when angle θ (Fig. 2.17a) is minimized, the small distortion from an off-axis mirror

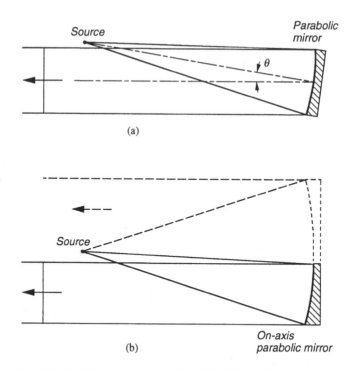

Fig. 2.17   A collimated beam can be produced by (a) an off-axis parabolic mirror and (b) an on-axis parabolic mirror.

is acceptable for most applications and it is the configuration that is used most often.

The alternative is a collimating lens. That option is usually the most expensive, with the on-axis mirror less expensive and the off-axis mirror least expensive. When parabolic mirrors are purchased from vendors that supply the amateur telescope market, mirrors of fractional wavelength surface accuracy are economical.

2.3 Coherent Superposition

2.3.1 Pure Two-beam Interference

Consider the scheme of Fig. 2.18a to achieve superposition of two beams. Let the source be a laser that emits light of wavelength λ and polarization perpendicular to the plane of the diagram. Let the light be collimated by a lens or parabolic mirror to produce a parallel beam with a plane wave front w_0. At a later time, wave front w_0 appears as wave fronts w_1 and w_2. After division by the beam splitter, one beam travels along path 1 and a portion of it is reflected at the second beam splitter toward the camera. Assuming perfect optical elements, the beam emerges with a plane wave front w_1. The second beam traverses path 2. Since it travels at a lower velocity through the medium of refractive index n, it emerges later than beam 1 and its wave front w_2 lags w_1 by a distance S.

A generic diagram of two-beam interference is shown in Fig. 2.18b. The shaded box represents any one of various two-beam interferometers. The optical system within the box divides the input beam into mutually coherent beams; it alters the optical path length of one or both beams to create a relative phase difference. Consequently, the beams emerge with one leading the other by a distance S.

Consider the number of wave cycles that occurred at an arbitrary point z_0 as the two beams (i.e., the two wave trains) propagate through the point. At a given instant, the head of the lagging wave train advanced beyond z_0 by a smaller distance. Therefore, the lagging beam experienced S/λ fewer wave cycles. Its phase is $2\pi S/\lambda$ less than that of the leading beam. This phase difference persists for all time during the passage of the two beams.

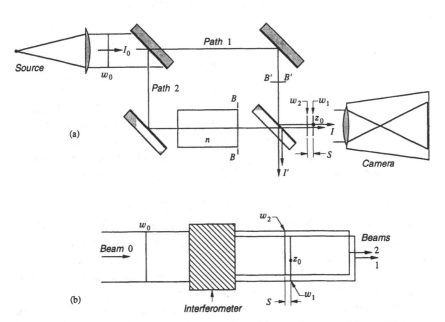

Fig. 2.18   (a) Superposition of mutually coherent beams with optical path difference S. Note: refractions that occur at the beam splitters are not shown. (b) Generic representation of two-beam interference.

If at z_0 the field strength of wave train 1 is assumed to vary with time as

$$A_1 = a \cos 2\pi \omega t \tag{2.18}$$

then the strength of wave train 2 is

$$A_2 = a \cos 2\pi \left(\omega t - \frac{S}{\lambda} \right) \tag{2.19}$$

These equations assume the two waves have the same peak strength a. Since the phase difference at z_0 is the same for all time, and since the sum of two harmonic functions having the same frequency is another harmonic of that frequency, the resultant field strength is

$$A = A_1 + A_2 = K \cos 2\pi (\omega t + \phi)$$

where

$$K = \left[2a^2 \left(1 + \cos 2\pi \frac{S}{\lambda} \right) \right]^{1/2}$$

By Eq. 2.6, the resultant intensity of the recombined wave trains is K^2, or

$$I = 2a^2 \left(1 + \cos 2\pi \frac{S}{\lambda} \right) = 4a^2 \cos^2 \pi \frac{S}{\lambda} \tag{2.20}$$

This is a fundamental relationship, called the intensity distribution of pure two-beam interference. It is derived in Sec. 2.3.8. The intensity of the combined beams is persistent, or independent of time. It varies cyclically from maximum to zero to maximum, and so on, as a function of the difference S of path lengths traveled by the two beams.

In the schematic arrangement of Fig. 2.18, S might have any value. If S is an integral number of wavelengths, i.e. $S/\lambda = 0, 1, 2, 3,$---, the intensity at arbitrary point z_0 is maximum and *constructive* interference occurs. If $S = \lambda/2$ plus an integral number of wavelengths, or $S/\lambda = 1/2, 3/2, 5/2,$---, the intensity reaches a minimum value of zero and *destructive* interference occurs.

This example is called *pure* two-beam interference. It occurs when two mutually coherent beams of equal amplitude (and intensity) travel different optical path lengths and then recombine. It is characterized by a maximum intensity that is equal to twice that obtained by incoherent superposition (Sec. 2.4), and by a minimum intensity of zero (i.e., complete destructive interference).

Since w_1 and w_2 are parallel and they travel with identical velocities, and since z_0 is an arbitrary point, the intensity—or the condition of constructive or destructive interference—is the same at every point in space where beams 1 and 2 coexist. The case of nonparallel wave fronts will be considered later.

2.3.2 Conservation of Energy

Referring again to Fig. 2.18a, assume that the beam splitters and mirrors are made with dielectric coatings of zero absorption. Let the beam splitters have 50% reflectance and 50% transmittance, which means that the intensity in each path is half the intensity of the incident light. Let the two full mirrors have 100% reflectance. Let the incident intensity be $4a^2$.

The intensity of beam 1 (i.e., the beam with wave front w_1) is halved at each beam splitter to become a^2; its maximum amplitude is a. Similarly the intensity of beam 2 (with wave front w_2) becomes a^2 and its maximum amplitude is a. These two emergent beams can be represented by Eqs. 12.18 and 2.19. Their resultant intensity is given by Eq. 2.20.

However, Eq. 2.20 shows that the output intensity of the combined horizontal beams can be as high as $I = 4a^2$. For certain values of S, the output intensity is equal to the input intensity. Then, can we extract energy from the output intensity I' in the perpendicular path?

Of course not! When $I = 4a^2$, interference in the perpendicular path must give $I' = 0$. In fact, I and I' are always complementary, that is,

$$I' = 4a^2 \sin^2 \pi \frac{S}{\lambda}$$

This shift is caused by the phase changes of π that occur upon external reflection (Sec. 2.1.3). The two beams that emerge in the direction of the camera experience two external reflections in each path, so there is no relative phase change induced by reflection. For output in the direction of I', path 1 encounters one external reflection (with its phase change of π) while path 2 encounters two external reflections (2π change) and one internal reflection (zero phase change). The relative phase change is π and therefore I' is the complement of I. Thus, $I + I' = I_0$ for every value of S. If not for the phase changes upon reflection, the output intensity could be double the input and we could create energy!

2.3.3 Impure Two-beam Interference

Impure interference occurs when beams 1 and 2 (Fig. 2.18) have unequal amplitudes a_1 and a_2 and thus unequal intensities I_1 and I_2. The resultant intensity at arbitrary point z_0 is

$$I = I_1 + I_2 + 2\sqrt{I_1 I_2}\cos 2\pi\frac{S}{\lambda} \qquad (2.21)$$

This is the general expression for interference, or superposition of any two coherent wave trains; the relationship is derived in Sec. 2.3.8. The resultant intensities are plotted in Fig. 2.19 as a function of path difference $N = S/\lambda$. Curve 1 is for pure two-beam interference and the other curves apply for the specified ratios of input beam intensities. Interference of beams of unequal strengths is *impure* because destructive interference is incomplete, or $I_{\min} \neq 0$.

2.3.4 Fringe Patterns and Walls of Interference

The apparatus represented by Fig. 2.20 is similar to that of Fig. 2.18, but here collimated beams 1 and 2 emerge with a relative angle 2θ. This would occur, for example, if wedge-shaped optical elements are inserted in paths 1 and 2 of Fig. 2.18a. As always (in free space and isotropic media) their wave fronts w_1 and w_2 are perpendicular to the beams and they, too, intersect at angle 2θ.

Wave fronts w_1 and w_2 originated from division of wave front w_0. Consequently, w_1 and w_2 have identical phase. By Eq. 2.20 or 2.21, their separation $S_{(x,y)}$ determines the local intensity at every x,y point. Where their separation S is an integral number of wavelengths, interference is constructive; midway between neighboring locations of constructive interference, the interference is destructive.

Visualization of the intensity distribution in the space where the two beams coexist is aided by Fig. 2.21. The figure illustrates the two wave trains and it depicts wave fronts w_1 and w_2 and neighboring wave fronts. The separation between wave fronts in each train is λ. The harmonic curves with ordinates labeled A_1 and A_2 represent the field strengths of the two wave trains in space at the given instant. These curves have equal amplitudes, indicating conditions for pure two-beam interference. Phases marked ϕ_1 and ϕ_2 are identical since wave fronts w_1 and w_2 have equal phase. Consequently, when w_1 reaches point a, w_2 has already advanced five wavelengths beyond a. Point b lies 2.5 wavelengths ahead of w_1 and 2.5 wavelengths behind w_2, so the path difference at b is 5 wavelengths. In fact, the path difference is 5 wavelengths at every point along ac; in Eq. 2.20, $S/\lambda = 5$

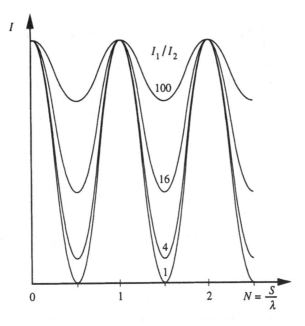

Fig. 2.19   Resultant intensity versus path difference, S/λ, for two-beam interference, where I_1 and I_2 are the intensities of the two input beams.

at these points. The result is constructive interference at every point along ac.

By the same argument, along de the difference S of optical path lengths traveled by the two wave trains is 4.5 wavelengths, or $S/\lambda = 4.5$. Along qr, $S/\lambda = 4$. This explains the formation of constructive and destructive interference at the given instant, but why is the effect persistent, or independent of time?

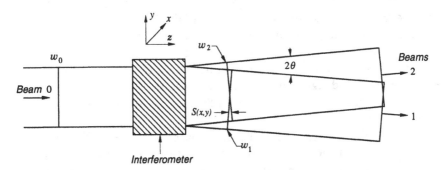

Fig. 2.20   In this case, the interferometer causes the two output beams to emerge with a relative angle 2θ.

At a later instant these two wave fronts have advanced, but their velocities are identical and the distances traveled are identical. Therefore, the phase difference at point a remains five cycles, and the phase difference at all other points remains unchanged. A steady-state condition of constructive and destructive interference is formed. It is clear from the figure that these lines of interference lie parallel to the bisector of the incoming beams.

Furthermore, the wave fronts of Fig. 2.21 are not merely the lines shown, but they are planes that stand perpendicular to the diagram. The optical interference occurs in the three-dimensional space where the two beams intersect. Constructive and destructive interference occurs in a series of parallel planes that stand perpendicular to the diagram and contain the lines labeled constructive and destructive interference.

The planes of constructive interference lie within thin volumes of space where the resultant intensity is relatively high. The planes of destructive interference lie within intermediate thin volumes of space where the resultant intensity is relatively low. One can visualize these thin volumes of space surrounding the planes of constructive interference as *walls* of constructive interference—walls in which light intensity exists. These walls are separated by voids that represent the (relative) absence of light intensity. The conceptual model is that of three-dimensional space occupied by parallel, uniformly spaced *walls of light intensity*. They will be designated by that name and, more simply, by *walls of interference*.

Such walls can be visualized wherever two beams of mutually coherent light intersect in space, regardless of the angle of intersection. Of course, the walls do not obstruct anything. Other beams of light can pass through them and emerge uninfluenced by their presence, just as two beams of light can cross in space and emerge from the intersection zone unimpaired.

A photographic plate that cuts this space absorbs energy from the walls of light intensity and records their presence. It receives no energy where it cuts through the voids. Thus, the photographic plate exhibits dark and light bands such as those in Fig. 2.22. The bands are called *fringes*, and the array of fringes is called an *interference fringe pattern*.

Let the photographic plate be interposed along BB in Fig. 2.21. A fundamental relationship defines the distance G between adjacent fringes on the plate, or between adjacent walls of interference in space. From the shaded triangle in Fig. 2.21, whose hypotenuse is G and short leg is $\lambda/2$, we find

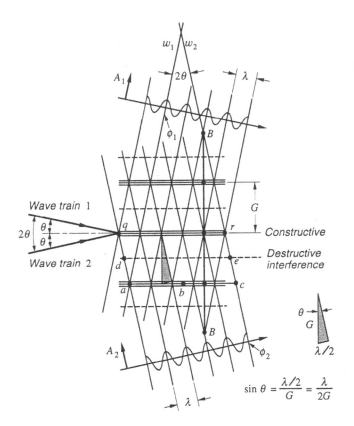

$$\sin \theta = \frac{\lambda/2}{G} = \frac{\lambda}{2G}$$

Fig. 2.21 Regions of constructive and destructive interference in space where two equal coherent beams intersect. $A_1 = A_2$ at points a, b, c, so constructive interference is produced. $A_1 = -A_2$ at d and e, resulting in destructive interference.

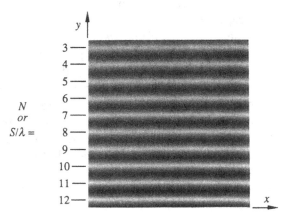

Fig. 2.22 Interference fringes recorded by a photographic plate installed in plane BB of Fig. 2.21.

$$\sin \theta = \frac{\lambda / 2}{G} \tag{2.22}$$

The frequency of the fringes, or the *fringe gradient* on the plate (e.g., fringes/mm), is F, where

$$F = \frac{1}{G} \tag{2.23}$$

The fringe gradient is determined by

$$F = \frac{2}{\lambda} \sin \theta \tag{2.24}$$

Like Eq. 2.20 or 2.21, the relationship of Eq. 2.24 pertains to every case of optical interference; it will arise repeatedly in our studies. The relationship applies for any value of θ between 0 and 90° and it applies for any separation S, except that S must be small compared to the length of the wave train or the coherence length of the light source.

If a photographic plate is interposed along BB, it would record a pattern similar to Fig 2.22; this represents a positive print and shows destructive interference as black, or the absence of light. The bright bands (or bright fringes) represent zones of constructive interference. The fringes are given numerical values according to the corresponding wave front separations, namely

$$N(x,y) = \frac{S(x,y)}{\lambda} \tag{2.25}$$

where N is called *fringe order*. In Fig. 2.22, the centers of bright fringes are loci of points of integral values of $N(x,y)$, i.e., points of $N(x,y) = 0,1,2,3,\text{---}$. At other points, N is a noninteger; every point in the field has a value of N. Equation 2.25 shows that the fringe pattern is a contour map of the separation $S(x,y)$ between wave fronts w_1 and w_2, where the contour interval is one wavelength per fringe.

The fringe pattern of Fig. 2.22 has a rather low frequency (fringes/mm). How small was the angle 2θ between the beams that created the pattern? What are the limits on fringe frequency and angle θ? Table 2.2 provides numerical values for fringe frequency F versus angle θ, according to Eq. 2.24. The red light of the helium-neon laser, $\lambda = 632.8$ nm, is assumed for the calculation. When θ is in the range of one part per 10,000 to one part per 1000 (0.006° to 0.06°), the pattern exhibits 0.3 to 3.2 fringes per mm (8 to 80 fringes per inch). That range of fringe frequencies is common in practical applications of optical interferometry for deformation and surface contour measurements. For such measurements, the angle between interfering beams is extremely small. On the other hand, high

TABLE 2.2 Fringe Frequency for Various Angles of Interfering Beams

θ	Angle of intersection (2θ)	F fringes/mm	F fringes/in.
1/10,000	2/10,000	0.3	8
1/1000	2/1000	3.2	80
1/100	2/100	31.6	803
1°	2°	55.2	1,400
5°	10°	275	7,000
20°	40°	1,080	27,500
40°	80°	2,030	51,600
60°	120°	2,740	69,500
80°	160°	3,110	79,000
90°	180°	3,160	80,300

for $\lambda = 632.8$ nm

frequency diffraction gratings are made by recording the interference pattern created by two beams that have large angular separation (Fig. 4.32). Practical values of angle θ can be as large as 80° to produce gratings of 3000 lines/mm or 75,000 lines/in.

Figure 2.22 illustrates a case of pure two-beam interference. If the two interfering beams had unequal intensities, the resultant intensity at each point (in plane B-B) would be governed by Eq. 2.21 and Fig. 2.19. The dark fringes would be photographed as shades of gray instead of black. Table 2.3 lists parameters relating to the visibility of the fringes for different relative intensities of the input beams. The output is given by the ratio of maximum to minimum intensities, I_{max}/I_{min}, for impure interference (Eq. 2.21), and also by contrast, where

$$\% \text{ contrast} = \frac{I_{max} - I_{min}}{I_{max}} \times 100\% \qquad (2.26)$$

It becomes clear that pure two-beam interference is not required for good fringe visibility. If the intensities of two interfering beams are in the ratio of 4:1, the resultant fringe contrast is good, namely 89%. Two beams with a 100:1 intensity ratio produce a discernible interference pattern with a contrast of 33%. (This, incidentally, is why extraneous interference patterns or "ghost patterns" are so commonly observed when coherent laser light is used. Small amounts of extraneous light that reflect accidentally from a local zone, e.g., from the frame of an optical element, combines with the main beam to produce low-contrast but noticeable interference fringes).

TABLE 2.3 Two-beam Interference for Coherent Beams
of Unequal Intensities

Two-beam input	Output: result of coherent summation	
$\dfrac{I_1}{I_2}$ or $\dfrac{a_1^2}{a_2^2}$	$\dfrac{I_{max}}{I_{min}}$	Contrast (%) Eq. 2.26
1	∞	100
2	34	97
4	9	89
9	4	75
16	2.8	64
25	2.2	56
100	1.5	33

2.3.5 Warped Wave Fronts

Referring again to Fig. 2.20, let wave front w_2 emerge as a surface
that is not flat, but deviates from flatness in a smooth continuous
manner. It is smoothly warped, e.g., in the manner sketched in Fig.
2.23. Wave front w_1 remains plane in this example. If a photographic
plate is installed in the two output beams, it would record the
nonuniform interference fringe pattern of Fig. 2.23b, which is a
contour map of separation $S_{(x,y)}$ between the two wave fronts.

The wave front separation S and the distance between neighboring
fringes both vary systematically with x and y. Along any continuous
bright contour, $S_{(x,y)}/\lambda$ has the constant value given by fringe order N,
where N is an integer. N is constant at all x,y points along any dark
contour; N is constant at all points along a continuous contour of any
intermediate intensity, so N assumes all values: integer, half-order,
and intermediate or decimal values.

Since w_2 is smoothly warped, fringe orders must vary in a smooth,
continuous manner. Fringes must be smooth contours and the
change of fringe order between adjacent fringes must be either 1 or 0.

In moiré interferometry, both beams usually have warped wave
fronts. However, the interpretation is the same—the fringe pattern
recorded on the photographic plate is a contour map of the separation
between wave fronts w_1 and w_2, that is, between wave fronts of equal
phase in the two output beams.

Two-beam interference with coherent light is said to be
nonlocalized. This means that the predictions of Eqs. 2.20 and 2.21

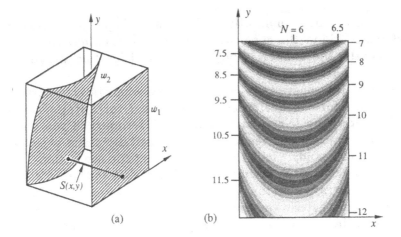

Fig. 2.23   When one or both of the wave fronts is smoothly warped, the interference fringes are continuous curves. The interference pattern is a contour map of the separation $S(x,y)$ between the wave fronts, where fringe orders $N(x,y) = S(x,y)/\lambda$.

remain effective at every point in the space where the two beams coexist. The contrast of interference fringes recorded on a photographic plate is essentially the same at every location in this space.

The concept of walls of interference applies in the three-dimensional space where any pair of coherent light beams coexist, including beams with warped wave fronts. The walls are flat, parallel, and uniformly spaced only for the special case of interference between beams with plane wave fronts. In the general case of warped wave fronts, the walls have irregular shapes. As a consequence of the wave front changes depicted in Fig. 2.2, each wall of interference has a continuous but warped shape, with its warpage systematically different from that of the neighboring wall. A photographic plate that cuts the wall at one location might record the fringe pattern of Fig. 2.23, but at a different location the fringe pattern would be somewhat different. Clearly, as these two wave fronts propagate, the warpage of w_2 changes, the wave front separation $S(x,y)$ changes, and the interference fringe pattern representing $S(x,y)$ changes.

2.3.6 The Camera

A predicament is arising. The fringe pattern is not unique—the pattern changes with the location where it is observed. Where in space must the fringe pattern be recorded?

The fringes are nonlocalized, i.e., the fringes of two-beam interference exhibit the same fringe contrast everywhere in space. Consequently, the answer is not some special location where the fringes are clearest.

Instead, the answer is this: in two-beam interferometric techniques for metrological purposes, including moiré interferometry, the fringe pattern must be recorded in the plane where the information beam (or beams) exit from the object or specimen under study. It is in this plane that the separation $S_{(x,y)}$ between emergent wave fronts is *coordinated with the object points* from which the rays (and thus the wave fronts) emerge. As a warped wave front propagates away from the object, its warpage changes and the x,y coordinates in the wave front no longer corresponds to x,y positions in the object. Thus, to capture the most meaningful interference map of wave front separation $S_{(x,y)}$, the two wave trains (beams) must be photographed by focusing the camera precisely on the plane of the object.

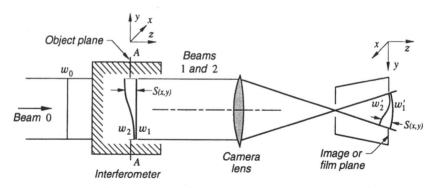

Fig. 2.24   A camera reproduces the wave front separation, $S_{(x,y)}$, in the image plane; $S_{(x,y)}$ is the same at corresponding object and image points, regardless of image magnification.

Figure 2.24 represents a generic interferometer in which AA is the plane of the object under study. The warped wave front emerging from the object is illustrated as w_2. Depending upon the type of interferometer under consideration, the wave front w_1 might not lie in the object space. In that case, the wave front shown in the figure is the virtual image of w_1. (For example, in Fig. 2.27, w_1' is the virtual image of w_1).

Wave front w_2 propagates toward the camera. It is warped; since light propagates normal to the wave front, the rays comprising beam 2 are inclined to the optical axis. However, its warpage is greatly exaggerated; it was noted in Table 2.2 and the related discussion that the largest off-axis angle of these rays is very small. Accordingly, a camera lens that is somewhat *larger than the object* will intercept all the rays and direct them to the image plane.

In Fig. 2.24, wave front w_1 remains plane as its beam propagates to the camera lens. The lens converges the beam to a node, where it diverges to the image plane. The wave front has a spherical surface as it crosses the image plane; it is labeled w_1'. But, what is the shape of w_2' as it crosses the image plane?

The *OPL* between any object point and its image point is constant, regardless of the paths of rays through the lens. In addition, the *OPL* between any pair of wave fronts in a wave train is always constant. It follows, therefore, that separation $S_{(x,y)}$ of wave fronts as they cross the image plane equals their separation as they emerge from the object plane. The shape of w_2' corresponds to the shape of w_2, plus an additional spherical warpage. In the image plane, the original shape of each wave front is modified by the same spherical warpage.

The key feature is the equality of wave front separation $S_{(x,y)}$ at corresponding points in the object and image planes. Thus, a camera reproduces in the image plane the phase relationships of the light that emerges from the object plane. The interference pattern representing $S_{(x,y)}$ is produced faithfully in the image plane when the camera is focused on the object.

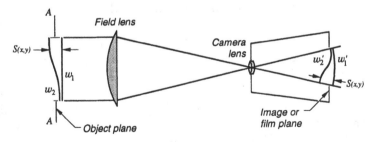

Fig. 2.25   A camera system suitable for large fields.

It is interesting to notice that the lateral magnification of the image does not influence the nature of the interference patterns. The wave fronts that are reconstructed in the image plane can be magnified in the x,y directions, but $S_{(x,y)}$—which lies in the z direction—is independent of magnification.

Figure 2.25 illustrates an alternative camera system that is useful for large fields of view. A large collecting lens, or *field lens*, directs the two interfering beams into a smaller camera lens. Again, the wave front separation $S_{(x,y)}$ is preserved and the interference pattern in the image plane is a faithful map of $S_{(x,y)}$. In practice, the collecting lens should be located as close to the object as possible to minimize the effect of lens aberrations.

Two mutually coherent beams with warped wave fronts produce full-contrast interference fringes wherever they coexist, including the space inside a camera. The finite camera lens aperture does not degrade the fringe contrast (Sec. 2.2.5). An excellent way to assure correct focus of the camera is to scratch a fine mark on the object and critically focus the camera on that mark. Then, the intensity distribution seen in the camera is given by Eq. 2.20 (or 2.21), and the fringe order at each point is given by Eq. 2.25, repeated here,

$$N_{(x,y)} = \frac{S_{(x,y)}}{\lambda}$$

Another practical matter is the camera shutter, which sometimes can vibrate the optical system. A convenient alternative is to locate a separate shutter just ahead of the laser to control the exposure time. If a mechanical shutter is used, it is best to mount it from a separate stand that rests on the floor and does not touch the optical table. Otherwise, an electronic shutter can be used, which does not introduce vibrations.

A question could arise on how to cope with low light levels when using a sheet-film camera. When modest laser power is used and larger image magnifications are demanded, the light coming from the ground-glass camera screen might be too weak for effective viewing. An easy modification circumvents the problem. Replace the ground-glass screen with a clear glass of the same thickness. Adjust a focusing magnifier to focus on a mark on the inside surface of the clear glass. Then, view the moiré pattern through the focusing magnifier. The image will be much brighter, because all the light that reaches the screen is directed into the observer's eye. With a ground-glass screen, the light is spread in all directions and only a small portion reaches the eye.

A video camera is convenient when fringes are not too closely spaced, i.e., when the image contains several pixels per fringe pitch G. Otherwise, a photographic film camera is advantageous.

2.3.7 Two-beam Interferometers

The optical arrangement of Fig. 2.18a is called a *Mach-Zehnder* interferometer. Light that passes through path 1 is called a *reference* beam and light that passes through path 2 is called an active or *information* beam. It has been used extensively for wind tunnel tests

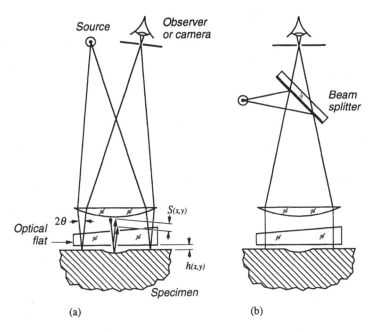

Fig. 2.26   Fizeau interferometer to measure surface topography.   (a) Small oblique incidence and (b) normal incidence arrangements.

of flow over objects. Visualize the shaded box in Fig. 2.18a as the cross-section of a wind tunnel, with glass windows normal to the optical axis on each side of the tunnel. When an object such as an airfoil is mounted in the tunnel, it induces a distribution of increased and decreased pressure in the surrounding air. The pressure is related systematically to the density and refractive index of the air, and thus to the optical path length across the tunnel. Interference between the reference and information beams creates a fringe pattern that provides the pressure distribution.

Surface topography can be measured by various methods, but one that involves very simple apparatus is the *Fizeau* interferometer illustrated in Fig. 2.26a. Suppose it is important to measure the topography of a slightly warped specular surface, e.g., a wear pattern in a polished steel plate. An uncoated optical flat can be placed on the plate. If it is illuminated by a collimated beam of monochromatic light as shown, interference fringes can be observed in the reflected light. About 4% of the incident intensity is reflected by the optical flat, and about 40% by the steel plate (reflectance $\approx$ 0.4). Impure two-beam interference (Eq. 2.21) would produce fringes of adequate contrast, about 73% (Eq. 2.26). The difference of optical path lengths of the reflected beams is

$$S(x,y) = 2h(x,y) \cos \theta$$

where $h(x,y)$ is the thickness of the gap at every x,y point between the specimen and optical flat. Accordingly, the fringe pattern is a contour map of $h(x,y)$ whereby, by Eq. 2.25,

$$h(x,y) = \frac{\lambda}{2 \cos \theta} N(x,y) \qquad (2.27)$$

and where $N(x,y)$ is the fringe order at each point. The contour interval is $\lambda/(2 \cos \theta)$ per fringe order. Angle θ would typically be small, whereby $\cos \theta \approx 1$. An alternative arrangement utilizes a beam splitter, to achieve $\theta = 0$, as illustrated in Fig. 2.26b.

When the optical flat rests on the specimen and h is small, a laboratory light source such as a mercury vapor lamp can be used. The requirement for monochromatic purity is greatly relaxed when $S(x,y)$ is small. Of course, a laser (and beam expander) which provides high monochromatic purity could be used instead. The Fizeau interferometer is not a true two-beam interferometer; multiple reflections occur between the two partially reflective surfaces, but they have relatively low intensities and may be neglected.

Figure 2.27 illustrates an interferometer that is frequently used for measurements of surface topography. It is called a *Twyman-Green* interferometer. The optical arrangement is the same as the well-known *Michelson* interferometer, except a collimated input beam is used in the Twyman-Green whereas an extended (or diffuse) source is used in the Michelson. This arrangement is more complicated than the Fizeau system, but it is especially useful when the reference plate cannot be located near the specimen because of size or shape limitations, environmental limitations, or otherwise.

In this case, an optically flat beam splitter directs half the light to the specimen and the other half to a flat reference mirror. After reflection from the specimen and reference surfaces, the beams meet again at the beam splitter and a portion of each propagates horizontally to be collected by the observer. When focused on the specimen, the interference pattern seen by the observer is the contour map of separation $S(x,y)$ between the warped wave front w_2 and the plane wave front w_1', which is the virtual image of w_1. The warpage of w_2 is twice the warpage of the specimen surface. Thus, the interference pattern is a contour map of the z coordinate of the specimen surface, where the contour interval is $\lambda/2$ per fringe order.

The surface topography can be represented by different patterns. For example, let Fig. 2.28a be the interference pattern of the surface of a body that was plastically deformed by pressing a punch into the

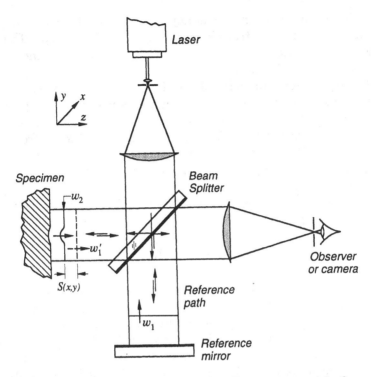

Fig. 2.27   The Twyman-Green interferometer.   The arrangement is the same as a Michelson interferometer, but the Michelson uses an extended source.

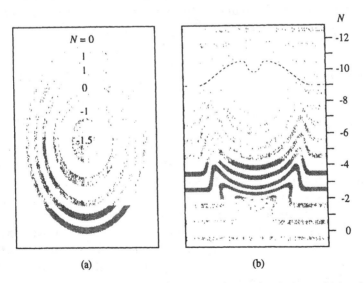

Fig. 2.28   (a) Interference pattern depicting a depression in an anisotropic plate caused by a spherical punch.   (b) Carrier fringes detect the ridge along the boundary of the depression; observe the optical illusion of three-dimensional shape by viewing the pattern at oblique incidence from the top or bottom.

body. If the reference mirror is adjusted to give it a slight tilt, the interference pattern is transformed to that of Fig. 2.28b. The new pattern is the contour map of the separation between w_2 and w_1', but now, w_1' is slightly inclined.

The adjustment introduced carrier fringes (sometimes called *wedge fringes*), which are straight, uniformly spaced fringes when the specimen surface is plane. These carry the added information when the specimen surface is deformed. Carrier fringes frequently provide details that are otherwise ambiguous. In this case, for example, the depression in the body is bounded by a ridge, but that fact cannot be extracted from Fig. 2.28a.

The fringe order N represents the out-of-plane position W of each point on the surface by

$$W_{(x,y)} = \frac{\lambda}{2} N_{(x,y)} \qquad (2.28)$$

where W is measured relative to a datum plane, which is the plane of the virtual image of the reference mirror. In Fig. 2.28a, the datum plane corresponds to the undeformed portion of the specimen surface; in (b) the datum plane is slightly inclined to the surface, with its lowermost edge nearly coincident with the undeformed surface. N and W are negative when the surface is in the $-z$ direction relative to the datum plane.

With the Fizeau interferometer (Fig. 2.26), the reference plane cannot lie inside the specimen. Nevertheless, the fringe patterns can be interpreted in the same way if $h_{(x,y)}$ is interpreted as

$$h_{(x,y)} = W_{(x,y)} + D$$

where D is the distance from the selected datum plane to the optical flat.

It is often necessary to inspect surface topography on a microscopic scale. Various interference microscopes and various microscope accessories are available. One rather simple attachment is suitable for measurement of scratch profiles, cleavage steps, or similar isolated details. Let the specimen illustrated in Fig. 2.29 have a fairly smooth reflective surface with a narrow groove or blemish. If a plane wave front is incident upon the surface, the reflected wave front will have the form depicted by w_2. The microscope attachment utilizes a doubly-refracting crystal to generate a second wave front w_1, which has the same shape as w_2 but is laterally displaced as shown. Interference then provides the contour map of separation

$S(x,y)$ between the wave fronts, which closely approximates the depth contours of the groove or blemish. The method is called *wave front shearing interferometry*, since the wave front is divided into two equal parts and one of these is displaced (or sheared) laterally with respect to the other.

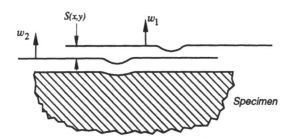

Fig. 2.29   A wave front shearing microscope produces two mutually displaced images of the emergent wave front. Interference reveals the separation, $S(x,y)$, which provides a depth measurement of isolated features.

2.3.8 Derivation of the Interference Equations

The following derivation is for two beams of mutually coherent light. Figure 2.24 is used as a model of the optical systems. Changes of path lengths are incurred in or near the object plane and the interference fringe pattern is revealed in the image plane. The information beam travels from a point source to each x,y point in the object, and then through the camera lens to the corresponding x,y points of the image. The reference beam travels from the same source, through the camera lens and to each x,y point of the image. The notation (x,y) is deleted in the equations, but it is to be understood that the analysis applies to every x,y point.

The two interfering beams have a common plane of polarization, e.g., the yz plane, where their field strengths are A_{y1} and A_{y2}. The common polarization does not change, and the subscript denoting the polarization is deleted in the following equations.

The fringe pattern is governed by the difference of optical path lengths from the source to the image. In the general case the two interfering beams reach each image point after traveling different optical path lengths δ_1 and δ_2 from the source. (Symbols δ and OPL represent the same quantity; δ is used here for convenience.) At the image plane the strengths of the two disturbances are, at a representative point,

$$A_1 = a_1 \cos 2\pi \left(\omega t - \frac{\delta_1}{\lambda} \right)$$

$$A_2 = a_2 \cos 2\pi \left(\omega t - \frac{\delta_2}{\lambda} \right) \tag{2.29}$$

and their intensities are, by Eq. 2.6,

$$I_1 = a_1^2$$

$$I_2 = a_2^2 \tag{2.30}$$

The resultant field strength is

$$A = A_1 + A_2$$

$$= a_1 \cos \left(2\pi \omega t - 2\pi \frac{\delta_1}{\lambda} \right) + a_2 \cos \left(2\pi \omega t - 2\pi \frac{\delta_2}{\lambda} \right)$$

The following trigonometric identities should be recalled (where α and β are unrelated to symbols used elsewhere)

$$\cos (\alpha - \beta) = \cos \alpha \cos \beta + \sin \alpha \sin \beta \tag{2.31a}$$

$$\sin^2 \alpha + \cos^2 \beta = 1 \tag{2.31b}$$

Then, using the identity of Eq. 2.31a,

$$A = a_1 \cos 2\pi \omega t \, \cos 2\pi \frac{\delta_1}{\lambda} + a_1 \sin 2\pi \omega t \, \sin 2\pi \frac{\delta_1}{\lambda}$$

$$+ a_2 \cos 2\pi \omega t \, \cos 2\pi \frac{\delta_2}{\lambda} + a_2 \sin 2\pi \omega t \, \sin 2\pi \frac{\delta_2}{\lambda}$$

Combining terms,

$$A = \left(a_1 \cos 2\pi \frac{\delta_1}{\lambda} + a_2 \cos 2\pi \frac{\delta_2}{\lambda} \right) \cos 2\pi \omega t$$

$$+ \left(a_1 \sin 2\pi \frac{\delta_1}{\lambda} + a_2 \sin 2\pi \frac{\delta_2}{\lambda} \right) \sin 2\pi \omega t \tag{2.32}$$

Since the terms in parentheses are constants for any x, y point, they can be equated to new constants as follows, provided that the equations can be satisfied simultaneously. Let

$$\left(a_1 \cos 2\pi \frac{\delta_1}{\lambda} + a_2 \cos 2\pi \frac{\delta_2}{\lambda} \right) = K \cos \phi \tag{2.33}$$

$$\left(a_1 \sin 2\pi \frac{\delta_1}{\lambda} + a_2 \sin 2\pi \frac{\delta_2}{\lambda} \right) = K \sin \phi \tag{2.34}$$

By squaring and adding these equations, and reducing them by Eqs. 2.31, we obtain

$$K^2 = a_1^2 + a_2^2 + 2a_1a_2 \cos \frac{2\pi}{\lambda} \left(\delta_1 - \delta_2 \right) \tag{2.35}$$

and by division,

$$\tan \phi = \frac{a_1 \sin 2\pi \dfrac{\delta_1}{\lambda} + a_2 \sin 2\pi \dfrac{\delta_2}{\lambda}}{a_1 \cos 2\pi \dfrac{\delta_1}{\lambda} + a_2 \cos 2\pi \dfrac{\delta_2}{\lambda}} \tag{2.36}$$

Thus, Eqs. 2.33 and 2.34 are satisfied simultaneously.

By substituting Eqs. 2.33 and 2.34 into Eq. 2.32,

$$A = K \left(\cos \phi \cos 2\pi\omega t + \sin \phi \sin 2\pi\omega t \right)$$

and by Eq. 2.31a

$$A = K \cos \left(2\pi\omega t - \phi \right) \tag{2.37}$$

The intensity of the combined beams is, by Eq. 2.6 and from Eq. 2.35

$$I = K^2 = a_1^2 + a_2^2 + 2a_1a_2 \cos \frac{2\pi}{\lambda}(\delta_1 - \delta_2) \tag{2.38}$$

Since the difference of optical path lengths is the separation S of the respective wave fronts of equal phase,

$$\delta_1 - \delta_2 = S \tag{2.39}$$

Therefore,

$$I = a_1^2 + a_2^2 + 2a_1a_2 \cos 2\pi \frac{S}{\lambda} \tag{2.40}$$

By substitution of Eqs. 2.30, and remembering that the analysis applies to the whole field,

$$I_{(x,y)} = I_1 + I_2 + 2\sqrt{I_1 I_2} \, \cos 2\pi \frac{S(x,y)}{\lambda} \tag{2.21}$$

which is the intensity equation of impure two-beam interference. If

$$a_1^2 = a_2^2 = a^2 = I_0$$

Eq. 2.40 becomes

$$I_{(x,y)} = 2I_0 \left(1 + \cos 2\pi \frac{S(x,y)}{\lambda} \right) = 4I_0 \cos^2 \pi \frac{S(x,y)}{\lambda} \tag{2.20}$$

which applies to pure two-beam interference; I_0 is the intensity of each of the two interfering beams.

These equations define the intensity distributions graphed in Fig. 2.19. By substituting Eq. 2.25 $(N = S/\lambda)$, they define the intensity distributions of two-beam interference patterns in terms of their fringe orders (Fig. 2.23b). For pure two-beam interference

$$I_{(x,y)} = 2I_0 \left(1 + \cos 2\pi N_{(x,y)}\right) = 4I_0 \cos^2 \pi N_{(x,y)} \qquad (2.41)$$

and for impure two-beam interference

$$I_{(x,y)} = I_1 + I_2 + 2\sqrt{I_1 I_2} \cos 2\pi N_{(x,y)} \qquad (2.42)$$

Complex Notation

The same intensity equations can be derived by use of complex notation. Recall the complex identities

$$i^2 = -1 \qquad (2.43a)$$

$$H = e^{i\theta} = \cos \theta + i \sin \theta \qquad (2.43b)$$

$$\overline{H} = e^{-i\theta} = \cos \theta - i \sin \theta \qquad (2.43c)$$

where $\overline{H}$ is the complex conjugate of H (i.e., its complex equivalent). Thus,

$$H + \overline{H} = e^{i\theta} + e^{-i\theta} = 2\cos \theta \qquad (2.43d)$$

A harmonic wave can be expressed as the real part of a complex quantity

$$Z = ae^{i\theta}$$

Its intensity $I = a^2$ can be determined by

$$I = Z\overline{Z} \qquad (2.44)$$

since

$$Z\overline{Z} = ae^{i\theta} \cdot ae^{-i\theta} = a^2 e^0 = a^2$$

which is a real quantity.

Accordingly, the two waves represented by Eqs. 2.29 can be expressed as the real parts of the following complex quantities

$$Z_1 = a_1 e^{i2\pi(\omega t - \delta_1/\lambda)} = a_1 \exp\left[i2\pi\left(\omega t - \frac{\delta_1}{\lambda}\right)\right]$$

$$Z_2 = a_2 \exp\left[i2\pi\left(\omega t - \frac{\delta_2}{\lambda}\right)\right] \qquad (2.45)$$

where $A_1 = \text{Re }[Z_1]$ and $A_2 = \text{Re }[Z_2]$. Their complex conjugates are

$$\overline{Z_1} = a_1 \exp\left[-i2\pi\left(\omega t - \frac{\delta_1}{\lambda}\right)\right]$$

$$\overline{Z_2} = a_2 \exp\left[-i2\pi\left(\omega t - \frac{\delta_2}{\lambda}\right)\right]$$

(2.46)

The resultant field strength is

$$A = A_1 + A_2$$

which is the real part of

$$Z = Z_1 + Z_2$$

$$= a_1 \exp\left[i2\pi\left(\omega t - \frac{\delta_1}{\lambda}\right)\right] + a_2 \exp\left[i2\pi\left(\omega t - \frac{\delta_2}{\lambda}\right)\right]$$

(2.47)

The intensity I is the product of Z and its complex conjugate $\overline{Z}$,

$$I = Z\overline{Z}$$

$$= \left\{a_1 \exp\left[i2\pi\left(\omega t - \frac{\delta_1}{\lambda}\right)\right] + a_2 \exp\left[i2\pi\left(\omega t - \frac{\delta_2}{\lambda}\right)\right]\right\}$$

$$\bullet \left\{a_1 \exp\left[-i2\pi\left(\omega t - \frac{\delta_1}{\lambda}\right)\right] + a_2 \exp\left[-i2\pi\left(\omega t - \frac{\delta_2}{\lambda}\right)\right]\right\}$$

(2.48)

which reduces to

$$I = a_1^2 + a_2^2 + a_1 a_2\left\{\exp\left[i\frac{2\pi}{\lambda}(\delta_1 - \delta_2)\right] + \exp\left[-i\frac{2\pi}{\lambda}(\delta_1 - \delta_2)\right]\right\}$$

Since

$$\delta_1 - \delta_2 = S$$

we obtain

$$I = a_1^2 + a_2^2 + a_1 a_2\left[e^{i2\pi S/\lambda} + e^{-i2\pi S/\lambda}\right]$$

and by Eq. 2.43d

$$I = a_1^2 + a_2^2 + 2a_1 a_2 \cos 2\pi\frac{S}{\lambda}$$

(2.49)

The individual intensities are

$$I_1 = Z_1 \overline{Z_1} = a_1^2$$

$$I_2 = Z_2 \overline{Z_2} = a_2^2$$

(2.50)

Thus, the equation of impure two-beam interference is derived as

$$I_{(x,y)} = I_1 + I_2 + 2\sqrt{I_1 I_2} \, \cos 2\pi \frac{S_{(x,y)}}{\lambda} \qquad (2.21)$$

2.3.9 Interference with Broad Spectrum Light and White Light

The interference phenomena also occurs with light that is not monochromatic, but useful data is obtained only when the path difference $S_{(x,y)}$ of interfering beams is a small number of wavelengths. Figure 2.30 illustrates what happens when an interferometer is used with light of a broad continuous spectrum. As indicated in (a), interference occurs for each individual wavelenth in the spectrum, producing a simple harmonic intensity distribution. For any given value of path difference S, the intensity varies with the wavelength; when S is large enough, the curves overlap and zones of constructive interference for one wavelength coincide with zones of destructive interference for another wavelength.

The resultant intensity distribution is obtained by *scalar* summation of the intensities contributed by each wavelength. That distribution is shown in Fig. 2.30b. A few cycles of constructive and destructive interference fringes of high contrast occur, where constructive interference corresponds closely to integral values of fringe order N_0 and destructive interference to half orders.

In the case illustrated here, the light source was assumed to emit a spectral band of constant intensity in the wavelength range of $\lambda_0 \pm$ 5%. Under these circumstances, only a few fringes are useful for most metrological purposes. As the spectral band is narrowed, good fringe contrast is maintained for increasingly large fringe orders.

The contrast or visibility at which interference fringes cease to be useful is a subjective issue. For most practical purposes the contrast is sufficiently good in regions where

$$N_1 = N_2 + \frac{1}{2} \qquad (2.51)$$

where N_1 and N_2 are fringe orders corresponding to wavelengths λ_1 and λ_2, respectively, at the half-intensity points of the spectrum (Fig. 2.7). We will use this criterion as a limit for interference fringes of good contrast. Fringe orders are related to the wave front separation S by

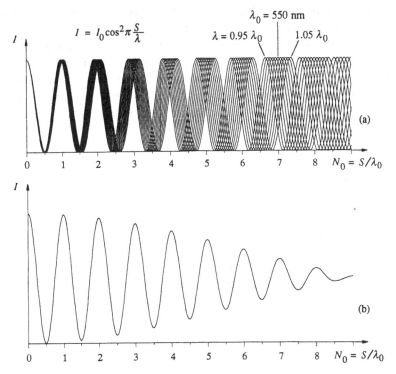

Fig. 2.30   Interference with broad spectrum light: $\lambda = 550$ nm $\pm$ 5%. (a) Intensity distribution for specific wavelengths. (b) Total intensity by scalar summation of intensities from all wavelengths in this range.

$$S \;=\; N_0\lambda_0 \;=\; N_1\lambda_1 \;=\; N_2\lambda_2$$

By Eq. 2.51, and approximating $\lambda_1\lambda_2$ by $\lambda_0{}^2$, we obtain

$$N_{0(max)} \;=\; \frac{\lambda_0}{2\left(\lambda_2 - \lambda_1\right)} \tag{2.52}$$

Thus, the fringe order at which good contrast ceases (by this criterion) can be calculated for different light sources. For high pressure mercury vapor lamps ($\lambda_0 = 546$ nm; $\lambda_2 - \lambda_1 \approx 5$ nm), N_0 is approximately 55; for medium pressure, $N_0 \approx 500$.

Equation 2.51 is an arbitrary criterion for good fringe contrast, but it is a suitable choice. Figure 2.30 can be used to corroborate the choice. For the conditions of Fig. 2.30, Eq. 2.52 yields $N_0 = 5$. Observe that at $N_0 = 5$ the curves for the shortest and longest wavelengths experience 5.25 and 4.75 cycles, respectively, which means $N_1 = 5.25$ and $N_2 = 4.75$. Equation 2.51 is satisfied. The contrast at $N_0 = 5$ is determined from Fig. 2.30b as 72%. We realize, furthermore, that the analysis corresponds to a rectangular spectral distribution instead of the peaked function of Fig. 2.7. A peaked function yields an appreciably higher contrast, since the intensity is concentrated near the central wavelength. Consequently, the criterion corresponds to a fringe contrast appreciably higher than 72%, the exact figure being dependent upon the shape of the spectral intensity curve of the light source.

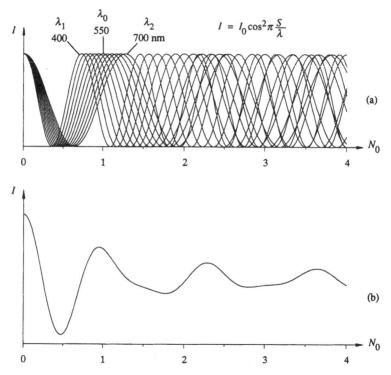

Fig. 2.31 Interference with white light. (a) Intensity contributions for several wavelengths between 400 and 700 nm. (b) Distribution of total intensity; this is the distribution perceived by a color-blind receiver.

The visible spectrum spans the wavelength range from 400 to 700 nm. Interference with *white* light, i.e., the entire visible spectrum, produces the intensity distributions illustrated in Fig. 2.31. In (a), the intensity versus fringe order curves are drawn for the limiting wavelengths and several intermediate wavelengths. The resultant intensity distribution is the scalar summation of all the individual intensities, which is represented by the intensity curve in (b). Only one interference cycle is visible before the intensity fluctuation becomes irregular.

Figure 2.31 illustrates the fringes as they would be perceived by a colorblind receiver, like a black and white video camera or an orthochromatic photographic film. However, the human eye and other color sensitive receivers discriminate between intensities produced by different wavelengths. A color sensitive receiver would see white light where $S = N = 0$, since constructive interference occurs there for all wavelengths. Where $S \sim 650$ nm, red light experiences its second cycle of constructive interference, while other colors contribute relatively little intensity. At about $S \sim 860$ nm, where yellow experiences its second cycle of destructive interference, the most luminous part of the spectrum is extinguished; however, red and blue combine here and form a narrow band of purple, called the *tint of passage*. For slightly higher values it becomes predominately blue. With higher values, predominate colors become green, orange, and red. The progression continues until the colors overlap so thoroughly that they merge into white light. About eight cycles of red fringes can usually be detected, the color becoming pale pink in the later cycles. A full-field interference pattern would exhibit continuous curved bands, as in Fig. 2.23, but they would appear as bands of different colors instead of bands of high and low intensities.

White light fringes have the special property that the zero order fringe is distinguishable from all the others, since it has no color. They can be used, for example, to achieve precision alignment of two mirror surfaces, as illustrated in Fig. 2.32. In the alignment procedure, the optical path lengths to each mirror are adjusted to equal the reference path length, whereupon the zero order fringe

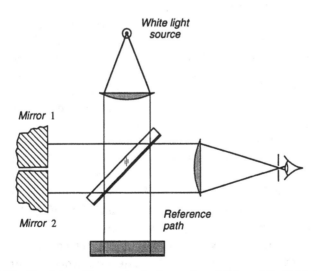

Fig. 2.32   Precision alignment of mirrors 1 and 2 by white light fringes.

appears across both mirrors. Then the reference mirror is tilted slightly to form a few wedge fringes or carrier fringes. Mirrors 1 and 2 are coplanar when the white light wedge fringes cross the boundary between them with no discontinuity of fringe position or angle.

Notice that the reference mirror is shown in Fig. 2.32 as a rear surface mirror. True white light fringes are produced when equal portions of refractive materials lie in the information and reference paths. This follows from the requirement for equal information and reference path lengths for every wavelength of the white light. Since refractive materials exhibit dispersion, i.e., variations of refractive index with wavelength (Sec. 2.1.1), the requirement is satisfied by equal amounts of refractive materials in the two paths.

2.4 Superposition of Incoherent Light Waves

Persistent optical interference—as distinguished from instantaneous effects—requires mutually coherent light. As implied in the derivation of the interference equations, the wave trains that interfere must have the same frequency and the same polarization. To achieve this, the two wave trains must come from the same source. Why? The answer lies in the nature of light. We remember that light is emitted as numerous wave trains of finite lengths. The individual wave trains are emitted randomly in time, so they are not synchronized; instead their phases vary randomly. (This applies for lasers, too. If the wave trains are 3 meters long, a single train passes a fixed point in only 10^{-8} seconds, which is too brief for persistence. The following wave train is not synchronized with the proceeding one, so again their phase relationship is random.) As a result of their random phases, the wave trains from two different sources interfere randomly. Some pairs of wave trains produce constructive interference while other pairs produce intermediate and destructive interference. Any interference effect that is present in a single pair of wave trains is washed out.

On the other hand, when light from a single source is divided into two beams, each original wave train is divided into two wave trains that are synchronized with respect to each other. Consider a beam from a source of monochromatic light. Let it be divided into two beams, let them experience a relative phase shift of $2\pi S/\lambda$, and let them subsequently be reunited into a single beam. Then, every original wave train is divided into a mutually coherent pair that experiences the same $2\pi S/\lambda$ shift. The equations of interference prescribe the same interference effect for all the mutually coherent pairs of wave trains. The output intensity is the sum of all the intensity contributions from these pairs. It is a function of $2\pi S/\lambda$ and it is persistent.

Thus, for mutually coherent light, i.e., light from a single source, the random phase of each original wave train relative to the others is irrelevant. Aside from the physical size of the source, the only limitations for interference are that the shift S must be small compared to the length of the wave trains and the light must be sufficiently monochromatic.

In the previous section, *Interference with Broad Spectrum Light*, we considered a case where some components of the light were mutually coherent while others were not. When the broad spectrum light passed through an interferometer, the reunited wave trains of each unique frequency produced interference. Then their intensities at each point were summed arithmetically.

The arithmetical summation might seem arbitrary, since we know that incoherent waves (of equal polarization) interact on an instantaneous basis. However, the rationale can be appreciated from the following argument.

Imagine two beams of different frequencies (and wavelengths) to occupy the same space as they propagate through a fixed point z_0. Consider only two wave trains of equal amplitudes that travel together through z_0, and let them have a common plane of polarization. Their electric field strengths at z_0 are graphed as functions of time in Fig. 2.33, (a) and (b). Their frequencies are denoted by ω_1 and ω_2. At any instant, the field strength at z_0 is their sum, which is graphed in (c). At some instants, the field strengths have the same sign and reinforce each other; at other times they have opposite signs and cancel each other. The result is the beat-like oscillation of field strength at z_0. The instantaneous intensity can be defined as the square of the amplitude of the envelope of $A_1 + A_2$; it is graphed in (d). Thus, the intensity of z_0 varies with time with the beat frequency of A_1 and A_2.

The instantaneous intensity varies from 0 to $4a^2$, and its average value over many cycles is $2a^2$. Normally,* the beat frequency is so high that *measurements* give the average intensity, $2a^2$. We note that the intensities of the individual beams are each a^2 and their sum is the same as the average or measured intensity. Thus, the intensity of

---

* So-called heterodyne methods extract phase information from the beat patterns. The heterodyne techniques use extremely small $\omega_1 - \omega_2$ (produced by Doppler shifting from a coherent beam) to obtain low frequency beats. Photoelectric detectors are used to sense the relative phase of beat patterns at various individual points in the field of view.

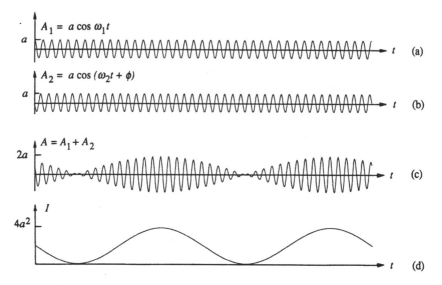

$A_1 = a \cos \omega_1 t$

a

t (a)

$A_2 = a \cos (\omega_2 t + \phi)$

a

t (b)

$A = A_1 + A_2$

$2a$

t (c)

I

$4a^2$

t (d)

Fig. 2.33 Superposition of two beams of unequal frequencies and wavelengths. The amplitudes and intensity at a fixed point z_0 vary with time.

(any number of) superimposed beams of unequal frequencies is equal to the scalar sum of the intensities of the individual beams.

The rule is more general. *The intensity of any number of superimposed incoherent beams is equal to the scalar sum of the individual intensities.*

Coherence involves polarization, too. Equal polarizations are required for persistent constructive and destructive interference. The field strength of a polarized beam is a vector quantity. In Fig. 2.34 vectors a_1 and a_2 represent two plane polarized beams of equal frequency that travel together in space. Vector a_2 can be resolved into its x and y components, as shown. Then beams with field strengths a_1 and a_{2y} combine by optical interference. The remaining light with amplitude a_{2x} cannot interfere; instead, it contributes a background intensity $(a_{2x})^2$ that adds arithmetically to the interference pattern.

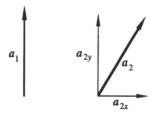

a_1

a_{2y}

a_2

a_{2x}

Fig. 2.34 Vector representation of polarized beams.

Again, incoherent superposition of two beams requires scalar addition of their intensities; the result is independent of the phase of the two beams. On the other hand, coherent superposition is controlled by the equations of optical interference, which prescribe constructive and destructive interference according to the relative phase of the two beams. In Fig. 2.34, a_{2y} is combined with a_1 by coherent superposition and a_{2x} is combined with that result by incoherent superposition.

2.5 Diffraction Gratings and Diffraction

Moiré interferometry uses diffraction gratings. The method depends on diffraction of light as well as interference. Diffraction gratings are illustrated in Fig. 2.35, where (a) indicates that a grating is a surface with regularly spaced bars or furrows. In the case of transmission gratings, the incident light passes through the grating; the incident and diffracted beams appear on opposite sides of the grating. With reflection gratings, they are on the same side and the substrate may be opaque. *Amplitude gratings* consist of opaque bars and transparent spaces, or else reflective bars and nonreflective spaces. These are usually low frequency gratings (e.g., 10 to 50 lines/mm or 250 to 1250 lines/in.) and they are used for geometric moiré. *Phase gratings* have furrowed or corrugated surfaces, with either symmetrical or nonsymmetrical furrow profiles. They are usually high frequency gratings (e.g., 300 to 2400 lines/mm or 7500 to 60,000 lines/in.) and they are used for moiré interferometry. The period or *pitch* of a grating is the distance g between corresponding features of adjacent bars or furrows. A *cross-line* grating has the same repeating arrangement of bars or furrows in two orthogonal directions. The SEM image of a cross-line phase grating shows that the surface consists of an orthogonal array of hills. In fact, a uniformly spaced arrangement of features of any shape produces diffraction and is considered to be a diffraction grating.

Frequency f of a grating is the number of bars or furrows per unit length, usually expressed as *lines* per millimeter, meaning repetitions or cycles per millimeter. Frequency and pitch are related by

$$f = \frac{1}{g} \qquad (2.53)$$

This notation applies for high frequency as well as low frequency gratings.

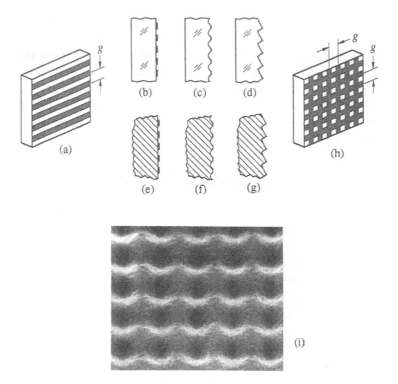

Fig. 2.35   Diffraction gratings are comprised of regularly spaced bars or furrows. Cross-sectional views (b), (c), and (d) illustrate transmission gratings, while (e), (f), and (g) illustrate reflection gratings; (b) and (e) represent bar and space gratings called amplitude gratings; (c) and (f) represent symmetrical phase gratings; (d) and (g) represent blazed phase gratings; (h) illustrates a cross-line grating, which can be either amplitude or phase type. (i) Scanning electron microscope image of the surface of a 1200 lines/mm cross-line symmetrical phase grating; courtesy of D. Mollenhauer (Wright-Patterson AFB) and J. P. Williams (UES, Inc.)

A grating divides every incident wave train into a multiplicity of wave trains of smaller intensities, and it causes these wave trains to emerge in certain preferred directions. This is indicated in Fig. 2.36a and (b) for transmission and reflection gratings, respectively. The incident and diffracted beams are represented by rays. When a parallel beam is incident at angle α, the grating divides it into a series of parallel beams which emerge at preferred angles: ---, β_{-1}, β_0, β_1, β_2, ---. These beams are called *diffraction orders* and are numbered in sequence beginning with the zero order. The zero

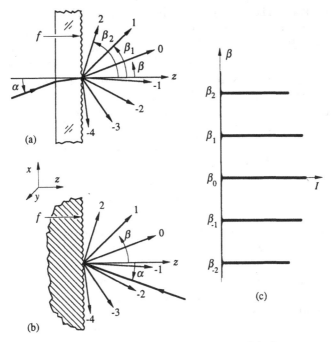

Fig. 2.36   A grating divides an incident beam of light into a number of diffracted beams, where the diffracted beams lie in a systematic array of preferred directions. The beams are represented by rays. (a) Transmission grating; (b) reflection grating; (c) the intensity of light diffracted into each preferred direction depends upon furrow shape (for phase gratings) or ratio of bar-to-space widths (for amplitude gratings).

diffraction order is an extension of the incident beam for transmission gratings and the mirror reflection of the incident beam for reflection gratings. In the zero order, a grating acts either as a window or a mirror.

The angles of diffraction are determined by the grating equation

$$\sin \beta_m = \sin \alpha + m\lambda f \qquad (2.54)$$

where m is the diffraction order, f is the grating frequency, α is the angle of incidence and β_m is the angle of the mth diffraction order.

Figure 2.36c is a graph of intensity versus diffraction angle β_m for the emergent light. Light is diffracted into beams with preferred directions, with very little light emerging in other directions. For high quality gratings the angular spread $\Delta\beta$ within a diffraction order is exceedingly narrow and can be assumed to be zero for our purposes.

Sign Convention: Diffraction orders whose angles are counterclockwise with respect to the *zero order* are considered positive for diffracted beams that propagate in a direction having a positive z component. Clockwise and counterclockwise are assessed by viewing from the positive direction of the axis parallel to the grating lines (from the $+y$ direction in Figs. 2.36–2.39). For diffracted beams that propagate in a direction having a negative z component, positive diffraction orders are clockwise with respect to the zero order. The angles α and β are measured from the normal to the grating surface. Counterclockwise angles are positive when the beams have a component in the $+z$ direction; clockwise angles are positive for beams with a component in the $-z$ direction.

Figure 2.36 illustrates diffraction in two dimensions, where the plane of incidence (i.e., the plane containing the incident ray and the grating normal) is perpendicular to the grating lines. This special case is violated if the grating is rotated by an angle ψ about the z axis, and three-dimensional diffraction must be considered. The general case is illustrated in Fig. 2.37, where A and B_m represent unit vectors that define the directions of incident and diffracted beams,

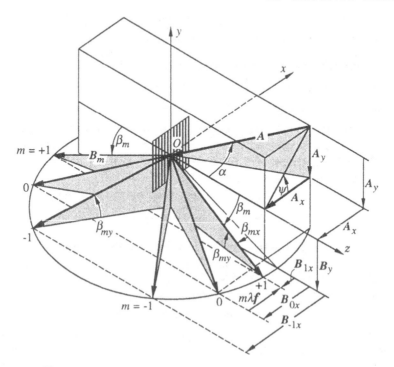

Fig. 2.37 The general case of three-dimensional diffraction is illustrated, where A and B_m are unit vectors that define the directions of incident and diffracted rays (beams), respectively.

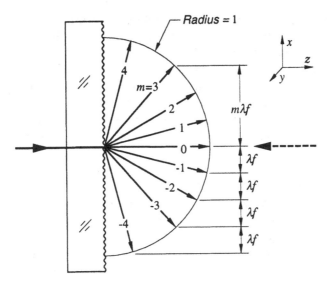

Fig. 2.38   The angles of diffraction vary systematically with fixed increments λf.

respectively.[1,2] For generality, both the reflected system of diffracted beams (in the +z space) and the transmitted system (in the −z space) are shown here. The x,y axes are assumed to rotate with the grating.

Consider the projections of unit vectors $\boldsymbol{A}$ and $\boldsymbol{B}_m$ that lie parallel to the xy plane. Their x and y components in that plane are $\boldsymbol{A}_x$, $\boldsymbol{A}_y$, $\boldsymbol{B}_{mx}$ and $\boldsymbol{B}_y$, respectively, as shown in the figure. Since $\boldsymbol{A}$ and $\boldsymbol{B}_m$ are unit vectors, their x,y components fully define the directions of incident and diffracted light.

In terms of these vector components, the three dimensional grating equations are very simple. They are

$$\boldsymbol{B}_{mx} = \boldsymbol{A}_x + m\lambda f$$
$$\boldsymbol{B}_y = \boldsymbol{A}_y$$

(2.55)

The frequency f of the grating is a vector quantity, too, as defined later in Sec. 3.1.6. The vector $m\lambda f$ is shown in the figure for $m = +1$ (on reflection side).

The equivalent scalar relationships can be expressed if the angle of diffraction β_m is divided into two components, β_{mx} and β_{my}; let β_{mx} be defined as the angle between the unit vector $\boldsymbol{B}_m$ and the yz plane and let β_{my} be the angle between $\boldsymbol{B}_m$ and the xz plane. Angle β_{my} is the same for all diffraction orders. The scalar relationships are

$$\sin \beta_{mx} = \sin \alpha \cos \psi + m\lambda f$$
$$\sin \beta_{my} = \sin \alpha \sin \psi$$

(2.56)

When the angle of in-plane rotation ψ of the grating with respect to the plane of incidence is zero, these equations reduce to the two-dimensional form, Eq. 2.54.

The diffraction angles illustrated in Fig. 2.37 can be demonstrated easily by a desk-top experiment. Place a grating (transmission or reflection type) upright on a white sheet of paper and illuminate it obliquely by an unexpanded laser beam. A small helium-neon laser that is held by hand is suitable. Narrow beams of diffracted light will intercept the paper as a set of bright dots that lie on a circle.

Notice in Fig. 2.37 that the x component of the distance between the tips of neighboring diffraction vectors is a constant, namely λf. Of course, this characteristic is implicit in the diffraction equations, too. Figure 2.38 emphasizes the point for a less complicated case, where a grating is illuminated at normal incidence. The diffracted rays are unit vectors (length = 1). The angles between neighboring diffraction orders are not constant. Instead, they vary systematically whereby the dimension λf is constant.

The angle of diffraction β_{mx} cannot exceed 90°. Equations 2.54 and 2.56 prescribe that the number of diffraction orders that emerge within the range ± 90° is a large number when f is small (i.e., for a coarse grating), and conversely, m_{max} is a small number when f is large. With normal incidence ($\alpha = 0$), the maximum diffraction order is the whole number calculated by $m_{max} = 1/(\lambda f)$. The number of diffraction orders in the 180° arc in the transmission or reflection space is given in Table 2.4 for various grating frequencies; normal incidence ($\alpha = 0$) and a red helium-neon laser ($\lambda = 632.8$ nm) are assumed. For the larger grating frequencies, only three orders exist: $m = -1, 0, +1$.

The angle between neighboring diffraction orders is small for a coarse grating and it is large for a fine grating. Table 2.4 also lists the angle of first-order diffraction for several grating frequencies.

A special case of interest is illustrated in Fig. 2.39. It is where the zeroth diffraction order and its neighbor, the −1 order, are symmetrical with respect to the grating normal. If the angle of incidence is α, we have $\beta_0 = \alpha$ and $\beta_{-1} = -\alpha$. Equation 2.54 reduces to

$$\sin \alpha = \frac{\lambda}{2} f \qquad (2.57)$$

This defines the angle of incidence α to achieve the special condition of symmetry.

When α is large enough (Fig. 2.39b), only two diffraction orders can exist. The others would be "over the horizon" (diffraction angles > 90° and they cannot exist). Under what conditions do only two

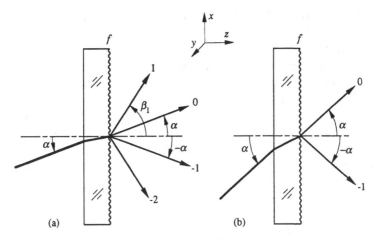

Fig. 2.39 The zero and −1 diffraction orders have symmetrical directions when
sin α = λf/2. When α is large, only two diffraction orders exist.

symmetrical diffractions emerge from the grating? The lower limit
is determined by setting $\beta_1 = 90°$ (Fig. 2.39a). By substituting this and
Eq. 2.57 into Eq. 2.54, we find $f = 2/(3\lambda)$, sin $\alpha = 1/3$ and $\alpha = 19.5°$.
Curiously, the limiting angle is independent of λ.

The upper limit is approached as α approaches 90°, and by Eq.
2.57, as f approaches $2/\lambda$. Thus, only two symmetrical orders emerge
when

$$\frac{2}{3\lambda} \leq f < \frac{2}{\lambda} \tag{2.58}$$

or

$$\frac{3}{2}\lambda \geq g > \frac{1}{2}\lambda \tag{2.58a}$$

The frequency range is about 1000 to 3000 lines/mm (25,000 to 75,000
lines/in.) for red light and 1300 to 4000 lines/mm (33,000 to 100,000
lines/in.) for blue-green light.

The grating equations prescribe the directions of the diffracted
beams, but not their intensities. For phase gratings, the contour of
the furrows determines the distribution of intensities of the diffracted
beams. Most of the energy can be concentrated into a given
diffraction order by prismatic furrows, as illustrated in Fig. 2.40 for
transmission and reflection gratings. When the angle of refraction
determined by Snell's law (Eq. 2.14) coincides with the angle of
diffraction (Eq. 2.54) of a given order, the beam of that diffraction
order is strongly favored. For reflection gratings, the favored

diffraction is the order that emerges in the direction of mirror reflection (Eq. 2.11).

These are called blazed gratings. They are made by precision ruling engines that form the furrows with a diamond stylus. The diamond burnishes the groove (by plastic deformation) into a soft metal. This forms a master grating. Submasters are made on glass blanks by replicating the surface contours with a plastic resin, usually an epoxy. For reflection gratings these are overcoated with evaporated aluminum or another metal. When excellent prismatic furrows are achieved, blazed gratings can direct nearly 100% of the light into the desired diffraction order.

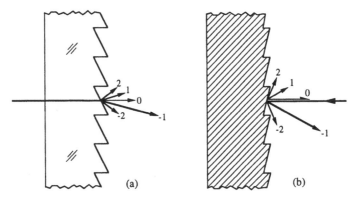

Fig. 2.40   Blazed gratings. The intensity of a specific diffraction order is maximized when the angle of diffraction coincides with (a) the angle of refraction, or (b) the angle of reflection.

Gratings with sinusoidal profiles distribute the light into various orders. In the special case of Fig. 2.39b, however, where only two diffraction orders emerge, all the energy is concentrated in the two orders and blazed gratings are not superior.[3] The ratio of intensities in the two orders can be controlled by the depth of the sinusoidal grooves.

Gratings with sinusoidal profiles are made by exposing a coating of photosensitive material to the virtual grating created by two intersecting beams of coherent light. Techniques for producing these gratings are discussed in Sec. 4.9.2. These gratings are called *holographic gratings*, because holograms are made in a similar way. Blazed gratings can be made by the so-called holographic techniques, too.

2.5.1 Wave Fronts and Optical Path Lengths

Huygen's construction gives useful insight into the formation of diffracted beams. Figure 2.41a shows a grating, illuminated at normal incidence, with Huygen's wavelets of equal phase radiating from each opening. The wavelet separation is one wavelength. A series of envelopes drawn as tangents to the wavelets represent wave fronts of the diffracted beams. The angles of diffraction are calculated in Fig. 2.41b, which corroborates Eq. 2.54. The construction is consistent with experimental evidence that beams propagate and carry energy in these directions.

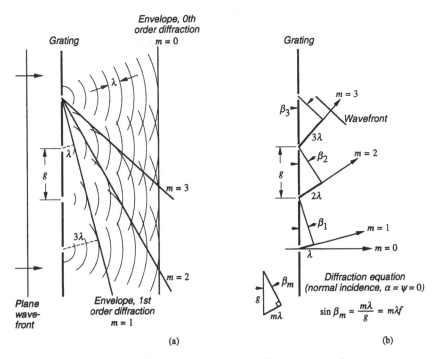

(a)   (b)

Fig. 2.41   Directions of diffraction orders from Huygen's wavelet construction.

Figure 2.41a illustrates graphically that the optical path length of a diffracted beam changes by $m\lambda$ in each increment g (pitch) of the grating, where m is an integer. As before, the waves in a diffracted beam all oscillate in synchronism, so the conceptual model of the wave train is retained. The diffracted wave train is still described by its electric field strength as

$$A \;=\; a \, \cos 2\pi \omega t \qquad\qquad (2.4)$$

where t is the time since an arbitrary starting time t_0 during the passage of the diffracted wave train. The phase of the disturbance remains defined by $2\pi\omega t$ and the wave front remains described as the locus of points where the phase is constant (Eq. 2.5).

Nevertheless, a beam of light that is bent (diffracted) by a grating has different characteristics than one bent (refracted) by a prism. The difference is illustrated in Fig. 2.42. In (a), the optical path lengths of all rays between the incident and emergent wave fronts are equal (Sec. 2.1.1), since the increased OPLs within the prism compensates for the decreased mechanical path lengths. In (b), however, the surface of constant optical path lengths does not coincide with the wave front. Wave fronts are no longer surfaces of equal optical path lengths from the source. They remain surfaces perpendicular to all the rays in the diffracted beam, and *within* the diffracted beam the optical path lengths between any pair of wave fronts is a constant.

Diffraction by two equal gratings, as in Fig. 2.42c, can compensate for the change of path lengths. Here, the OPL between w_0 and w_2 is constant for all rays. This compensation scheme is utilized in the design of achromatic moiré interferometers (Sec. 4.7.8).

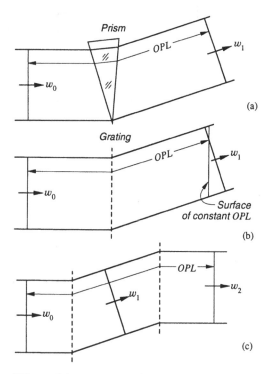

Fig. 2.42   For diffracted beams, wave fronts do not coincide with surfaces of constant optical path lengths.

2.5.2 Equivalency and Virtual Gratings

Equation 2.24 relates the fringe gradient (or frequency of fringes F), formed when two coherent beams intersect, to the half-angle θ. Equation 2.57 relates the angle of emergence α, of the zero and -1 diffraction orders from a grating, to its frequency f. The equations are the same. If we let $\theta = \alpha$, we find $F = f$.

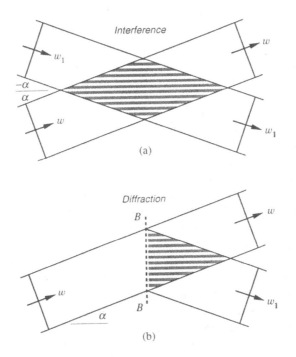

Fig. 2.43   Virtual gratings formed (a) by two coherent beams and (b) by a real grating.

There exists a kinship—actually an equivalency—between fringe patterns and gratings. Figure 2.43 illustrates the relationship. In the volume of space where two coherent beams coexist, a steady-state formation of walls of constructive and destructive interference is generated; their separation is g. On the other hand, a real grating of the same pitch produces two coherent beams that coexist in space and they create the same walls of interference in their region of intersection.

Remember the photographic plate BB exposed to the interference pattern in Fig. 2.21. The exposed and developed plate is a pattern of parallel, equally spaced bars governed by Eq. 2.24; it is a diffraction grating. If this diffraction grating is illuminated at a suitable angle

$\alpha = \theta$, it will generate zero and -1 order beams equal to those that originally formed the grating. This is an equivalency relationship.

The three-dimensional zone of regularly spaced walls of interference represented in Fig. 2.43 will be called a *virtual grating*. Regardless of how it is created, this light possesses the same properties as light emerging from a real grating. The significant concept is that a virtual grating can be formed either by a real grating or by intersection of two coherent beams.

The equivalency is a powerful concept.* Diffraction grating lines and regularly spaced interference fringes can be interchangeable entities. In moiré techniques, a real reference grating can be replaced by a virtual grating.

2.5.3 Self-imaging of Gratings

A fascinating effect of diffraction, easily seen with coarse amplitude gratings, is the grating self-imaging phenomenon. When a grating of frequency f is illuminated by monochromatic collimated light, the grating divides the light into a multiplicity of diffracted, collimated beams. These mutually coherent diffracted beams coexist in space and they combine by optical interference. At preferred distances from the grating, the interference pattern is that of alternating dark and bright bars, repeating with the same frequency as the grating. These are virtual gratings, formed in this case by multiple-beam interference. The virtual gratings are good images of the real amplitude grating, hence the name *self-imaging*.

Figure 2.44 illustrates the phenomenon. The grating images occur at successive distances z from the grating, given by

$$z = n D_T \qquad n = 1, 2, 3, \cdots$$

$$D_T = \frac{g^2}{\lambda} = \frac{1}{\lambda f^2} \qquad (2.59)$$

where g and f are the pitch and frequency, respectively, of the grating and virtual gratings. Numerous successive images are produced without significant degradation. Virtual gratings of lower contrast are visible near these preferred planes; the contrast improves and becomes excellent at the preferred distances. Midway between preferred planes, the contrast is zero.

The phenomenon differs from that of Fig. 2.43, which involves *two-beam* interference. Self-imaging at discrete distances is produced by

---

* This equivalency is the basis of holography.

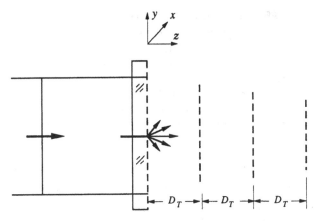

Fig. 2.44   Virtual images of a grating are produced at regularly spaced distances from the grating.

multiple-beam interference.   Numerous diffraction orders must emerge from the grating and therefore coarse *bar-and-space* gratings are effective. Table 2.4 shows that a grating of 10 lines/mm produces hundreds of diffraction orders, which can interact systematically in multiple-beam interference.

The interval D_T is called the Talbot distance,[4] after H. Talbot, who reported the phenomenon in 1836; it has also been called the preferred distance[5] and the self-image or Fourier distance.[6] Numerical values are given in Table 2.5. In the case of shadow moiré (Sec. 3.3.1), the reference grating should be placed within about $0.05D_T$ from the object to obtain clear shadows. Alternatively, it could be placed at a distance $0.95D_T$ to $1.05D_T$, but then, collimated monochromatic light should be used.

TABLE 2.4 Diffraction by Gratings of Various Frequencies

f lines / mm	f lines / in. (approximate)	Total no. of diffraction orders	First order diffraction angle, β_1
10	250	317	0.36°
40	1000	79	1.45°
100	2,500	31	3.63°
1000	25,000	3	39.3°
1,200	30,000	3	49.4°
1,500	38,000	3	71.7°

normal incidence ($\alpha = \psi = 0$), $\lambda = 632.8$ nm

TABLE 2.5 Talbot Image Distances

f, lines/mm	40	10	1
D_T	1 mm	1.6 cm	1.6 m
f, lines/in.	1000	250	25
D_T	0.04 in.	0.64 in.	5.4 ft.

for λ = 632.8 nm

The Talbot effect has been used in a moiré method called *deflectometry* by Kafri and coworkers[7] to measure wave front warpage in a large variety of applications, ranging from combustion studies to testing of optical elements.

2.6 References

1. A. Livnat and D. Post, "The Governing Equations for Moiré Interferometry and Their Identity to Equations of Geometrical Moiré," *Experimental Mechanics*, Vol. 25, No. 4, pp. 360-366 (1985).

2. G. H. Spencer and M. V. R. K. Murty, "General Ray-Tracing Procedure," *J. Optical Society of America*, Vol. 52, No. 6, pp. 672-678 (1962).

3. M. C. Hutley, *Diffraction Gratings*, Academic Press, New York (1982).

4. H. Talbot, "On Facts Relating to Optical Science," *Phil. Mag.*, Vol. 9, pp. 401-407 (1836).

5. A. J. Durelli and V. J. Parks, *Moiré Analysis of Strain,* Prentice-Hall, Englewood Cliffs, New Jersey (1970).

6. E. Keren and O. Kafri, "Diffraction Effects in Moiré Deflectometry," *J. Optical Society of America A,* Vol. 2, No. 2, pp. 111-120 (1985).

7. O. Kafri and I. Glat, *The Physics of Moiré Metrology*, John Wiley & Sons, New York (1990).

2.7 Key Relationships

electric field strength

$$A = a \cos 2\pi \omega t \qquad (2.4)$$

intensity

$$I = a^2 \qquad (2.6)$$

optical path length

$$OPL = \sum_{1}^{i=q} n_i D_i \qquad i = 1, 2, 3, \cdots, q \qquad (2.9)$$

reflection

$$\alpha' = \alpha \qquad (2.11)$$

refraction

$$\frac{\sin \alpha''}{\sin \alpha} = \frac{n_1}{n_2} \qquad (2.14)$$

thin lens

$$\frac{1}{D_1} + \frac{1}{D_1} = \frac{1}{FL} \qquad (2.16)$$

$$M = \frac{D_2}{D_1} \qquad (2.17)$$

pure two-beam interference

$$I = 2a^2 \left(1 + \cos 2\pi \frac{S}{\lambda} \right) = 4a^2 \cos^2 \pi \frac{S}{\lambda} \qquad (2.20)$$

impure two-beam interference

$$I = I_1 + I_2 + 2\sqrt{I_1 I_2} \cos 2\pi \frac{S}{\lambda} \qquad (2.21)$$

fringe gradient, fringe frequency

$$F = \frac{2}{\lambda} \sin \theta \qquad (2.24)$$

fringe orders

$$N(x, y) = \frac{S(x, y)}{\lambda} \qquad (2.25)$$

fringe contrast

$$\% \text{ contrast} = \frac{I_{max} - I_{min}}{I_{max}} \times 100\% \qquad (2.26)$$

Twyman-Green (Michelson) interferometer

$$W_{(x,y)} = \frac{\lambda}{2} N_{(x,y)} \tag{2.28}$$

diffraction gratings

$$f = \frac{1}{g} \tag{2.53}$$

diffraction angles

$$\sin \beta_m = \sin \alpha + m\lambda f \tag{2.54}$$

symmetrical diffraction angle

$$\sin \alpha = \frac{\lambda}{2} f \tag{2.57}$$

sign conventions

    diffraction order   (Sec. 2.5)

    diffraction angle   (Sec. 2.5)

equivalency

    interference of two beams creates a virtual grating ↔ a real grating creates the same two beams

Talbot distance

$$D_T = \frac{g^2}{\lambda} = \frac{1}{\lambda f^2} \tag{2.59}$$

2.8 Exercises

2.1 Derive Snell's law from Fig. 2.6b.

2.2 Let unpolarized white light of intensity I_0 strike a window at 45°. Determine the total intensity of reflected light. (Use Fig. 2.5; neglect multiple reflections.)

2.3 Details that lie in the object plane are focused by two thin lenses. Determine the distance D to the image and the magnification M of the image. (Answer: $D = 93$ mm, $M = 0.5\overset{..}{3}$)

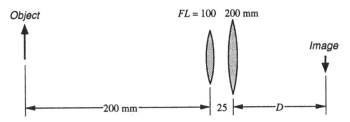

2.4 (a) Determine the frequency of carrier fringes in Fig. 2.28b. (b) Calculate the angle of rotation of the reference mirror in Fig. 2.27 to produce the carrier fringes. Assume the magnification of the pattern in Fig. 2.28 is $M = 5$. Let $\lambda = 633$ nm.

2.5 (a) Sketch the interference pattern generated in plane AA. (b) Determine the frequency of the interference pattern. (c) Sketch the pattern in plane BB and determine the frequency of the fringes; BB is midway between AA and the double wedge. (d) What assumptions were made in the answers? Let $\Theta = 10°$, $n = 1.5$, $h = 30$ mm, $\lambda = 514$ nm.

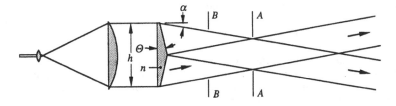

2.6   A shallow cylindrical surface is inspected in a Fizeau interferometer. (a) Sketch the fringe pattern seen by the observer. (b) If 10.0 fringes appear between A and B (and also between B and C), determine the radius of the cylindrical surface. What assumptions were made? (Answer: 20.0 m)

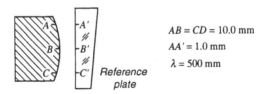

$$AB = CD = 10.0 \text{ mm}$$
$$AA' = 1.0 \text{ mm}$$
$$\lambda = 500 \text{ mm}$$

2.7   In Exercise 2.6, derive an expression for the fringe contrast in terms of reflectance R_s of the specimen and reflectance $R_r = 4\%$ of the reference plate. (Neglect multiple reflections.) Plot contrast vs. R_s.

2.8 Show the algebraic manipulations needed to derive Eq. 2.52.

2.9   In Exercise 2.6, determine the maximum spectral band width of the light source, $\lambda_2 - \lambda_1$, to maintain good fringe contrast. What types of light source would be satisfactory? What assumptions were made?

2.10   For quality control of parts with precision curved surfaces, the surface can be compared to a master gage that has the same surface contour as the required part, or else the negative (or opposite) surface contour. Design a practical interferometric apparatus to measure the deviations of surface contour of the test parts relative to the master gage. Estimate the smallest deviation that can be measured with the apparatus? What assumptions are made?

2.11   In Exercise 2.5, α can be very small (near $\alpha = 0°$) and it can be moderate (near $\alpha = 10°$), but it cannot be large (near $\alpha = 90°$). (a) Sketch an optical arrangement using full mirrors in which the collimated beam is divided into two intersecting beams, where the angle of intersection 2α can approach 180°. (b) Sketch another arrangement using full mirrors and one partial mirror (beam splitter). (c) Discuss the relative merits of each arrangement.

2.12   Design an optical system to cause two mutually coherent beams to intersect with an included angle of 90°, (a) using one diffraction grating plus full mirrors, and (b) using only diffraction gratings. Specify the frequency of each grating. What assumptions were made?

2.13   Certain optical techniques require that a transparent solid must be immersed in a liquid of the same refractive index. Assume you are given a transparent block and two liquids; let their refractive indices be $n_b \approx 1.55$, $n_1 \approx 1.5$ and $n_2 \approx 1.6$, respectively. The liquids are easily mixed to produce any intermediate refractive index. The block can be modified to any shape desired.
  Design an experimental apparatus and procedure to obtain an index matching liquid. The match is to be optimum at wavelength $\lambda = 633$ nm. An interferometer should *not* be used. Estimate the uncertainty of the index match.

2.14 Design an experimental apparatus and procedure for Exercise 2.13 using an interferometric method. Estimate the uncertainty of the index match.

2.15   The copper rod on the left is heated by a torch.   (a) Design an interferometric method to determine its change of length. Sketch the apparatus and paths of light beams. (b) Briefly describe the method. (c) Sketch the fringe pattern obtained. (d) Derive the relationship between the fringes and the change of length. (e) List any assumptions you made.

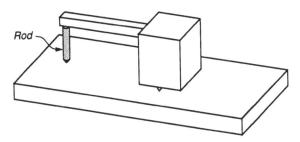

2.16   Design a practical experiment to measure the coefficient of thermal expansion of a new alloy steel in the temperature range of 20° to 120°. Discuss the sources of potential error. Estimate the overall accuracy of the coefficient.

2.17   Design another practical experiment for Exercise 2.16 that circumvents or reduces the most significant potential errors.

3
Geometric Moiré

3.1 Basic Features of Moiré

Our goal in this chapter is to build a foundation for easy assimilation of the concepts of high-sensitivity moiré interferometry. An additional goal is to review many ideas of geometric moiré that are usually addressed in the study of optical methods of experimental mechanics and engineering practice.

Emphasis is on the measurement of deformations that occur when a body is subjected to applied forces. Moiré methods can measure the in-plane and out-of-plane components of displacement. Moreover, the data are revealed as contour maps of displacements, which means that displacements are measured simultaneously over the whole field of view.

We will concentrate on displacements in this chapter. Although strains can be extracted from in-plane displacement fields, the sensitivity of geometric moiré is not adequate in most cases for determination of strain distributions. Strain analysis is treated in Chapter 4, in conjunction with the high-sensitivity measurement of in-plane displacements. However, it is useful to appreciate at the outset that the equations relating moiré fringes to in-plane displacements are identical for geometric moiré, moiré interferometry, and microscopic moiré interferometry.[1]

3.1.1 Gratings, Fringes, Visibility

Moiré patterns are common. In Fig. 3.1a, two combs produce the moiré effect. In (b), two wire mesh bowls used in food preparation produce moiré patterns. The broad dark and light bands are called *moiré fringes*. They are formed by the superposition of two amplitude gratings (Sec. 2.5), each comprised of opaque bars and clear spaces. Whereas gratings with straight bars and spaces were described in

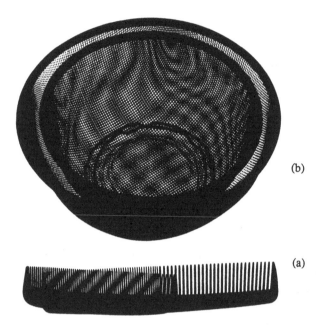

(b)

(a)

Fig. 3.1 Moiré patterns.

Chapter 2, the bars and spaces can be *curvilinear*. The visibility of the moiré fringes is good in (a), but poor in (b).

Moiré fringes of excellent visibility occur when these conditions are present:

- the widths of the bars and spaces are equal
- the two gratings are well defined, i.e., the distances between the bars and the curvatures of the bars in each grating are smoothly varying functions of the coordinates x and y
- the angle of intersection of the two gratings is small, e.g., 3° or less
- the ratio of the pitches of the two gratings is small, e.g., 1.05:1 or less.

A corollary to the fourth condition is that the distance between adjacent moiré fringes is large compared to the distance between adjacent bars of the gratings, e.g., in the ratio 20:1 or larger. Of course, excellent (and good) fringe visibility is a subjective assessment and less stringent conditions are frequently acceptable.

The *pitch* of each grating is the distance between corresponding points in adjacent bars. The reciprocal of pitch is *frequency*, or the number of bars per unit length. The bars are often called *lines*, so the grating pitch, g, is the distance between adjacent lines and the grating frequency, f, is the number of lines per unit length; 10 lines/mm (250 lines/in.) is a common example of grating frequency for geometric moiré. For most applications, grating frequencies do not exceed 40 lines/mm (1000 lines/in.). The pitch, G, of moiré fringes is the smallest distance between corresponding features on neighboring fringes. Their frequency, F, is the number of fringes per unit length, e.g., 2 fringes/cm or 5 fringes/in. Usually we are interested in the fringe frequencies along orthogonal x and y axes, denoted F_x and F_y; they are the number of fringes that cut lines that are parallel to the x and y axes, respectively, per unit length along these lines.

3.1.2 Intensity Distribution

The observer sees the moiré fringes as a cyclic function of the intensity of light; the fringes appear as alternating bands of brightness and darkness. This is especially true in the usual case where the bars of the superimposed gratings are closely spaced and the observer cannot resolve them. Ideally, the observer perceives the *average* of the light energy that emerges from each small region, where the dimensions of the region are equal to the pitch of the coarser grating. This perceived intensity distribution is illustrated by graphs in Fig. 3.2 for uniform linear gratings each with bars and spaces of equal width. Pure rotation is depicted on the left, where equal gratings are superimposed with an angular displacement θ. Pure extension is illustrated on the right, where two gratings of slightly different pitches are superimposed. In both cases, some of the light that would otherwise emerge from the first grating is obstructed by the superimposed grating. At the centers of dark fringes, the bar of one grating covers the space of the other and no light comes through; the emergent intensity, I, is zero. Proceeding from there toward the next dark fringe, the amount of obstruction diminishes linearly and the amount of light (averaged across the grating pitch) increases linearly until the bar of one grating falls above the bar of the other. There, the maximum amount of light passes through the gratings. Then the obstruction increases again and the intensity (averaged) decreases again. This progression is indicated in the graphs, in which coordinate x lies perpendicular to the moiré fringes.

The triangular variation of intensity with x is obvious from the diamond-shaped spaces formed in the pure rotation case. For the pure extension case, the unobstructed spaces become wider in a linear progression, and then narrower, as seen in the figure. This linear progression matches the triangular distribution when the intensity is averaged over the pitch of the coarser grating.

When several fringes cross x, the intensity distribution is depicted in Fig. 3.2c. Fringe spacings are typically one to two orders of magnitude larger than grating line spacings. Since the moiré fringes are much further apart, they can be resolved even when the grating lines cannot. The limited resolution of the observer (or camera) influences the perceived intensity distribution. With *reduced resolution*, the triangular intensity distributions become rounded near maxima and minima points. The rounding is the result of intensity averaging in regions larger than the grating pitch, for example, when averaging occurs over a few grating pitches. The effect is illustrated in Fig. 3.2d. Since the averaging distance is fixed for a given observation, the departure from a triangular distribution is more severe for fringes that are more closely spaced than for fringes that are further apart.

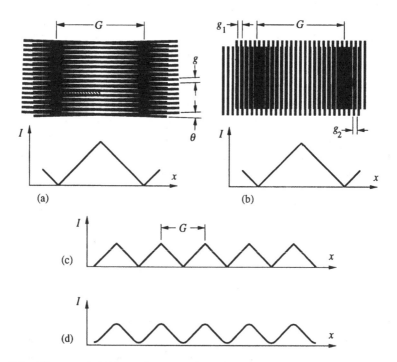

Fig. 3.2   These moiré fringes have a triangular intensity distribution when the emergent light is averaged over the pitch of the coarser grating. In (d), the rounding is caused by averaging over several pitches.

3.1.3 Multiplicative and Additive Intensities

Multiplicative Intensities.

Figure 3.3 illustrates alternative means to superimpose gratings. In (a) the gratings are on transparent substrates and they are in contact, or nearly in contact. They are illuminated by a broad diffuse source of light. In (b), one grating is printed in black ink on a diffusely reflecting surface, (e.g., a white paper or a white matte painted surface) and the other grating is on glass. In both cases, light that would otherwise reach the observer is partly obstructed by the gratings.

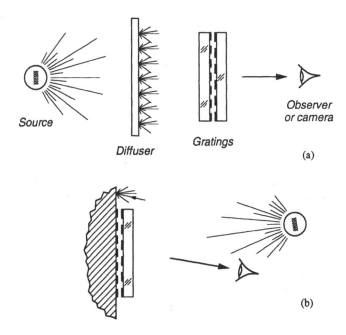

Fig. 3.3   Superposition of gratings in (a) a transmission system and (b) a reflection system.

In the transmission system (Fig. 3.3a), the intensity of light that emerges from the first grating is the incident intensity I_0 times the transmittance $T_1(x,y)$ at each point of the first grating, or $I_0 T_1(x,y)$. The intensity of light that emerges from the second grating is

$$I = I_0 T_1(x,y) T_2(x,y) \tag{3.1}$$

For the bar and space gratings that are commonly used in geometric moiré, $T \approx 0$ at the opaque bars and $T \approx 1$ at the transparent spaces. Thus, there are two levels of intensity: zero in obstructed regions and I_0 elsewhere. When the emergent intensities are averaged over the grating pitch, the conditions of Fig. 3.2 prevail.

In the reflective system (Fig. 3.3b), light passes twice through the transmission grating and it is reflected once by the reflection grating. Accordingly, the intensity of the emergent light is

$$I = I_0 \left(T(x,y) \right)^2 R(x,y)$$

(3.2)

where $R(x,y)$ is the reflectance at each x,y point of the reflective grating. At the reflective bars, R is some finite reflectance, perhaps 0.8 in many cases; at nonreflective spaces $R \approx 0$. Again, T is approximately zero or one. Thus, the emergent intensity is zero in the obstructed regions and a constant finite value elsewhere. When averaged, the conditions of Fig. 3.2 prevail here, too.

In view of Eqs. 3.1 and 3.2, the patterns developed when two real gratings are in contact are called moiré patterns of *multiplicative intensities*; sometimes they are called *multiplicative moiré* patterns. They are characterized by moiré fringes of good visibility.

What is meant by contact? The two gratings may be pressed tightly together, but that is not required. Essentially no deterioration of the moiré pattern will be seen when the gap between the two gratings is small compared to the distance g^2/λ, e.g., 5% of g^2/λ (Sec. 2.5.3), where λ is the wavelength of the light.

Additive Intensities.

A different optical system is illustrated in Fig. 3.4, where (a) represents projection of a bar and space grating onto a diffuse screen. A lantern slide projector could be used. The observer sees a grating on the screen, comprised of dark and bright bars, or lines. For the sake of explanation, visualize two projected gratings superimposed on the screen, as illustrated schematically in Fig. 3.4b. Let the two projected gratings be slightly different from each other so that moiré fringes are formed on the screen. As in the previous case, the difference might be a small rotation, a small change of pitch, or both.

Referring to Fig. 3.4a, with the single projector, let the observer have limited resolution. Let the smallest detail that can be resolved be equal in size to the pitch of the projected grating (or a small multiple of the pitch). Then the observer would perceive each and every pitch as having the *average* intensity of the dark bar and bright bar. The screen would appear to be illuminated by a uniform intensity. The second projector would produce the same effect. None

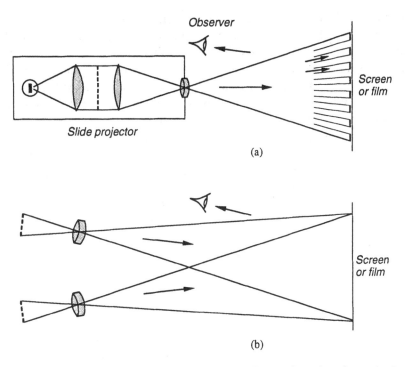

Fig. 3.4   (a) Projection of a grating.   (b) Two superimposed gratings by projection.

of the dark and bright bars that actually exist on the screen would be perceived. If the observer cannot resolve the individual lines of the grating, only the average intensity is detected and no moiré fringes are seen. The effect of limited resolution is totally different from that found in Fig. 3.2 for multiplicative intensities.

If the observer or camera can resolve the individual grating lines, a periodic structure is seen and perceived as moiré fringes. In regions where the bright bars of the two gratings fall directly on top of each other, bright and dark grating lines are visible. In regions where the bright bars of one grating fall midway between those of the other, a gray band is formed. The periodic structure of gray bands alternating with bands of grating lines comprises the moiré fringes. An example is shown in Fig. 3.5a, with the corresponding graph of averaged intensity in (b).

The visibility of fringes of additive intensity can be improved by nonlinear recording, e.g., as in Fig. 3.5c. A strong overexposure was used to saturate the photographic film. Thus, all the bright zones have the same brightness level; the nonlinear recording converted the gray levels of (a) to maximum brightness. When the photograph is viewed and the intensity is averaged across the grating pitch, the perceived intensity (idealized) is that graphed in (d). Accordingly, the grating lines must be resolved on the screen or film, but then the moiré fringes can be discriminated by their variation of (averaged) intensity. When nonlinear recording is used, it is not necessary for the observer to resolve the individual lines to see the moiré, even though the grating lines must be resolved on the screen or film. Still, the contrast of fringes is low compared to moiré fringes of multiplicative intensities.

Patterns formed by projected gratings are designated by the term *moiré of additive intensities*. The term *additive moiré* is reserved for another phenomenon addressed in Sec. 3.1.5.

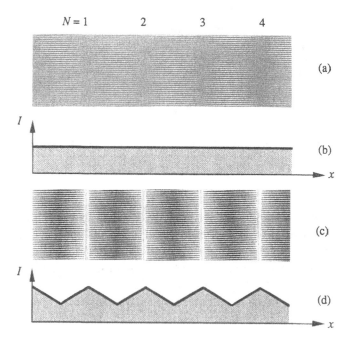

Fig. 3.5   (a) Moiré of two projected gratings recorded with linear photographic sensitivity.   (b) Averaged intensity for linear recording.   (c) Moiré with nonlinear recording.   (d) Averaged intensity for nonlinear recording.

3.1.4 Pure Rotation and Extension

A beautiful feature of geometric moiré is that the relationships between the moiré fringes and grating lines can be determined simply by geometry. For the case of pure rotation, the shaded triangle in Fig. 3.2a gives, for small angles, $g/2 = \theta G/2$, where the grating pitch g and the fringe pitch G are defined in the figure. Thus, for small pure rotations,

$$G = \frac{g}{\theta} \qquad (3.3)$$

In terms of frequency, the relationship is

$$F = f\theta \qquad (3.4)$$

Note that the fringes of pure rotation lie perpendicular to the bisector of angle θ, or nearly perpendicular to the lines of the gratings.

For pure extension (Fig. 3.2b) there is one more line in the finer grating than in the coarser one, for each moiré fringe. Thus, the pitch of the moiré fringes is given by

$$G = ng_2 = (n + 1)g_1$$

where n is the number of pitches of the coarser grating that fall within G. By eliminating n this reduces to

$$\frac{1}{G} = \frac{1}{g_1} - \frac{1}{g_2} \qquad (3.5)$$

and in terms of frequency, the relationship is

$$F = f_1 - f_2 \qquad (3.6)$$

where f_1 and f_2 are the frequencies of the finer and coarser gratings, respectively. An example would be $f_1 = 200$ lines/cm, $f_2 = 196$ lines/cm, and $F = 4$ fringes/cm. The fringes of pure extension lie parallel to the grating lines.

3.1.5 Moiré Fringes as Parametric Curves

A very nice treatment of the relationship between moiré fringes and the gratings that create them was presented by Durelli and Parks.[2] That approach will be introduced here. Moiré patterns feature three families of curves. The two superimposed gratings are two families of curves, which we will call families L and M. These families interact in such a way that they form a third family, the family of moiré fringes, which we will call family N. Figure 3.6 illustrates the three families, but for simplicity the grating lines and fringes are shown straight instead of curved.

The centerlines of the bars of the L and M families are extended in the figure. Each member of each family is given an index number, representing the sequential order of the parametric family. A specific number applies to every point on one curve of the family. The figure shows that the centerlines of the bright moiré fringes pass through the intersections of the centerlines of the bars. Inspection shows that the centers of the moiré fringes comprise the locus of points where $L - M$ is a constant integral number. These moiré fringes are a parametric system of curves defined by $N = L - M$, where N is the index number assigned to each point in the moiré pattern. Consistent with practice, N is called the moiré fringe order, and more specifically, it is called the *subtractive moiré* fringe order.

The encircled region in Fig. 3.6 indicates that all points between curves of integral index numbers have intermediate index numbers. Every point in the field has index numbers L and M and fringe order N.

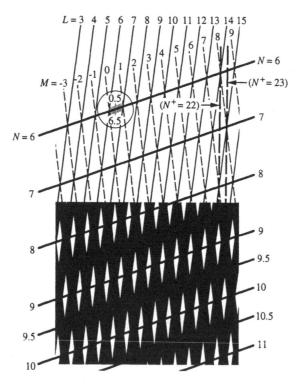

Fig. 3.6 The L and M parametric families of grating lines creates the N family of moiré fringes, where $N = L - M$ at every point.

Moiré as a Whole-field Subtraction Process

Moiré can be used to subtract data, and it can do the subtraction simultaneously everywhere in an extended field. Figure 3.7 is an example. The specimen was cut from a plastic sheet, polymethylmethacrylate, of 3 mm (1/8 in.) thickness. The objective was to map the changes of thickness near the notch caused by an applied tensile load; subwavelength sensitivity was required.

Employing the experimental method of Ref. 3, Fizeau interferometry (Sec. 2.3.7) was used to record the contour maps of the specimen thickness before and after the tensile load was applied. They are shown in Fig. 3.7 as (a) and (b), respectively. The initial (or zero-load) pattern is the M family of curves, in this case, and it was

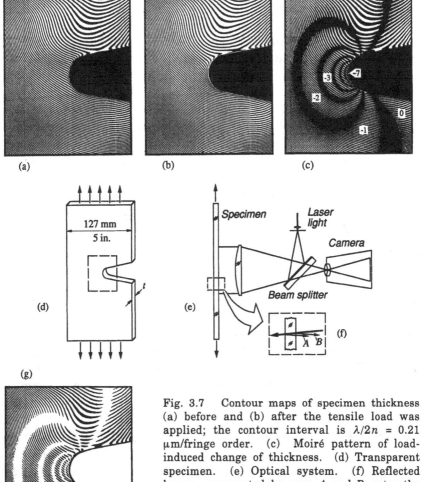

(a)   (b)   (c)

(d)   (e)   (f)

(g)

Fig. 3.7   Contour maps of specimen thickness (a) before and (b) after the tensile load was applied; the contour interval is $\lambda/2n = 0.21$ µm/fringe order. (c) Moiré pattern of load-induced change of thickness. (d) Transparent specimen. (e) Optical system. (f) Reflected beams represented by rays A and B enter the camera and interfere to produce the maps of specimen thickness. (g) Photographic print of superimposed films of (a) and (b). Note: The patterns represent the region in the dashed box of (d).

subtracted by superposition of (a) and (b). The resultant N family of parametric curves shown in (c) is the moiré pattern, where the fringes represent contours of constant change of thickness, specifically, contours of the load-induced changes of thickness. The numbers attached to the moiré fringes are their fringe orders; when multiplied by the contour interval, 0.21 μm, the moiré pattern becomes a map of the thickness changes throughout the field.

In practice, the photographic negatives of contour maps (a) and (b) are superimposed and inserted into a photographic enlarger to make a print. The result is shown in Fig. 3.7g, which is a negative image of (c). This format is sometimes preferred, since it is easier to sketch the centerlines of integral and half-order fringes directly on the print. The positive image (c) can be produced by using auto-positive printing paper in the photographic enlargement step; if the negatives are large enough, a xerographic copying machine or a computer scanner can be used instead.

This concept of moiré as a whole-field subtraction process is powerful. In moiré interferometry, for example, it can be used to cancel distortions of the initial field; it can be used to determine deformations incurred between two different load levels in order to evaluate nonlinear effects; and it can be used to perform whole-field differentiation, to determine derivatives of displacements.

Whole-field Additive Moiré

Figure 3.6 raises a question. The centerlines of the L and M families of curves intersect to form diamond-shaped quadrilaterals, and the lines joining the minor axes of the diamonds lie on the moiré fringes. Is it logical that the major axes of the diamonds exhibit analogous properties? Yes, inspection shows that lines joining the major axes comprise the locus of points where $L + M$ is constant. These lines are the centerlines of moiré fringes representing the additive family $L + M = N^+$. The dashed lines labeled $N^+ = 22$ and 23 are examples; N^+ is the symbol used to denote the fringe orders of additive moiré fringes. Thus, the interaction of two parametric families of curves, L and M creates two families of moiré fringes, N and N^+.

However, it is evident that only one family of moiré fringes has good visibility. A condition for visibility is that the distance between moiré fringes must be large compared to the distance between grating lines (Sec. 3.1.1). Thus, the subtractive moiré is visible in Fig. 3.6 and the additive moiré is not.

Figure 3.8 illustrates remarkable features of moiré fringes. The L family of parametric curves exhibits a rare condition where the index numbers are increasing with coordinate x in one region and

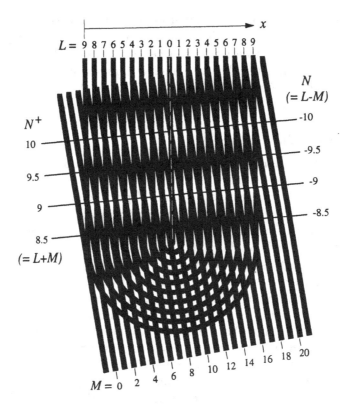

Fig. 3.8 On the right side of the dashed line the moiré fringes represent the subtractive parameter $L - M$. On the left side they represent $L + M$.

decreasing in the other. The M family exhibits the usual condition where index numbers increase monotonically with x. On the right side of the dashed line, examination proves that the clearly visible moiré fringes are the subtractive family, the $N = L - M$ family. Proceeding downwards on the right side, these moiré fringes lose visibility where the angles between the L and M families becomes large. Continuing up the left side, the angles become small again and moiré fringes gain visibility again. On the left side, however, examination proves that the clearly visible moiré fringes are the additive family, the $N^+ = L + M$ family. The rules of visibility apply equally to the subtractive and additive families of moiré fringes. The parametric system of curves carries both the $L - M$ and $L + M$ information in all cases, but only one of these families appears in the form of visible moiré fringes, either the subtractive or the additive family.

In most applications of moiré, the two superimposed gratings are very similar to each other, with one exhibiting only small perturbations of grating lines relative to the other. Figure 3.7 is an example. Consequently, the moiré fringes of good visibility encountered in normal practice are subtractive fringes. Yet, experiments can be configured to generate the additive fringes.

Example of Additive Moiré

Again, the challenge was to determine load-induced or stress-induced thickness changes, but unlike Fig. 3.7 the specimen was opaque.[4] The specimen was a graphite/epoxy composite plate, which was very strong and very stiff. The objective was to map the changes of plate thickness near a 19 mm (0.75 in.) diameter hole as a function of compressive loads on the plate.

Subwavelength sensitivity was required. Contour maps of out-of-plane displacements could be obtained from both surfaces of the specimen by means of holographic interferometry. Then, point-by-point addition of the two displacements could be used to obtain the thickness change, but that operation would be tedious and prone to error. Instead, the experiment was configured to utilize additive moiré to sum the two out-of-plane displacement fields.

Reflection holographic interferometry[4] was used on both sides of the specimen, as illustrated schematically in Fig. 3.9a. This is the technique in which the light first passes through the holographic plate and then reflects from the specimen to pass through the holographic plate in the opposite direction. The incoming light is the reference beam and the light returning from the specimen is the information beam. Normally, two exposures are made on the holographic plate, one exposure before the compressive load is applied and one after, while maintaining the holographic plate in a fixed position. In the reconstruction process, the hologram displays a contour map of the change of gap, S_1 (or S_2) between the specimen surface and the holographic plate.

For this experiment, the two holographic plates were *linked together* in a frame that could pivot about an axis through point O in the specimen. The first exposure was made with the specimen under a very small load and with the holographic plates parallel to the specimen surface. Then the holographic plates were inclined so that the gaps become wedge-shaped; the inclination was tiny, but it caused variations of S_1 and S_2 of many wavelengths. The specimen was loaded in compression and the second holographic exposures were made. The two reconstructed images showed contour maps of the changes of gaps S_1 and S_2. Thus the maps represented the out-of-

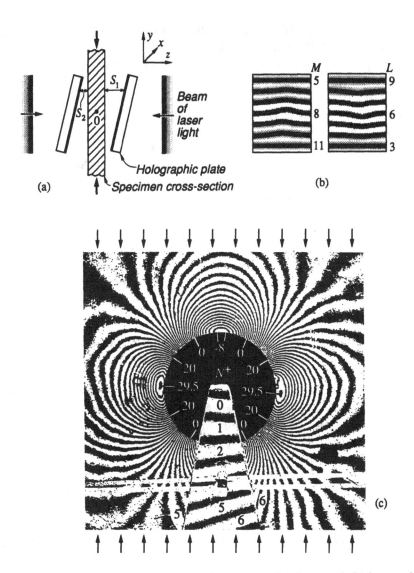

Fig. 3.9   Demonstration of additive moiré to map the change of thickness of a graphite/epoxy plate with a central 19 mm (3/4 in.) hole.   The compressive load increment is 66.7 kN (15,000 pounds).

plane (W) displacement fields of the specimen plus the change of inclination of the holographic plates.

The reconstructed contour maps are depicted schematically in Fig. 3.9b, in which the families of contours are given the designations L and M. Family L represents the right side, where S_1 increases with y and the fringe orders (of holographic interferometry) increase with y. On the left side, S_2 decreases with y and M decreases with y. Thus, the conditions for high visibility additive moiré prevail, just as in the left side of Fig. 3.8. The two holographic images were photographed and the photos were superimposed to create the moiré pattern of the additive parameter.

The result is Fig. 3.9c, which is a contour map of the *relative* changes of thickness of the specimen. The device inside the hole was used to establish the absolute change of thickness at one point in the specimen.[4] The numbers represent the fringe orders N^+, which denote the thickness change when they are multiplied by the contour interval of $\lambda/2$, or 0.316 µm. Accordingly, additive moiré was responsible for the production of whole-field contour maps of the submicron thickness changes.

The pattern, Fig. 3.9c, reveals a significant feature of composite laminates. The maximum change of thickness occurs at a finite distance from the hole boundary (along the horizontal diameter). With isotropic materials, it occurs on the boundary.

Specular Surfaces

The experimentalist will be interested in another aspect of the work described above. Optimum conditions for reflective holography include specular specimen surfaces—highly reflective, smooth surfaces. The graphite/epoxy specimen had rough black surfaces, but they were treated as follows. An epoxy cement was used to bond flat acrylic plates to the specimen surfaces; the liquid epoxy was squeezed to thin films by applying weights to the sandwich. Since epoxy does not bond well to acrylics, the plates were easily pried off after the epoxy hardened, leaving the specimen with thin, smooth surfaces of epoxy. The surfaces were mirrorized by evaporated aluminum (Sec. 2.2.3) to provide the desired high reflectance. Of course, these thin coatings had negligible influence on the deformation of the specimen and they revealed the W displacements with high fidelity.

Generalization of Parametric Curves

The principle of parametric curves is not restricted to systems where moiré fringes become visible. Figure 3.10 consists of two families of parametric curves, labeled L and M. These contour lines represent the two velocities of light at each point inside a stressed photoelastic body,[5] i.e., a birefringent body. The parametric interpretation of the pattern is the same as in Fig. 3.6, but here the angles of intersection are larger and moiré fringes are not visible. Nevertheless, the diagonals of the quadrilaterals connect systematically to form the subtractive and additive families of parametric curves. A few members of these derived families are sketched on the figure, with labels N and N^+, respectively.

Additive and subtractive contour maps can always be drawn in this way. The principle was used effectively in Ref. 6, where contour maps of shear strains required the sum of two components of shear.

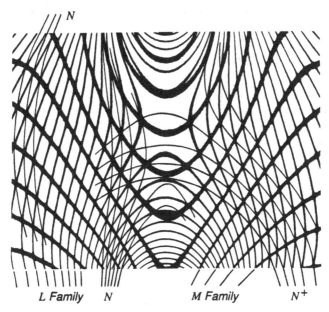

Fig. 3.10   The additive and subtractive contours can be drawn for any pair of parametric families of curves.

3.1.6 Fringe Vectors

Parametric curves have vector properties. In any local region surrounding a point O, Fig. 3.11, each of the four families ($L, M, N,$ N^+) has a direction, magnitude and sense. The direction is defined as the normal to the curve through point O. The magnitude is the spatial frequency of curves near O, which is the reciprocal of the local pitch of the curves measured along the shortest distance between the curves. The units of frequency are grating lines per unit length for L and M, and moiré fringes per unit length for N and N^+. The sense of the vector is the direction of increasing index number or increasing fringe order.

The vector relationships are[7]

$$\boldsymbol{F}_N = \boldsymbol{f}_L - \boldsymbol{f}_M \tag{3.7}$$

and

$$F_{N^+} = f_L + f_M \qquad (3.8)$$

where vectors f and F represent grating frequency and moiré fringe frequency, respectively, and the subscripts denote each of the four parametric families. The relationships are illustrated in Fig. 3.11 for the parametric families near the general point O. The frequency vectors lie normal to the curves, their lengths are proportional to the spatial frequencies of the curves and their positive direction is the direction of increasing index number or fringe order. The vector diagrams depict the subtractive and additive fringe vectors, F_N and F_{N^+}. The usual case is illustrated, where F_N is smaller and F_{N^+} is larger than the grating vectors. The corresponding moiré patterns near O are sketched at the bottom of the figure, emphasizing that the subtractive moiré has much greater visibility.

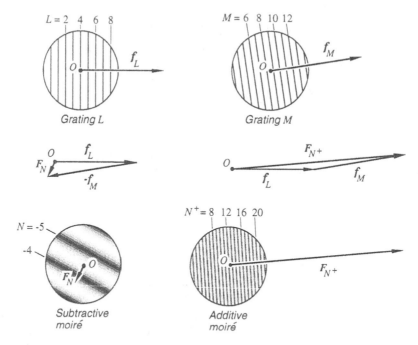

Fig. 3.11   Vector relationships for parametric curves. The vectors represent grating frequencies for L and M and moiré fringe frequencies for N and N^+.

It is useful to resolve the moiré fringe vector into its x and y components, as illustrated in Fig. 3.12. The corresponding fringe pitches are labeled as the reciprocals of the frequencies. The components represent the fringe gradients in the x and y directions,

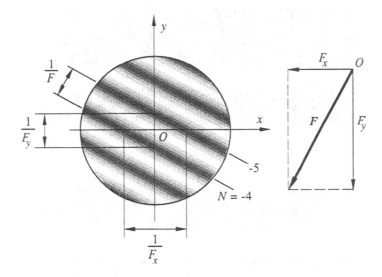

Fig. 3.12   Fringe vectors depict the local moiré fringe direction, frequency (i.e., fringe gradient), and sense. The fringe vector at any point can be resolved into its x and y components.

i.e., $F_x = \partial N / \partial x$ and $F_y = \partial N / \partial y$. These x and y components are important in strain analysis, as treated in Chapter 4.

3.1.7 Amplification by Moiré

When two uniform gratings of slightly different pitches are superimposed, they form moiré fringes governed by Eqs. 3.5 and 3.6. Two such gratings and their superpositions are shown in Fig. 3.13. The moiré pattern of Fig. 3.13c represents the initial condition, before the *grating of pitch* g_1 is moved horizontally by a distance δ. When it is moved by some integral multiple of δ, ($\delta = g_1, 2g_1, 3g_1, ---$), the resulting fringe pattern appears the same as in Fig. 3.13c. This is logical because the stationary grating is obstructed by the moveable grating in exactly the same way.

When $\delta = g_1/4$, the moiré fringes move to new positions, displaced by $\Delta = G/4$ from their original positions, as shown in (d); in (e), $\Delta = G/2$. Generalizing

$$\Delta = \delta \left(\frac{G}{g_1} \right) \tag{3.9}$$

where G/g_1 is an amplification factor. The fringe movement Δ amplifies the grating movement δ by this factor. When $g_1 < g_2$, the

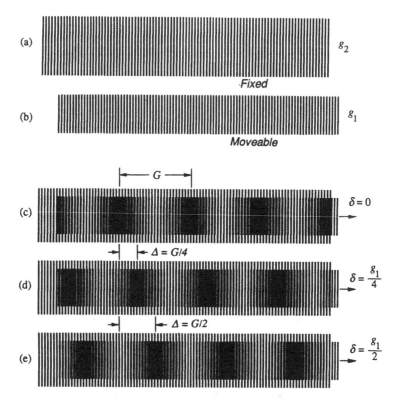

Fig. 3.13   Amplification: the fringe movement Δ is G/g_1 times the grating movement δ. As an instructive exercise, the reader can make transparencies of (a) and (b) on a copying machine and manipulate the superimposed gratings.

fringes move in the same direction as the moveable grating. When $g_1 > g_2$, the directions are opposite.

This principle is used widely for measurement and control of movements of machine tools. On a milling machine, for example, a moiré device can be used to measure (and control) the position of the workpiece relative to the cutting tool. A long grating, called a scale grating, is attached to the moveable table of the machine and a short reference grating is attached to the stationary base of the machine. A photodiode is used to sense the intensity of moiré fringes that pass a stationary point as the scale grating moves by. The photodiode produces oscillations of voltage in harmony with the oscillations of the moiré fringe intensity. The oscillations are counted electronically, and the fringe count, multiplied by the pitch of the scale grating (e.g., 1/50 mm), specifies the position of the table.

A single sensor cannot discriminate the direction of movement, i.e., whether δ is positive or negative. Accordingly, a second photodiode is positioned at a point $G/4$ away from the first. A logic circuit interprets the two signals and outputs the correct position of the table.

3.1.8 Fringe Sharpening and Multiplication in Geometric Moiré

The triangular distributions of intensities illustrated in Fig. 3.2 can be altered by changing the ratio R of bar-width to space-width in each grating; R is sometimes called the black-to-white ratio. The moiré fringes can be narrowed, or *sharpened*, by using complementary gratings of $R \neq 1$. For complementary gratings, $R_2 = 1/R_1$. Figure 3.14 illustrates the superposition of two complementary gratings for the case of $R_1 = 2$ and $R_2 = 1/2$. The enlarged view shows that the unobstructed openings between bars provides an intensity distribution with sharpened intensity minima. Fringe sharpening can be defined by the ratio h/G, where h is the width of the dark fringe at the half-intensity level. The relationship between fringe width and grating characteristics is

$$\frac{h}{G} = \frac{1}{R_1 + 1} \tag{3.10}$$

where R_1 pertains to the darker grating. As before, the usual condition is assumed, where the moiré fringes are well resolved but the individual grating lines are not. Sharpened fringes offer the advantage that the location of minimum fringe intensity can be determined more accurately, thus enhancing the accuracy of measurements.

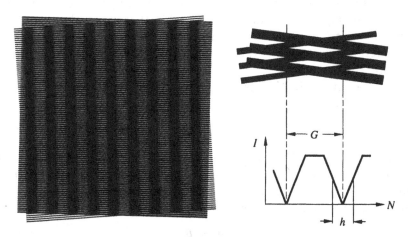

Fig. 3.14   Moiré fringe sharpening using complementary gratings.

The use of complementary gratings presents an opportunity for *geometric fringe multiplication*.[8] As illustrated in Fig. 3.15, an extra bar can be inserted in the space between the widely separated bars. As a result, an extra moiré fringe is formed between each pair of original fringes. Fringes are multiplied without any reduction of fringe sharpening. The intensity maxima are reduced, but that can be counteracted by the use of a stronger light source or by longer photographic exposures.

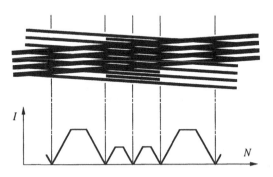

Fig. 3.15   Moiré fringe multiplication by a factor $\beta = 2$ is achieved by doubling the frequency of the lighter grating. For $\beta = 3$, its frequency is tripled.

When geometric moiré is used for whole-field in-plane displacement measurements (Sec. 3.2) it is difficult in some circumstances to apply a high frequency grating to the specimen. With a pair of low frequency gratings, an insufficient number of moiré fringes might be formed for an analysis of acceptable accuracy. However, a higher frequency reference grating could be used with a lower frequency specimen grating to increase the number of fringes. The term fringe multiplication is applied in that sense; the number of moiré fringes is multiplied in comparison to the usual case in which the reference grating has the same initial frequency as the specimen grating.

Fringe multiplication is produced when the reference grating frequency, f, is

$$f = \beta f_s \qquad (3.11)$$

where β is an integral number representing the fringe multiplication factor and f_s is the initial frequency of the specimen grating. Optimum conditions of combined sharpening and multiplication are produced when both gratings have bar-to-space ratios of

$$R = \beta \qquad (3.12)$$

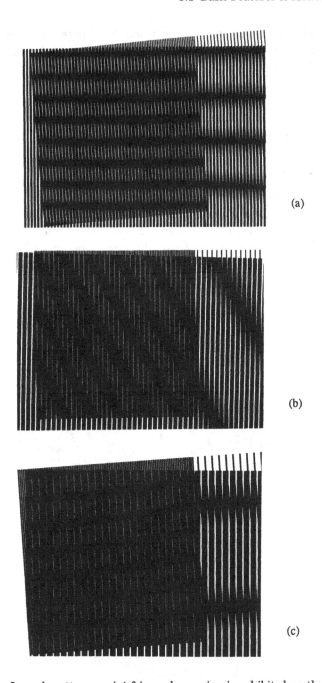

Fig. 3.16   In each pattern, moiré fringe sharpening is exhibited on the right side and moiré fringe multiplication on the left side. Multiplication factors are $\beta = 2$, 3, and 4.   (a) and (c) show fringes of rotation, and (b) shows rotation plus extension.

Figure 3.16 illustrates these conditions for fringe sharpening with complementary gratings and also for fringe multiplication of $\beta = 2$, 3, and 4. With geometric moiré, multiplication factors of $\beta > 5$ are impractical since the black-to-white ratios of the grating become large and the spaces where light penetrates the two gratings become very small. With interferometric moiré, fringe multiplication exceeding $\beta = 60$ is possible (Sec. 4.21).

3.1.9 Circular Gratings

Circular gratings are among the systems of parametric curves that produce interesting and potentially useful moiré fringes. Theocaris studied them in detail.[9] When the circular grating lines are deformed by a uniform strain field, the circles become ellipses and a new system of parametric curves is formed. Superposition with the

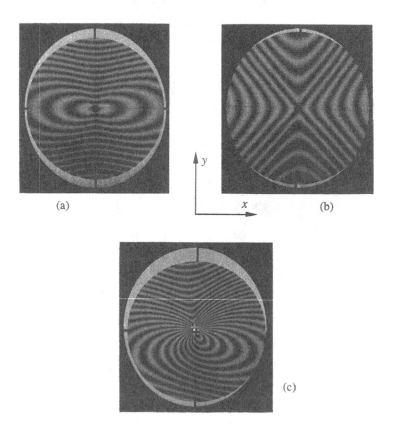

Fig. 3.17   Moiré of deformed and undeformed circular grating pairs.   (a) Biaxial tension where $\varepsilon_y > \varepsilon_x$. (b) Biaxial strain where $\varepsilon_x > 0 > \varepsilon_y$. (c) Improper centering.

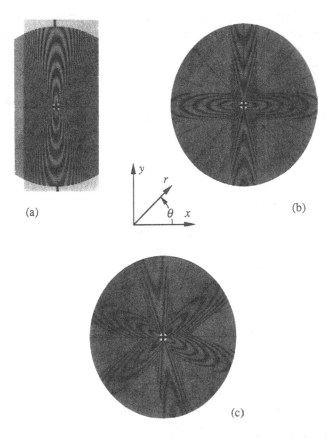

Fig. 3.18   Linear gratings superimposed upon (a) circular and (b, c) elliptical gratings.

original circular grating produces moiré fringes that reveal the *radial components* of displacement. Figure 3.17 shows the moiré patterns for uniform biaxial strains parallel to x and y coordinates. The fringe gradient in any radial direction determines the magnitude of the radial strain, ε_r, by

$$\varepsilon_r = g \frac{\Delta N}{\Delta r} \tag{3.13}$$

where ΔN is the number of fringes that cross a radial line of length Δr and g is the pitch of the undeformed grating. The two gratings must be superimposed with a common center point; otherwise the pattern of Fig. 3.17a degenerates to that of (c).

Figure 3.18a shows the moiré fringes produced by superposition of a circular grating and a linear grating, where the pitch of the circular grating is slightly smaller. In (b), the superimposed reference grating is comprised of two linear gratings with their lines perpendicular to orthogonal axes. Their pitch is g, the same as that of the undeformed circular grating. After deformation the circular grating corresponds to the case of $\varepsilon_y > 0 > \varepsilon_x$. Tensile strains are recognized by hyperbolic moiré fringes and compressive strains by loops. The strain magnitudes are determined by Eq. 3.13, but now the fringe count must be made along the axes of the hyperbolas or loops. The reference grating can be rotated to locate the principal strains, which corresponds to the angular position where the largest fringe gradient, $|\Delta N / \Delta r|$, is found. An advantage of linear reference gratings is that careful centering on the circular grating is not required.

Three linear gratings are used in Fig. 3.18c, with their lines perpendicular to three axes that are 60° apart. The configuration can be employed as a strain rosette. Conceptually, these moiré arrangements could be miniaturized and used as strain gages, but they lack the sensitivity needed for measurements of small strains.

3.2 In-plane Displacements of a Deformed Body

Concepts relating displacements and moiré are developed in this section. The focus here and in subsequent chapters is the application of moiré methods to deformation measurements, especially deformations of engineering materials and structural elements.

3.2.1 Physical Concepts, Absolute Displacements

Referring to Fig. 3.19a, let a point P on the surface of a solid body be displaced to point P'. P and P' represent the same physical point on the body; the displacement is caused by deformation of the body, or by rigid-body movement, or by a combination of both. The displacement of P is represented in (b) by vector δ, and its components U,V,W in the x,y,z directions, respectively. The displacement is usually described by scalars U,V,W, which are the magnitudes of the corresponding vectors, with + or − signs for vector components that point in the positive or negative direction of the coordinate axes, respectively.

U and V are called *in-plane* displacements since U and V lie in the original plane of the surface. W is perpendicular to the surface and it is called the *out-of-plane* displacement. U and V can be measured independently by geometric moiré, moiré interferometry and microscopic moiré interferometry. W can be measured

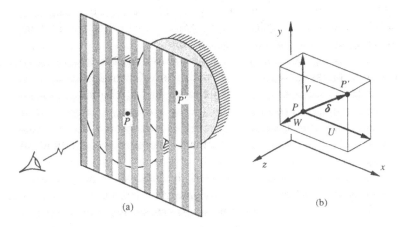

Fig. 3.19   (a) Point P moves to P' when the body is deformed. The x component of the displacement can be measured by means of the grating.   (b) A three-dimensional displacement is resolved into its U, V, W components.

independently by classical interferometry (e.g., with a Twyman-Green interferometer), holographic interferometry, shadow moiré and projection moiré.

Before discussing whole-field moiré measurements, consider the following concept.   Let point P on the specimen act as a tiny light source that glows continuously.   Let a reference grating on a glass substrate be fixed in front of the specimen, as illustrated in Fig. 3.19a; let the reference grating pitch and frequency be g and f, respectively. Let the observer be far away, such that every point is viewed at (nearly) normal incidence.   When the specimen is deformed or moved, P moves to P' while the reference grating remains stationary. The tiny light source appears to blink on and off as it is repeatedly obstructed by the opaque reference grating bars.   In fact, the number of intensity cycles or blinks, N, seen by the observer is

$$N = \frac{U}{g}$$

Since the number of blinks depends upon the number of opaque bars crossed by the bright point, the number is independent of V and W; it depends upon U alone.

Thus, the displacement U is the number of intensity cycles times the reference grating pitch, or

$$U = gN = \frac{1}{f}N \qquad (3.14)$$

U is the *absolute* displacement of point P, in the x direction, since N is the total number of cycles, independent of whether the movement is caused by deformation or rigid-body motion.

Next, let a grating of pitch g be printed on the specimen with its bars and spaces in registration with the reference grating. Let light be emitted from every point in the spaces of the specimen grating. The two gratings are originally in registration, so none of the light emitted from the specimen grating is obstructed by the reference grating. The observer sees a uniform (averaged) bright field and $N = 0$ everywhere.

If the reference grating is fixed, but the specimen experiences a rigid-body translation δ, the light emerging from the pair experiences N cycles of brightness/darkness. In this case, the instantaneous intensity is the same everywhere in the field; the number of whole-field intensity oscillations is given by Eq. (3.14). N is the moiré fringe order, and here it represents the absolute or total U displacement from the original position.

If the specimen and specimen grating deforms, displacement δ will vary from point-to-point in the field and (unless cracks are produced) the U displacements will vary as a continuous function of the x,y coordinates of the body. Point P cited earlier represents any and all light emitting points in the specimen and Eq. 3.14 applies for every point. Accordingly, N varies as a continuous function throughout the field. The absolute fringe order is determined by making the fringe count *while* the body is being deformed.

Of course, it is usually impractical to count the fringes while the body is being loaded or deformed. Fortunately, only *relative* displacements are needed for deformation analyses, and these can be determined without recourse to counting fringes during the loading operation. The following example illustrates the experimental analysis.

3.2.2 Practical Example, Relative Displacements

The demonstration illustrated in Fig. 3.20 investigates in-plane deformation. In (a), a deeply notched tensile specimen is imprinted with a uniform cross-line grating on its surface. The specimen is transparent and illuminated from behind, as in (b). A linear reference grating of the same frequency is superimposed in registration with the specimen grating; let the reference grating be oriented with its lines perpendicular to the y axis, so that it interacts with the corresponding set of lines in the cross-line specimen grating. The observer sees a *null field*, a field devoid of moiré fringes,

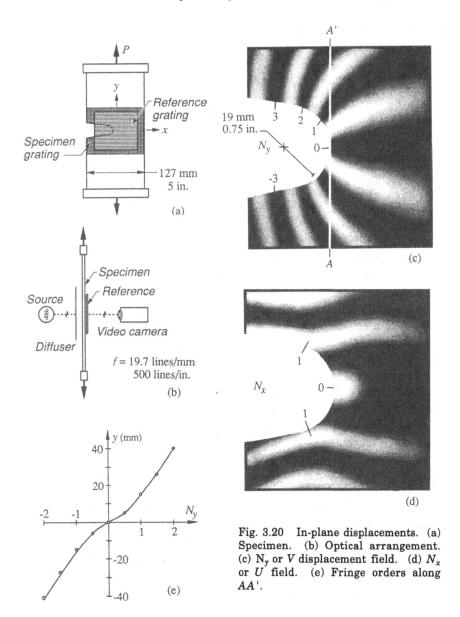

Fig. 3.20 In-plane displacements. (a) Specimen. (b) Optical arrangement. (c) N_y or V displacement field. (d) N_x or U field. (e) Fringe orders along AA'.

corresponding to the zero-load (zero deformation) condition of the specimen.

When tensile loads are applied, the specimen deforms, and the cross-line specimen grating deforms with it. Now, interaction of the specimen and reference grating produces the moiré pattern of Fig. 3.20c, with fringe orders N_y. The moiré pattern of Fig. 3.20d is formed by rotating the linear reference grating 90°, to interact with the orthogonal set of specimen grating lines; the fringe orders are designated N_x. As before, each pattern represents displacement components perpendicular to the lines of the reference grating. The displacements are

$$U(x,y) = gN_x(x,y) = \frac{1}{f}N_x(x,y)$$

$$V(x,y) = gN_y(x,y) = \frac{1}{f}N_y(x,y)$$

(3.15)

where the fringe orders are taken at the corresponding x,y points and where g and f represent the reference grating.

However, the fringe orders numbered in Fig. 3.20 are not absolute fringe orders. Instead, they were numbered by first assigning a zero-order fringe at a convenient location where the displacement was assumed to be zero. Thus, the moiré fringes in the figure represent displacements relative to an arbitrary zero-displacement datum. The free choice of a zero datum is equivalent to adding a freely chosen rigid-body translation to the absolute displacements; this approach is permissible because rigid-body motions have no importance for deformation studies. Only relative displacements are needed.

Accordingly, the normal procedure utilizes Eq. 3.15 together with an arbitrarily chosen zero-fringe-order location. The in-plane displacements of every x,y specimen point are determined from the fringe orders at the point, and the displacements are relative to the chosen zero-displacement datum.

The remaining question is how the non-zero fringe orders are assigned. They must change systematically as a parametric family of curves. In many cases, the question of whether the order of an adjacent fringe changes by +1 or −1 (i.e, whether the fringe orders are increasing or decreasing) is answered by the physics of the problem. For example, the specimen here is stretched in the y direction and therefore the relative displacements and fringe orders increase as y increases. The procedures for assignment of fringe orders in geometric moiré and moiré interferometry are identical. The question is treated in Sec. 4.10 in conjunction with moiré interferometry. In complex problems where the fringe ordering is not obvious, the question is answered by a simple experimental procedure described in Sec. 4.10.1.

Figure 3.20e is a graph of fringe orders along line AA', where the data points are taken at the centerlines of integral and half fringe orders. The graph emphasizes the fact that a fringe order exists at every point on the specimen. The intermediate fringe order of any x,y point can be estimated by visual interpolation, or else it can be determined by interpolation from a graph such as that in Fig. 3.20e.

3.2.3 Mismatch Fringes (Carrier Fringes)

With the relatively coarse gratings used in geometric moiré, the number of fringes in the field is usually small. The data for determination of displacements is not abundant, but the moiré mismatch method can be used to increase the number of data points.

The method is demonstrated in Fig. 3.21. A mismatch reference grating is used to produce a system of uniformly spaced fringes before the specimen is deformed. In this case the specimen and reference grating frequencies were 500 and 503 lines/in., respectively, so the frequency of mismatch fringes was 3 fringes/in. (0.12 fringes/mm). When the tensile load was applied, the mismatch pattern changed to that in (a). Fringe orders along AA' are plotted in (b).

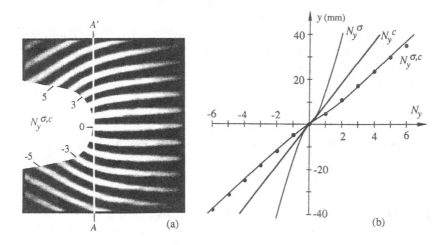

Fig. 3.21 (a) The N_y field with mismatch fringes; the tensile load is the same as in Fig. 3.20. (b) Graphs of N_y along AA'.

The superscripts c and σ in the figure signify mismatch (or carrier) fringes and stress-induced fringes, respectively. The $N_y^{\sigma,c}$ curve is plotted from the fringe pattern; N_y^c is known from the initial mismatched frequencies. The stress-induced fringe orders—those introduced by the tensile load—are the difference

$$N_y^\sigma = N_y^{\sigma,c} - N_y^c \qquad (3.16)$$

which is plotted in Fig. 3.21b. Of course, the result matches that of Fig. 3.20e.

Although additional data points become available to plot the curves, the accuracy of each data point is not improved. The ability to estimate the positions of points of maximum and minimum fringe intensity remains the same. By visual interpretation, N can be determined reliably within ±1/5 of a fringe order—and some investigators claim ±1/10—but the potential error of fringe order is the same with and without mismatch fringes. Nevertheless, the benefits are real. For smoothly varying displacement fields, the advantage of mismatch lies in the presence of more data points to improve the reliability of a smooth curve drawn through the points. Statistically, the improvement is realized because the errors are randomly distributed about the smooth curve. For displacement fields that contain an abrupt change of trends, the mismatch fringes also help to locate the region and magnitude of the abrupt change.

In geometric moiré, the initial fringes have been called *mismatch* fringes, consistent with the mismatch of specimen and reference grating frequencies. In moiré interferometry, the initial fringes are called *carrier* fringes; they are analogous to FM (frequency modulated) radio waves, where the information is the irregular part of an otherwise constant carrier frequency. Mismatch fringes and carrier fringes are names that represent the same phenomenon.

Specimen or Space Coordinates

The displacements caused by the tensile load in Fig. 3.21 are

$$V_{(x,y)} = \frac{1}{f}N_y^\sigma \qquad (3.17)$$

where the fringe order is taken at the corresponding x,y points. Of course, Eqs. 3.16 and 3.17 apply for every point in the field, and equivalent equations apply for the U displacement field and N_x fringes.

Equation 3.17 raises the question of which frequency to use, that of the specimen grating or reference grating. The answer depends upon the definition of the x,y coordinate system, and two different definitions should be considered.

One is when the x,y coordinates are attached to material points on the specimen and move as the specimen is deformed. By applying the concepts of Sec. 3.2.1, where any point P moves to P' and crosses N bars of the reference grating, we find that the number of intensity oscillations depends upon the pitch or frequency of the *reference* grating. Accordingly, f represents the reference grating; $f = 503$ lines/in. would be used for Fig. 3.21 when specimen coordinates are chosen.

The alternate situation appears when the x,y coordinates are fixed in space. They do not move as the specimen deforms. Then, a fixed point P on the *reference* grating is crossed by N lines of the specimen grating as the specimen deforms. The specimen point corresponding to P (in the undeformed specimen configuration) experienced a displacement of N pitches, i.e., N times the original pitch of the specimen grating. Consequently f represents the *specimen* grating in this case; $f = 500$ lines/in. would apply when fixed coordinates are chosen.

In summary, the value of f in Eq. 3.17 is the frequency of the reference grating when the x,y coordinates represent the deformed configuration of the specimen; f is the original frequency of the specimen grating when the x,y coordinates represent the initial configuration. Note that the issue does not arise in Eq. 3.15, since the reference grating frequency and the initial specimen grating frequency are equal. The issue appears only when mismatch fringes are present. Note, too, that the choice of specimen or space coordinates influences the results by only 0.6% in this case.

3.2.4 Fringe Shifting

Another method for increasing the number of data points is *fringe shifting*. It is accomplished by shifting or moving the reference grating relative to the specimen grating. A shift of δ (perpendicular to the grating lines) causes every fringe to move; the change of fringe order at every point in the field is

$$\Delta N = \frac{\delta}{g} = f\delta \qquad (3.18)$$

where g and f are the pitch and frequency of the reference grating. Fringe shifting is demonstrated in Fig. 3.13 for a homogeneous moiré pattern and in Fig. 3.22 for a nonhomogeneous displacement field. In Fig. 3.22 the mechanical shifts are $\delta = 0, g/3$ and $2g/3$, which produce fringe shifts of 0, 1/3 and 2/3 of a fringe order, respectively.

Fringe orders are marked on the patterns. In this case, fringe shifting provides three times the usual number of data points for plotting fringe orders along line AA', or along any other line. Again, the accuracy of each data point is not improved, but the increased number of data points provides a graph with statistically improved accuracy.

This technique does not substantially increase the resolution of displacement measurements. A large improvement would be achieved if the shifted fringes were also highly sharpened. Then, the locations of points of known fringe orders would be ascertained with greater accuracy, and many fringe shifts could be used to produce a displacement map with many contours. Such a method has been developed. It is called the *optical/digital fringe multiplication method* and its application to geometric moiré is presented in Appendix 1.

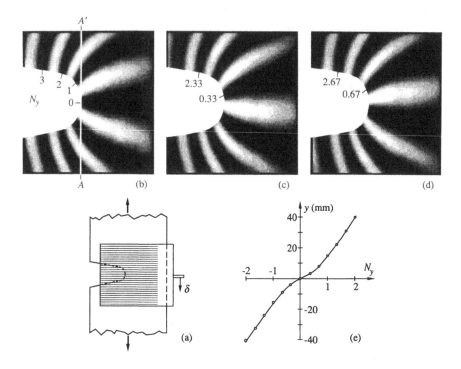

Fig. 3.22   (a) Fringe shifting is accomplished by translating the reference grating by δ. In (b), (c) and (d), the shifts are $\delta = 0$, $g/3$ and $2g/3$, respectively. (e) Fringe order along line AA'.

3.3 Out-of-plane Displacements

Two techniques for measuring out-of-plane displacements, W, by geometric moiré are described. They are the *shadow moiré* and *projection moiré* methods. Both measure W uniquely. Although U and V may also be present, the methods sense and measure only W. Both methods determine the topography of the specimen surface, i.e., its deviation from a plane surface. In mechanics, they are used to measure the topography of initially flat surfaces to evaluate warpage caused by load, temperature, humidity, age, or other variables. In metrology, they are used to document the shape of three-dimensional bodies.

3.3.1 Shadow Moiré

Figure 3.23 illustrates the basic concepts of shadow moiré. The specimen surface is prepared by spraying it with a matte white paint. A linear reference grating of pitch g is fixed adjacent to the surface. The grating is comprised of black bars and clear spaces on a flat glass plate. A light source illuminates the grating and specimen, and the observer (or camera) receives the light that is scattered in its direction by the matte specimen.

The explanation assumes rectilinear propagation of light; the assumption is reasonable when the gap W is small compared to the grating self-imaging distance g^2/λ, e.g., within 5% of g^2/λ (Sec. 2.5.3). Then, a clear shadow of the reference grating is cast upon the specimen surface. This shadow, consisting of dark bars and bright bars is itself a grating. The observer sees the shadow grating and the superimposed reference grating, which interact to form a moiré pattern.

In the cross-sectional view of Fig. 3.23, light that strikes points A, C and E on the specimen passes through clear spaces in the reference grating to reach the observer. Light that strikes points B

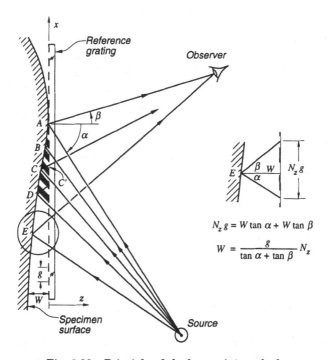

$$N_z g = W \tan \alpha + W \tan \beta$$

$$W = \frac{g}{\tan \alpha + \tan \beta} N_z$$

Fig. 3.23   Principle of shadow moiré method.

and D are obstructed by opaque bars of the reference grating. Thus, bright moiré fringes are formed in the regions near A, C, and E, while black fringes are formed near B and D.

Let the fringe order at A be $N_z = 0$, since the specimen and reference grating are in contact and $W = 0$. From the viewpoint of the observer, the number of shadow grating lines between A and C is compared to the number of reference grating lines between A and C'. The number of shadow grating lines is larger by one, and therefore the fringe order at C is $N_z = 1$. Similarly at E, the number of shadow grating lines is larger by two and the moiré fringe order at E is $N_z = 2$.

The relationship between the fringe order and the gap W is derived in the insert, which is an enlargement of the region around point E. Of course, W represents the out-of-plane displacement of point E. The angles of incoming and outgoing rays are α and β; the angles are both positive as shown, since clockwise angles are positive when the light travels with a component in the $-z$ direction (Sec. 2.5). The incoming and outgoing rays intercept the reference grating at points gN_z apart. Thus, the geometry provides the relationship shown in the figure,

$$W = \frac{g}{\tan\alpha + \tan\beta} N_z \tag{3.19}$$

Notice that angles α and β are variables that change with coordinate x. Therefore, the sensitivity of the measurement is not constant, i.e., the displacements are not directly proportional to fringe orders of the moiré pattern. Two schemes can be used to assure the constant proportionality, and it is usually sensible to apply one of them.

One is the scheme of Fig. 3.24, which prescribes that the light source and observer are at the same distance L from the plane of the specimen. With that constraint,

$$\tan\alpha + \tan\beta = \frac{D}{L} = K \tag{3.20}$$

where K is a constant, and

$$W = \frac{g}{K} N_z \tag{3.21}$$

Thus, W is directly proportional to the moiré fringe order. Actually, Eq. 3.20 is an approximation, since the distance to the specimen is $L + W$; however $W \ll L$ in practice and W can be neglected. This scheme is very practical and it is usually the arrangement of choice.

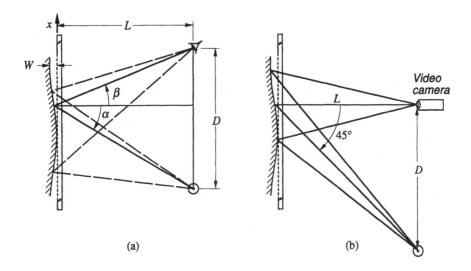

Fig. 3.24   Shadow moiré arrangements in which $\tan \alpha + \tan \beta$ = constant.

In the case of Fig. 3.24b, $D = L$ and the constant $K = 1$. Thus, the contour interval of the shadow moiré pattern becomes equal to the reference grating pitch g; the displacement becomes g per fringe order. An especially desirable feature of this arrangement is normal viewing, which results in a distortion-free view of the specimen.

The collimated light arrangements of Fig. 3.25 also provide constant sensitivity, since α and β are constants in Eq. 3.19. They are practical for small fields of view.

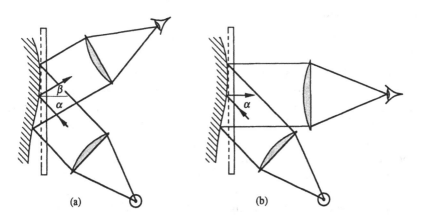

Fig. 3.25   Shadow moiré with collimated beams.

Gap Effect

A gap effect occurs in the shadow moiré arrangement of Fig. 3.23. Notice that the shadow of the reference grating is magnified by a factor of $1 + W/L$, where L is the distance from the source to the plane of the reference grating. If W is a uniform, finite gap, the magnification causes a pattern of mismatch fringes and these extraneous fringes are a source of error in the shadow moiré pattern.

If the camera and source are located at the same distance from the plane of the specimen, as is Fig. 3.24, the error is canceled. The cancellation occurs because the projection of the shadow on the reference grating is demagnified by the same factor when it is viewed from the same distance. The gap effect is another reason why the configurations of Fig. 3.24 are preferred.

Practical Considerations

In all these arrangements, the light source should be small, approaching either a point or a slit (oriented parallel to the reference grating lines). Monochromatic purity is not important and an incandescent lamp with a line filament or small filament is a good choice. When a camera is used to record the pattern, the aperture should be small or narrow, e.g., a narrow slit in front of the camera lens. Otherwise, when a broad source or large camera aperture is used, fringe contrast degrades with increasing gap W; then W should be restricted to a smaller range, perhaps to within 1% of g^2/λ.

It is often convenient to record the fringe pattern with a video camera. In most cases, only a few shadow moiré fringes appear in the field. Therefore, the pattern is comprised of several pixels per fringe, and a video camera is an excellent choice.

Fringe Ordering

To correctly interpret the results, it is important to know whether the moiré fringe pattern represents a bulge or depression of the specimen surface, i.e., whether a fringe order is larger or smaller than that of the neighboring fringe. Sometimes the answer is obvious from the nature of the experiment, but when it is not, a simple procedure can be used. While observing the fringes, gently press the reference grating toward the specimen so as to decrease the gap W. The fringes will move in the direction of *higher* fringe orders. For example, a second order fringe in the original pattern will move toward the position of the original third order fringe when the gap is reduced; the fringes will move toward the depression. Again, the zero-order fringe can be assigned arbitrarily for shape or deformation measurements.

Fringe Visibility

The moiré pattern formed by the reference grating and its shadow has the character illustrated in Fig. 3.2. The bright fringes are crossed with the lines of the reference grating. These extraneous lines can be made to disappear by a few different methods.

In an elegant method,[10,11] the reference grating is in motion during the time interval of the photographic exposure. The grating is translated in its plane, perpendicular to the direction of the grating lines. At points of destructive interference, as at point D of Fig. 3.23, either the incoming or outgoing ray is always obscured by the reference grating; destructive interference persists at D even as the grating translates. At points of constructive interference such as point E, light passes through the reference grating spaces during half of the exposure time and it is obstructed by the grating bars during the other half. Thus, translation of the reference grating has no influence on the positions of the moiré fringes, but the grating bars are smeared and become invisible because of their motion. Referring to Fig. 3.2b, the triangular intensity distribution is recorded even when the pattern is viewed with a high-resolution optical system. Averaging produced by reduced resolution is not needed.

An equivalent scheme is to record the pattern with a double exposure, wherein the reference grating is translated by half its pitch (by $g/2$) between exposures. This method assumes the usual type of reference grating, with equal widths of bars and spaces.

When a video camera is used, the reference grating lines disappear when the camera magnification is adjusted to provide a whole number of grating lines in the active width of the pixel. The intensity across each pixel is averaged, thus eliminating the grating lines.

Of course, the averaging method already mentioned in connection with in-plane measurements is effective here, too. The camera aperture can be decreased to prevent resolution of the reference grating lines, while the coarser moiré fringes remain well resolved. A useful technique is to use a slit aperture in front of the camera lens (or inside it), with the length of the slit parallel to the reference grating lines. With a slit, more light is admitted into the camera, and resolution in the direction parallel to the grating lines remains high.

Demonstration

The experiment depicted in Fig. 3.26 was performed by shadow moiré. The specimen was an aluminum channel subjected to compressive loading in a mechanical testing machine. It was painted matte white. The reference grating frequency was 300 lines/in. ($\approx$ 12 lines/mm). It was mounted on a separate fixture about 1/2 mm (0.02 in.) from the specimen surface; the gap was about 3.6% of g^2/λ.

The optical arrangement of Fig. 3.24b was used. Distances L and D were 762 mm (30 in.), providing $K = 1$ (Eq. 3.21) and a contour interval of 0.085 mm (0.0033 in.) per fringe order. The light source was a 250 watt mercury vapor arc lamp with an arc length of about 5 mm (0.2 in.). A video camera was used with a lens of about 30 mm (1.2 in.) focal length and 3 mm (0.12 in.) aperture.

The same specimen was used to demonstrate fringe multiplication in Appendix A.

3.3.2 Projection Moiré

One Projector, Deformations

Variations of projection moiré will be considered. The first is illustrated in Fig. 3.27, where a specimen (with a matte white surface) is illuminated by an array of uniformly spaced walls of light. Let the shaded bars represent the presence of light and the blank spaces represent the absence of light. The camera sees bright bars on the surface of the specimen, separated by dark spaces from which no light is emitted. The illuminated specimen is photographed first in its original condition; later it is photographed on another film in its deformed condition, after loads are applied. After development, the two films are bar-and-space gratings, and they create a moiré pattern when they are superimposed. By inspection of Fig. 3.27, it is seen that the band of light that intercepted point b when the specimen was in its original condition is three pitches removed from the light that strikes B, which is the same point after deformation. Thus, the moiré fringe order at B is 3. Similarly, it is 2.5 at A and 4 at C.

From triangle abB, we find that the out-of-plane displacement at B is

$$W = \frac{gN_z}{\tan \alpha}$$

where N_z is the fringe order at B. The relationship is the same as that of shadow moiré.

In fact, when the optical configuration is generalized to that of Fig. 3.28, the patterns of projection moiré and shadow moiré are

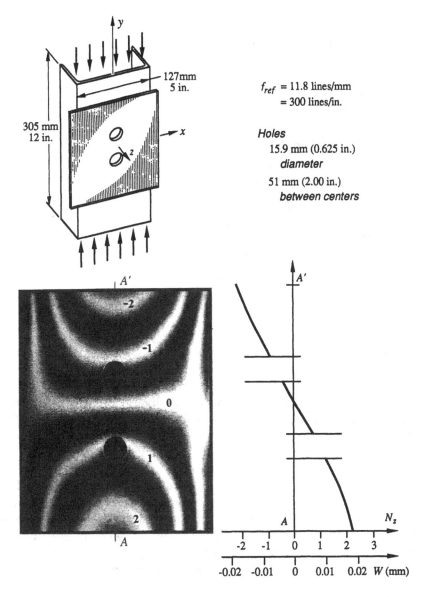

f_{ref} = 11.8 lines/mm
    = 300 lines/in.

Holes
    15.9 mm (0.625 in.)
        diameter
    51 mm (2.00 in.)
        between centers

Fig. 3.26   Shadow moiré applied to an aluminum structural member in compression. The optical arrangement of Fig. 3.24b was used.

identical. The gratings projected onto the original and deformed specimen surfaces are the same as the reference grating and its shadow in the corresponding shadow moiré system. The equations are identical; Eqs. 3.19 and 3.21 apply for projection moiré when collimated light is used.

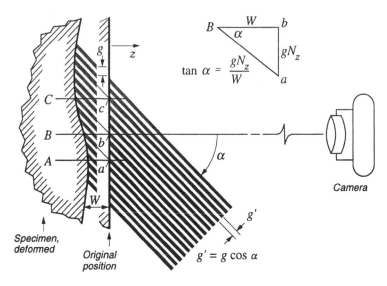

Fig. 3.27   Schematic of double-exposure projection moiré for deformation measurements.

The projector of Fig. 3.28a uses a fine grating to project a comparatively coarse grating. An alternative system for creating the projected grating is sketched in Fig. 3.28b. The input laser beam is divided into two beams which appear to diverge from two adjacent points in space. Their separation is adjusted by the inclination of the mirrors. After passing the lens, two collimated beams intersect at an angle 2θ to produce a projected grating of pitch g'. The frequency F of this projected grating is determined by Eq. 2.24 and numerical examples are listed in Table 2.2. In practice, θ is very small. The assembly of Fig. 3.28b and its use for creating two intersecting beams is well known; the arrangement is sometimes called a Michelson interferometer (Sec. 2.3.7). In principle, projected gratings of excellent quality can be produced, but in practice, speckle noise associated with coherent illumination of a matte specimen surface causes some deterioration.

One Projector, Topography

When three-dimensional shapes must be documented, rather than *changes* of shape, the advantage of before-and-after grating images is lost. It is not possible to superimpose images that are recorded before and after the changes. Then, the scheme of Fig. 3.29 can be applied. A real bar-and-space grating is laminated onto the ground-glass screen in the observation leg. Camera lens A is focused to image the specimen surface on the ground-glass screen. Thus, an image of the distorted grating from the curved surface is superimposed upon the real grating to form a moiré. The pitch and alignment of the real grating corresponds to the image that would be formed by placing a flat reference plane in the specimen space. Therefore, the moiré pattern is a contour map of the separation W between the specimen surface and the flat reference plane. The moiré pattern is recorded by a camera located behind the screen.

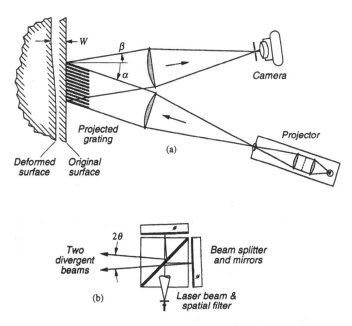

Fig. 3.28   (a) Projection moiré using a slide projector. (b) Alternate means to produce the projected grating by interference of light; this assembly is used instead of the projector.

The result is multiplicative moiré, generally characterized by fringes of good visibility. However, the distorted grating lines on the specimen must be resolved sharply by the camera lens A, and this means that very fine gratings cannot be used. When deeply warped surfaces are observed, the frequency of moiré fringes can become similar to the frequency of the grating lines, thus reducing the fringe visibility. As with shadow moiré, an elegant technique to wash out the grating lines uses moving gratings.[12] The technique can produce moiré fringes of excellent visibility. Reference 12 also cites numerous publications on projection moiré.

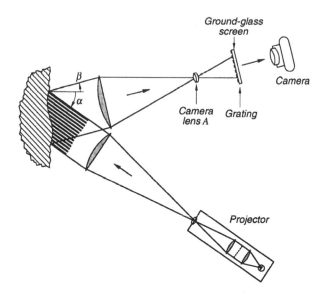

Fig. 3.29   Projection moiré for surface topography.

Two Projector Moiré

Another variation is illustrated in Fig. 3.30. In this case two sets of parallel walls of light are incident from symmetrical angles and they cross in space as depicted in Fig. 3.30b. The specimen intercepts the light, which casts two sets of bright bars on the matte specimen surface. They interact to produce a moiré pattern of additive intensity (Sec. 3.1.3).

Let the shaded walls represent the presence of light. The patterns of light received locally on the deformed specimen surface depends upon where the array of Fig. 3.30b is cut by the specimen surface. If some region of the specimen surface is represented by line a, that region is illuminated uniformly; in regions where the specimen surface coincides with lines b,d, and f, the surface is illuminated by alternating bright bars and dark spaces. The pattern on the specimen is characterized by moiré fringes of additive intensities, as depicted in Fig. 3.5a. To record the moiré fringes in the camera, the individual grating lines must be coarse enough to be individually resolved.

The out-of-plane displacement of points on one moiré fringe relative to the neighboring fringe is A, where A is the distance between planes a and c in Fig. 3.30b. From the triangle sketched in Fig. 3.30c, $A = g/(2 \tan \alpha)$. The out-of-plane displacement of any point is $W = A N_z$ and therefore

$$W(x,y) = \frac{g}{2\tan\alpha}N_z(x,y) \qquad (3.22)$$

where W is relative to an arbitrary datum where the moiré fringe order is assigned $N_z = 0$.

Note that the relationship is the same as that for shadow moiré when $|\beta| = \alpha$. This method has the advantage that the specimen can always be viewed at normal incidence. It has the disadvantages of additive intensity moiré, which requires resolution of the individual grating lines. Reference 12 provides means to transform the pattern to multiplicative moiré fringes and to wash out the grating lines; however, resolution of the individual grating lines is still required in the observation leg.

3.3.3 Comparison

Compared to shadow moiré, projection moiré has the advantage that no element of the apparatus is required to be close to the specimen. It has the disadvantage that the optical systems are more complicated, and the individual lines of the projected gratings must be resolved in the observation leg. With shadow moiré, only the moiré fringes must be resolved.

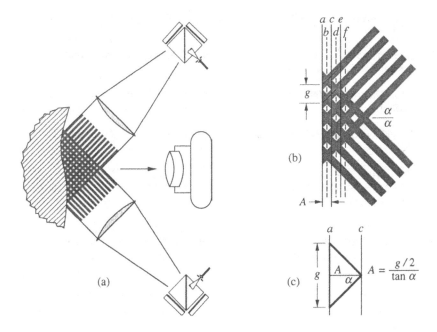

Fig. 3.30   (a) Two projector method. (b) The geometry of intersecting walls of light. (c) The geometry within one pitch of the projected specimen grating.

3.4 References

1. A. Livnat and D. Post, "The Grating Equations for Moiré Interferometry and Their Identity to Equations of Geometrical Moiré," *Experimental Mechanics*, Vol. 25, No. 4, pp. 360-366 (1985).

2. A. J. Durelli and V. J. Parks, *Moiré Analysis of Strain*, Prentice-Hall, Englewood Cliffs, New Jersey (1970).

3. G. Mesmer, "The Interference Screen Method for Isopachic Patterns," *Proc. Society for Experimental Stress Analysis*, Vol. XIII, No. 2, pp. 21-26 (1956).

4. D. Post, R. Czarnek and A. Asundi, "Isopachic Contouring of Opaque Plates," *Experimental Mechanics*, Vol. 24, No. 3, pp. 169-176 (1984). Also: J. D. Duke, Jr., D. Post, R. Czarnek and A. Asundi, "Measurement of Displacement Around Holes in Composite Plates Subjected to Quasi-Static Compression," *NASA Contractor Report 3989* (1986).

5. D. Post, "Holography and Interferometry in Photoelasticity," *Experimental Mechanics*, Vol. 12, No. 3, pp. 113-123 (1972).

6. D. Post, R. Czarnek, D. Joh and J. Wood, "Deformation Measurements of Composite Multi-span Beam Shear Specimens by Moiré Interferometry," *NASA Contractor Report 3844* (1984). Also: D. Post, R. Czarnek and D. Joh, "Shear Strain Contours from Moiré Interferometry," *Experimental Mechanics*, Vol. 25, No. 3, pp. 282-287 (1985).

7. G. L. Rogers, "A Geometrical Approach to Moiré Pattern Calculations," *Opt. Acta*, Vol. 24, No. 1, pp. 1-13 (1977).

8. D. Post, "Sharpening and Multiplication of Moiré Fringes," *Experimental Mechanics*, Vol. 7, No. 4, pp. 154-159 (1967).

9. P. S. Theocaris, *Moiré Fringes in Strain Analysis*, Pergamon Press, New York (1969).

10. H. Takasaki, "Moiré Topography," *Applied Optics*, Vol. 9, No. 6, pp. 1467-1472 (1970).

11. J. B. Allen and D. M. Meadows, "Removal of Unwanted Patterns from Moiré Contour Maps by Grid Translation Techniques," *Applied Optics*, Vol. 10, No. 1, pp. 210-212 (1971).

12. M. Halioua, R. S. Krishnamurthy, H. Liu and F. P. Chiang, "Projection Moiré with Moving Gratings for Automated 3-D Topography," *Applied Optics*, Vol. 22, No. 6, pp. 850-855 (1983).

3.5 Key Relationships

pure rotation

$$G = \frac{g}{\theta} \tag{3.3}$$

$$F = f\theta \tag{3.4}$$

pure extension

$$F = f_1 - f_2 \tag{3.6}$$

parametric curves, fringe vectors

$$\boldsymbol{F}_N = \boldsymbol{f}_L - \boldsymbol{f}_M \tag{3.7}$$

$$\boldsymbol{F}_{N^+} = \boldsymbol{f}_L + \boldsymbol{f}_M \tag{3.8}$$

fringe multiplication

$$f = \beta f_s \tag{3.11}$$

in-plane displacements

$$U(x,y) = gN_x(x,y) = \frac{1}{f}N_x(x,y)$$
$$V(x,y) = gN_y(x,y) = \frac{1}{f}N_y(x,y) \tag{3.15}$$

mismatch or carrier fringes

$$N_y^\sigma = N_y^{\sigma,c} - N_y^c \tag{3.16}$$

shadow moiré

$$W = \frac{g}{\tan\alpha + \tan\beta} N_z \tag{3.19}$$

$$W = \frac{g}{K} N_z \tag{3.21}$$

3.6 Exercises

3.1 Let the two equal gratings of Fig. 3.2a be oriented with angles +5° and –5° with respect to the x axis. (a) Derive the exact relationship for the pitch (G) of the moiré fringes in terms of the grating pitch (g) and the included angle of $\theta = 10°$. (b) What is the percent error when G is determined by the approximation given by Eq. 3.3? Answers: $G = g/[2 \tan (\theta/2)]$; 0.25%.

3.2 Using an office-type copying machine, make enlargements of Fig. 3.7, (a) and (b). Designate the as-loaded and no-load patterns as L and M families, respectively. (a) Assign fringe orders to the L and M patterns, beginning with 100 at the upper-right corner of the notch. (b) Mark a few points on the L pattern where $L - M = -5$. (c) What assumptions were made in assigning the fringe orders?

3.3 (a) Given the experimental arrangement of Fig. 3.7e, explain why the contour interval in (a) and (b) is $\lambda/2n$. (b) Explain why the contour interval for moiré fringes in (c) or (g) is the same as the contour interval for (a) and (b)?

3.4 (a) Using the data in Fig. 3.9c, plot the curve of thickness change ($2W$) versus position (x) along the horizontal centerline. (b) Briefly describe the trends or features exhibited by the graph.

3.5 (a) Assign index numbers (fringe orders) to the L and M families in Fig. 3.10; choose an arbitrary index number for one curve in each of these families. (b) Determine the corresponding index numbers for the curves in the N and N^+ parametric families.

3.6 The copper rod on the left side is heated by a torch. (a) Design an experiment using *shadow moiré* to determine its change of length. Show the optical arrangement and specify the parts. (b) Sketch the fringe pattern. (c) Derive the relationship between the fringe pattern and the change of length of the rod. (d) Estimate the uncertainty of the change of length measurement. (e) List the assumptions that were made.

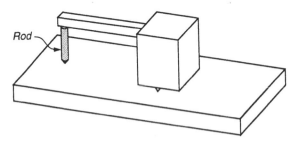

3.7 Same as Éxercise 3.6, except in-plane geometric moiré is to be used.

3.8 Consider the experiment of Fig. 3.9. Initially, the specimen is not perfectly flat, i.e., its surfaces are slightly warped. When the compressive load is applied, the general warpage of the plate increases and its plane at mid-thickness becomes increasingly warped. Explain with words and diagrams why the experiment measures the change of thickness uniquely, i.e., why the pattern is independent of the change of mid-plane warpage.

3.9 In the experiment of Fig. 3.9, let the two holographic plates be given opposite angular inclinations for the second exposure (as shown exaggerated, below) instead of the equal inclinations of Fig. 3.9a.   (a) What would the resultant fringe pattern—corresponding to Fig. 3.9c—represent in this case?   (b) Explain why with words and diagrams.   Answer: It is a contour map of the change of mid-plane warpage. (See Exercise 3.8.)

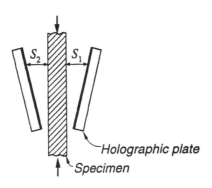

4
Moiré Interferometry

4.1 Introduction

Moiré interferometry combines the concepts and techniques of geometrical moiré and optical interferometry. In his definitive monograph, Guild[1] shows that all moiré phenomena can be treated as cases of optical interference, although moiré generated by low-frequency bar-and-space gratings can also be explained by obstruction or mechanical interference. Moiré interferometry is capable of measuring in-plane displacements with very high sensitivity. A sensitivity of 2.4 fringes/μm displacement is demonstrated for all the patterns in this chapter except Figs. 4.7 and 4.45, where the sensitivity is 4.0 and 1.2 fringes/μm displacement, respectively. In Chapter 5, the sensitivity is even higher.

Moiré interferometry provides whole-field patterns of high spatial resolution and excellent clarity. Figure 4.1 is an example;[2] the fringes are well defined, even where they are very closely spaced.

The relationship between fringe order and displacement is the same as in geometric moiré, namely

$$U_{(x,y)} = \frac{1}{f} N_x(x,y)$$

$$V_{(x,y)} = \frac{1}{f} N_y(x,y)$$

(4.1)

where f is the frequency of the (virtual) reference grating; $f = 2400$ lines/mm (60,960 lines/in.) is typical. Experimental determinations of displacement fields are important in themselves. Where the objective is to evaluate theoretical or numerical analyses by experimental measurements, there is no need to reduce the experimental displacement fields to strains or stresses. Instead, the verification can be made directly by comparison of experimental and theoretical (or numerical) displacement fields.

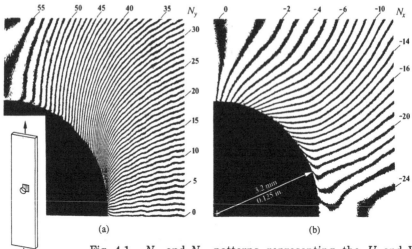

Fig. 4.1 N_x and N_y patterns representing the U and V displacement fields in a multi-ply graphite-epoxy composite specimen. The ply stacking sequence is $[0_2/\pm45/90]_S$. $f = 2400$ lines/mm (60,960 lines/in.).

Where strains are required, they can be extracted from the displacement fields by the relationships for engineering strains

$$\varepsilon_x = \frac{\partial U}{\partial x} = \frac{1}{f}\left[\frac{\partial N_x}{\partial x}\right]$$

$$\varepsilon_y = \frac{\partial V}{\partial y} = \frac{1}{f}\left[\frac{\partial N_y}{\partial y}\right]$$
(4.2)

$$\gamma_{xy} = \frac{\partial U}{\partial y} + \frac{\partial V}{\partial x} = \frac{1}{f}\left[\frac{\partial N_x}{\partial y} + \frac{\partial N_y}{\partial x}\right]$$
(4.3)

Of course, they apply to every x,y point in the field. These are the equations for small strains and small rotations. In view of the high sensitivity of moiré interferometry, the problems for which it is suited fall into this category. For example, in the case of pure rotation of 1° and grating frequency of $f = 2400$ lines/mm, the moiré pattern consists of 42 fringes/mm or 1060 fringes/in. (Eq. 3.4). Such dense fringe patterns are seldom encountered. In rare instances, the finite strain relationships[3] are appropriate.

The in-plane rotation ψ of any element in the field is defined by

$$\psi = \frac{\partial U}{\partial y} - \frac{\partial V}{\partial x} = \frac{1}{f}\left[\frac{\partial N_x}{\partial y} - \frac{\partial N_y}{\partial x}\right]$$
(4.3a)

Differentiation is required, but differentiation of experimental data always involves a loss of accuracy. A reasonable estimate is this: if displacements are known to 99% accuracy, the derivatives of displacement can be established to 90% accuracy. Careful procedures can improve the results, but a loss of an order of magnitude seems to be normal. The loss occurs regardless of the method used for differentiation. Nevertheless, because of the high sensitivity and abundance of displacement data from moiré interferometry, reliable strains can be extracted from the displacement fields.

Figure 4.2 is a schematic description of the moiré interferometry method we will study. A cross-line diffraction grating is produced on the specimen. This is a symmetrical, phase-type grating of the sort shown in Fig. 2.35, parts (f) and (i). The specimen grating is firmly bonded to the specimen. When loads are applied to the specimen, the surface deformation is transferred to the grating with high fidelity.

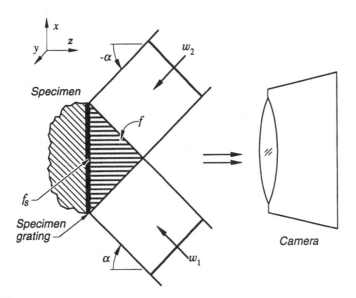

Fig. 4.2   Schematic diagram of moiré interferometry. The deformed specimen grating interacts with the virtual reference grating to form a moiré pattern. The phenomenon is analogous to geometric moiré.

Two beams of coherent light illuminate the specimen grating obliquely at angles $+\alpha$ and $-\alpha$. The two beams generate walls of constructive and destructive interference, i.e., a virtual grating, in the zone of their intersection (Fig. 2.43). The virtual grating is cut by the specimen, and thus an array of very closely spaced bright and dark bars appear on the specimen surface. These very closely spaced bars act like the reference grating of geometrical moiré. Their frequency is f, where

$$f = \frac{2}{\lambda}\sin \alpha \qquad (4.4)$$

This is a restatement of Eq. 2.24 in terms of f and α, which are symbols that represent the frequency of the virtual reference grating and the angle of the beams that create it.

The deformed specimen grating and the virtual reference grating interact to form a moiré pattern, which is viewed and photographed in the camera. This casual description repeats the essential description of geometric moiré and serves to emphasize their common qualities. A more rigorous description of moiré interferometry will follow. However, the casual interpretation faithfully predicts the moiré patterns, since the governing equations of geometric moiré and moiré interferometry are identical.[4]

After gaining an understanding of alternate and more rigorous descriptions, the reader is encouraged to view moiré interferometry as the interaction between a deformed specimen grating and an undeformed reference grating, analogous to geometric moiré. Such a treatment encourages a visual construction of the optical event, an interpretation that relates the deformation of the specimen to the moiré pattern.

In moiré interferometry, the initial frequency of the specimen grating, f_s, is half that of the virtual reference grating, f, i.e.,[*]

$$f_s = \frac{f}{2} \qquad (4.5)$$

The sensitivity of the measurements is controlled by the reference grating (Sec. 4.6). This condition is directly analogous to geometric moiré with a fringe multiplication factor of $\beta = 2$.

Generally, both the U and V displacement fields are desired. Together they fully define the in-plane deformation of the surface, and they are sufficient to determine the full state of strain at the surface, i.e., to find ε_x, ε_y and γ_{xy}. The optical arrangement sketched in Fig. 4.3 provides both fields. The moiré pattern depicting the U displacement field is formed by interaction of the x family of specimen grating lines (lines perpendicular to x) with the virtual reference grating formed by beams B_1 and B_2. The V field is formed

---

[*] Factors other than 1/2 are discussed in Sec. 4.21.3.

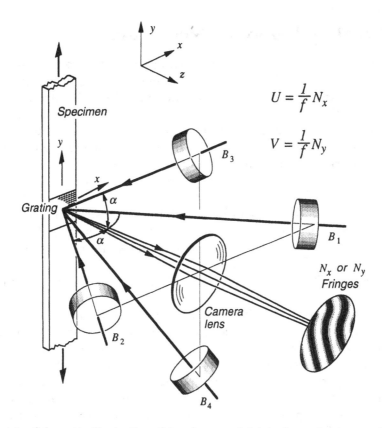

$$U = \frac{1}{f} N_x$$

$$V = \frac{1}{f} N_y$$

Fig. 4.3   Schematic illustration of four-beam moiré interferometry to record the N_x and N_y fringe patterns, which depict the U and V displacement fields.

by interaction of the y family of grating lines with the virtual grating from B_3 and B_4. For static problems, the two moiré patterns are photographed separately by blocking alternate pairs of incoming beams for each photograph.

Moiré interferometry is characterized by a list of excellent qualities, including

- real-time technique: the displacement fields can be viewed as loads are applied.
- high sensitivity to in-plane displacements U and V: typically 2.4 fringes/µm displacement, and higher for microscopic analyses; virtually no sensitivity to out-of-plane displacements W.
- high spatial resolution: measurements can be made in tiny zones.
- high signal-to-noise ratio: the fringe patterns have high contrast, excellent visibility.
- large dynamic range: the method is compatible with a large range of displacements, strains, and strain gradients.

4.2 A More Rigorous Explanation

The basic elements of a moiré-interferometry system are illustrated again in Fig. 4.4. The phase-type specimen grating of frequency f_s intercepts two beams of mutually coherent light which are incident at symmetrical angles α. One scheme for creating these two beams, beams 1 and 2, is illustrated in (a), where the beams are represented by corresponding rays. The beams are separated by the beam splitter and they are redirected by mirrors to meet at the specimen surface. The light should be polarized with its plane of polarization either perpendicular or parallel to the plane of the diagram. The specimen grating diffracts the incident light such that beams represented by 1' and 2' emerge from each specimen point and propagate to the camera. These beams interfere and generate an interference pattern on the film plane of the camera.

Of course, Fig. 4.4a illustrates an interferometer. It is represented by Fig. 2.24, where the beam splitter is the entrance to the box and the specimen grating is the object plane. The qualities discussed in Sec. 2.3 for two-beam interference apply fully to moiré interferometry. Various optical arrangements can be used to produce incident beams 1 and 2 but in each case the two beams are divided from a common beam, they travel different paths, and they meet again at the specimen. Two mutually coherent beams emerge from the specimen grating with warped wave fronts; they coexist in space and generate optical interference.

Figure 4.4b illustrates additional details. At the top and bottom of the figure, diffracted beams stemming from incident beams 1 and 2 are illustrated by ray diagrams. These diffraction directions are prescribed by the grating equations, Eqs. 2.54–2.56.

When $f_s = f/2$ and the lines of the specimen and reference gratings are exactly parallel, light from beam 1, diffracted in the first order of the specimen grating, emerges with angle $\beta_1 = 0$. Light from beam 2 that is diffracted in the −1 order emerges along $\beta_{-1} = 0$. The reason for the difference of grating frequencies becomes clear: when $f_s = f/2$, light is diffracted normal to the specimen. The specimen is viewed at normal incidence and the image of the specimen remains undistorted. If $f_s = f$, no diffraction order would emerge normal to the specimen and viewing at normal incidence would be impossible. Thus, the choice of $f_s = f/2$ is governed by the direction of diffracted light.

To show this relationship, note that the diffraction equation (Eq. 2.54) applied to the specimen grating is

$$\sin \beta_m = \sin \alpha + m\lambda f_s$$

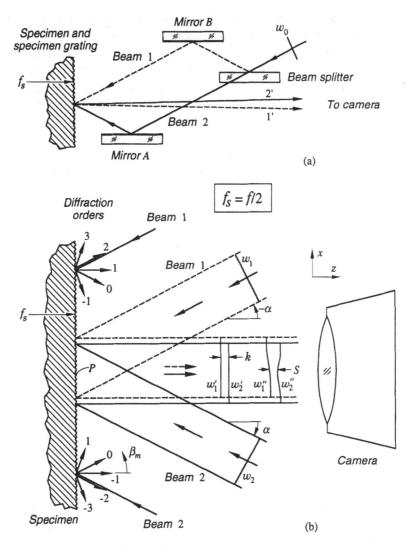

Fig. 4.4   (a) Schematic diagram of an interferometer for moiré interferometry. (b) Diffraction by the specimen grating produces output beams with plane wave fronts w_1' and w_2' for the initial no-load condition. Warped wave fronts w_1'' and w_2'' result from nonhomogeneous deformation of the specimen.

Substituting the prevailing conditions for beam 1:

$$m = 1,\ \sin(-\alpha) = -\lambda f/2 \ \text{(by Eq. 4.4), and } f_s = f/2$$

we obtain the result

$$\sin \beta_1 = 0$$

The same direction is found for the -1 diffraction order of beam 2.

Thus, beams with plane wave fronts w_1' and w_2' emerge from the undeformed specimen. These two coherent beams coexist in space, but their angle of intersection is zero. Their mutual interference produces a uniform intensity throughout the field, i.e., a pattern of zero frequency—zero fringes/mm. This is called a *null* field; in the literature it is also called an *infinite fringe*, meaning that the fringe pitch is infinite.

Now let the specimen (Fig. 4.4) be subjected to forces that stretch it uniformly in the x direction such that the uniform normal strain ε_x is a constant. The frequency of the specimen grating is thereby decreased to

$$f_s = \frac{f/2}{1 + \varepsilon_x}$$

Light from the first diffraction order of beam 1 does not emerge perpendicular to the grating. Instead, it emerges at an angle β_1 given by the grating equation. The conditions for beam 1 are the same as above except for the new grating frequency f_s. Now we find

$$\beta_1 = -\frac{\lambda f \varepsilon_x}{2}$$

Similarly, light from the −1 diffraction order of beam 2 emerges in the direction

$$\beta_{-1} = \frac{\lambda f \varepsilon_x}{2}$$

These two coherent beams propagate toward the camera with an angular separation of $2\beta_1$. The conditions of Fig. 2.20 are reproduced and again the result is an interference pattern with uniformly spaced parallel fringes (parallel to the y axis) just as in Figs. 2.21 and 2.22. The fringe gradient F_{xx} (i.e., $\partial N_x / \partial x$) is proportional to the angle of intersection $2\beta_1$. This angle is very small; by substituting $\sin \theta = |\beta_1|$ in Eq. 2.24, the magnitude of the fringe gradient is

$$F_{xx} = f \varepsilon_x \tag{4.6}$$

The fringe gradient (fringes/mm) is proportional to the reference grating frequency and the strain. The subscripts defining the fringe gradient refer first to the moiré pattern under consideration (the N_x or N_y pattern) and second to the direction of fringe gradient. For example, F_{xy} is the rate of change of fringe order in the y direction of the N_x pattern; it represents $\partial N_x / \partial y$.

Numerical evaluations of β_1 and F_{xx} are instructive. For light of the helium-neon laser (λ = 632.8 nm), a virtual reference grating of 2400 lines/mm and a normal strain of 0.1% (0.001 m/m), we find

$$\beta_1 = -0.00076 \, \text{rad} = -0.043°$$

and

$$F_{xx} = 2.4 \, \text{fringes/mm} = 61.0 \, \text{fringes/in.}$$

Angle β is tiny and, of course, the fringe frequency F_{xx} is 0.1% of the reference grating frequency f.

Next, let the specimen be subjected to nonhomogeneous deformation. This distorts the specimen grating such that its local frequency and orientation vary from point to point. They vary as continuous functions of x,y, so the directions of light diffracted from each point in +1 and −1 diffraction orders also vary as continuous functions and create the warped but continuous wave fronts depicted by w_1'' and w_2'' in Fig. 4.4b. The camera, which is focused on the specimen surface, collects these two beams of light and images w_1'' and w_2'' in its film plane with the same relative phase (or relative wave front separation) as they had when they emerged from the specimen grating.

The two beams produce an interference pattern in the camera which is a contour map of the separation, S, between these warped wave fronts (Sec. 2.3.6). The fringe orders in the moiré interference pattern are, from Eq. 2.25,

$$N(x,y) = \frac{S(x,y)}{\lambda} \qquad (4.7)$$

The mathematical analysis in the next section proves that the pattern represents in-plane displacements according to Eqs. 4.1.

4.3 Mathematical Analysis

The analysis follows the approach of Ref. 5. Consider any point P on the unloaded, undeformed specimen of Fig. 4.4. Beams 1 and 2 have path lengths from the source to P that differ by k. Their angle of intersection 2α satisfies Eq. 4.4, rewritten here as

$$\sin \alpha = \frac{\lambda f}{2} = \lambda f_s \qquad (4.8)$$

The +1 and −1 diffraction orders of beams 1 and 2, respectively, emerge along the normal to the specimen. Thus the *emergent* light from P can be represented by the electric field strengths

$$A_1' = a \cos 2\pi (\omega t)$$
$$A_2' = a \cos 2\pi (\omega t - k/\lambda) \tag{4.9}$$

where the phase difference is k/λ, amplitude a is equal for both beams, and time t is measured from an arbitrary datum.

Now consider the emergent light from the whole field. With the specimen still undeformed, the diffracted wave fronts w_1' and w_2' in Fig. 4.4b are plane and parallel. Thus, their phase difference is constant, equal at every x,y point in the field and equal to the phase difference at P. The interference pattern seen in the camera image plane is a null field, with the same intensity at every x,y point. The electric field strengths of emergent light at every x,y point on the specimen, and at every corresponding x,y point in the image plane, can be expressed by Eqs. 4.9.

What changes of phase occur when the specimen is deformed? They are $2\pi/\lambda$ times the changes of optical path lengths of rays to each point. The optical path lengths under consideration extend from the light source to the image plane. Figure 4.5 illustrates the details.

When the specimen is deformed, point P moves to a new location P', where U and V are the in-plane components of the displacement and W is the out-of-plane component. Coordinate y and displacement V lie perpendicular to the plane of the diagram. In general, the pitch and slope of the grating surrounding P' will differ from the initial condition at P.

In the figure, w_1 is a plane wave front in incident Beam 1; it extends as a plane normal to the diagram. The distance between the source and w_1 is the same for every ray in the incident beam. The same explanation applies for plane wave front w_2. Furthermore, $P'B$ lies in the object plane, so the optical path lengths from P' and B through the camera lens to the image plane is the same. [Note that the optical path length through the camera lens is independent of the diffraction angles at P' (Sec. 2.2.5).] Therefore, the changes of optical path lengths between the source and the image plane reduces to the changes of path lengths between w_1 and $P'B$ and between w_2 and $P'B$.

The changes of path lengths are analyzed in the figure, with the result

$$\Delta OPL_1(x,y) = W(x,y) (1 + \cos \alpha) + U(x,y) \sin \alpha$$
$$\Delta OPL_2(x,y) = W(x,y) (1 + \cos \alpha) - U(x,y) \sin \alpha \tag{4.10}$$

Observe that any V component of displacement does not change the distance between wave front w_1 and P', nor the distance between w_2 and P'. Therefore, the changes of optical path lengths are independent of V.

When the specimen is deformed, Eqs. 4.9 must be modified to take account of these changes of path lengths. The electric field strengths

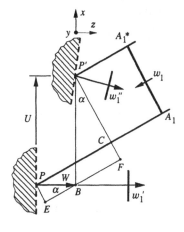

$$\Delta OPL_1 = A_1\,PB - A_1^*\,P'$$
$$= CP + PB$$
$$= (FB + BE) + PB$$
$$= (U \sin \alpha + W \cos \alpha) + W$$
$$= W(1 + \cos \alpha) + U \sin \alpha$$

$$\Delta OPL_2 = A_2\,PB - A_2^*\,P'$$
$$= A_2\,P - A_2^*P' + PB$$
$$= -(CP') + PB$$
$$= -(EP' - BF) + PB$$
$$= -(U \sin \alpha - W \cos \alpha) + W$$
$$= W(1 + \cos \alpha) - U \sin \alpha$$

Fig. 4.5   Changes of optical path lengths when deformation causes point P to move to P'.

of emergent light at each x,y point on the specimen, and corresponding x,y points in the image plane, become

$$A_1'' = a \cos 2\pi \left(\omega t - \frac{\Delta OPL_1(x,y)}{\lambda} \right)$$

$$A_2'' = a \cos 2\pi \left(\omega t - \frac{k}{\lambda} - \frac{\Delta OPL_2(x,y)}{\lambda} \right)$$

$$(4.11)$$

These equations have the same form as Eqs. 2.29. The parameters in Eqs. 2.29 and 4.11 are related by

$$a_1 = a_2 = a \qquad (4.12a)$$

$$\delta_1 = \Delta OPL_1 \qquad (4.12b)$$

$$\delta_2 = \Delta OPL_2 + k \qquad (4.12c)$$

$$S = \Delta OPL_1 - \Delta OPL_2 - k \qquad (4.12d)$$

Equations 4.10 and 4.8 provide for each x,y point

$$\Delta OPL_1 - \Delta OPL_2 = 2U(x,y) \sin \alpha = \lambda f U(x,y)$$

and thus,

$$S(x,y) = \lambda f U(x,y) - k \tag{4.12e}$$

Therefore, by substitution into Eq. 2.20, the intensity resulting from interference of A_1'' and A_2'' is

$$\begin{aligned}
I(x,y) &= 2a^2 \left[1 + \cos 2\pi \left(f U(x,y) - k/\lambda\right)\right] \\
&= 4a^2 \cos^2 \pi \left(f U(x,y) - k/\lambda\right)
\end{aligned} \tag{4.13}$$

In terms of fringe orders N_x, the intensity distribution in this fringe pattern of the U displacement field is

$$\begin{aligned}
I(x,y) &= 2a^2 \left(1 + \cos 2\pi N_x\right) \\
&= 4a^2 \cos^2 \pi N_x
\end{aligned} \tag{4.14}$$

where, by Eq. 2.25,

$$N_x = f U(x,y) - k/\lambda$$

Accordingly, the displacement is determined from the interference pattern by

$$U(x,y) = \frac{1}{f}\left(N_x + \frac{k}{\lambda}\right) \tag{4.15}$$

The constant $k/\lambda f$ is equivalent to a uniform displacement throughout the field, or a rigid body translation. When studying deformations we are not interested in rigid body motions. We can disregard the constant and interpret the pattern by

$$U(x,y) = \frac{1}{f} N_x \tag{4.16}$$

where $U(x,y)$ represents the displacement of every x,y point relative to an arbitrarily selected reference point of $N_x = 0$. The relationships for both displacement fields, U and V, is given by Eq. 4.1.

Note the strategic approach taken here. The strategy was applied because of the nature of diffraction. After diffraction, the OPL of a wave front changes by λ for each pitch of the grating (Fig. 2.41), and a wave front is not a surface of constant OPL from the source (Fig. 2.42b). However, the diffraction of only one ray was considered, not a wave front. The representative point P was fixed to the specimen, and thus it was fixed to a particular line on the grating. The change of OPL of the ray through P was caused only by specimen motion, i.e., none was caused by movement of P relative to the specimen grating lines.

Therefore, the changes illustrated in Figs. 2.41 and 2.42b did not influence its *OPL*. Subsequently, consideration of wave fronts was restricted to the space beyond the specimen grating—and *there* all rays between wave fronts travel equal optical path lengths.

4.3.1 Implications of the Analysis

Different Explanations

The theory of moiré interferometry, i.e., the interrelation of fringes and displacements, has been developed in stages:

(1) by analogy to geometric moiré, where the incident beams form a (virtual) reference grating which interacts with the specimen grating to form a moiré.
(2) by a warped wave front explanation, where interference between two beams with warped wave fronts generates the moiré pattern. It is quantified here for a case of uniform strain.
(3) by rigorous mathematical analysis, where the wave front warpages are quantified by the optical path lengths through each x,y point of the specimen.

Stages (2) and (3) construct a warped wave front model of moiré interferometry. It might appear that stages (2) and (3) are independent of each other since in (3) the optical path lengths traveled *prior* to the diffraction defines the moiré pattern. Actually, they are interrelated. The displacements control the optical path lengths; they also control the deformation of the specimen grating, which prescribes the directions of the diffracted light. (See Sec. 6.5 for additional discussion.)

The warped wave front model of moiré interferometry matches the wave front model of Chapter 2 for other types of two-beam interferometry. The model takes full account of the in-plane and out-of-plane displacements of the specimen surface, and the slope of the surface. The moiré patterns are quantitative maps of in-plane displacements at all x,y points in the field.

The mathematical analysis derives Eqs. 4.1, and in so doing, it confirms the identity to the equations of geometric moiré. The identity is derived in Ref. 4, too, but by an alternate mathematical approach based on the angles of diffraction at each x,y point. The fact that geometric moiré is governed by the same equations gives credibility to the analogy in stage (1) above. Thus, the analogy offers a powerful tool for visualizing the moiré interferometry phenomenon. For example, we can visualize the two incident beams in Fig. 4.2 as creating walls of interference parallel to the z axis. Then, it is clear that out-of-plane displacements of any point P are parallel to the walls; the W displacements do not cut the walls and therefore the moiré pattern is insensitive to W. In the same way, V displacements are parallel to the walls and the moiré pattern is independent of V.

W Displacements

The out-of-plane displacements, W, disappear in the analysis and the moiré fringe pattern is not influenced by W. This is a consequence of the symmetry of the two input beams (beams 1 and 2 in Fig. 4.4). Equations 4.10 show that the terms involving W are equal when both beams have the same angle of incidence, and they cancel when the two equations are subtracted. They show, too, that a small deviation from symmetry of the two incoming beams is not serious, i.e., the influence of W is small, except in rare cases where W is large compared to U.

The interference pattern depicts the U displacements faithfully when deformations, including out-of-plane slopes, are induced by the applied loads. However, additional out-of-plane slopes might occur as a result of accidental rigid body motion of the deformed specimen. This is treated in Sec. 4.19, where it is shown that the influence of accidental slopes is a second-order effect that is usually negligible, but when it is not negligible its effect can be nullified by subtraction.

Fringe Shifting

In Eq. 4.15, the term k/λ indicates that the fringes can be shifted and still represent the same displacement field, just like fringe shifting in geometric moiré (Sec. 3.2.4). Fringe shifting is accomplished by changing the phase of the virtual reference grating uniformly throughout the field, e.g., by translating mirror A (in Fig. 4.4a) in the x direction. When the phase is changed by a fraction of a cycle, k is changed by the same fraction of a wavelength, and the positions of intensity maxima and minima are shifted by the same fraction of a fringe order. Physically, the virtual reference grating is shifted with respect to the specimen grating. The fractional pitch shifts are tiny, but equipment to produce such tiny movements is available (Sec. 5.3.1).

4.4 The Virtual Reference Grating Concept; Analogous Gratings

Figure 4.6a illustrates a virtual reference grating with its walls of constructive and destructive interference parallel to the yz plane. The specimen grating is illuminated by dark and bright bars that act as a reference grating. The concept of the virtual grating connects moiré interferometry and geometric moiré. However, the mathematical analysis did not postulate a virtual reference grating.

Moiré interferometry does not require the existence of a virtual grating.

In Fig. 4.6a, the two incident beams are polarized perpendicular to the page, as indicated by the dots attached to rays. The wave motion is perpendicular to the page. The two beams that coexist in the region of intersection have a common direction of polarization and they interfere to create the walls of interference called a virtual grating. Light that emerges in the direction of the camera generates a full-contrast interference pattern of U displacements in the film plane.

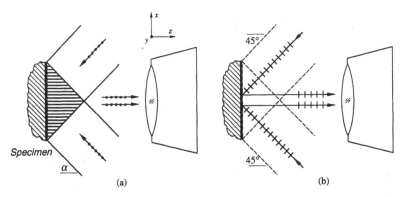

Fig. 4.6   (a) A virtual reference grating is formed when the incident light is polarized perpendicular to the plane of the page. (b) The incident beams have perpendicular polarization, but the diffracted beams have the same polarization.

In Fig. 4.6b, everything is the same except the incident light is polarized parallel to the page. Where the beams coexist, their polarizations are perpendicular to each other. The beams are not mutually coherent on account of their polarizations and they cannot interfere. *No virtual grating is formed.* Nevertheless, the beams are each diffracted by the specimen grating and propagate into the camera. The diffracted beams remain polarized in the plane of the page, so now they have a common direction of polarization and they interfere to form the same full-contrast interference pattern of U displacements. The results are the same as in Fig. 4.6a, although no virtual reference grating existed.

The conditions of Fig. 4.6b destroy the useful connection between interferometric and geometric moiré. The authors propose an artifice to rebuild the connection, so the analogy between the two can be maintained. The artifice is an *analogous reference grating*, which is a conceptualized equivalent of a virtual reference grating, employed when the latter does not exist. Its frequency is determined by the angles of incidence, $\pm\alpha$, and the wavelength, λ, by the same relationship, $f = (2/\lambda) \sin \alpha$. The analogous grating is assumed to interact with the deformed specimen grating to form the moiré pattern. In summation, it is not necessary to conceptualize a reference grating, but the concept is retained here to preserve the analogy between moiré interferometry and geometric moiré.

It is to be understood that the fringe patterns are truly patterns of optical interference. Nonetheless, they will also be called moiré patterns (which connotes the action of mechanical interference) in recognition of the analogy.

4.5 Theoretical Limit

It is clear from physical reasoning and also from Eq. 4.4 that the theoretical upper limit for reference grating frequency is approached as α (Fig. 4.4) approaches 90°. The theoretical limit is $f = 2/\lambda$. The corresponding theoretical upper limit of sensitivity is $2/\lambda$ fringes per unit displacement, which corresponds to a contour interval of $\lambda/2$ displacement per fringe order.

An experiment was conducted in which the specimen illustrated in Fig. 4.7 was prepared with a grating of 2000 lines/mm (50,800 lines/in.).[6] The optical system represented by Fig. 4.4b was used, where α was 77.4° and λ was 488.0 nm. This produced a virtual reference grating of 4000 lines/mm (101,600 lines/in.). For this wavelength, the theoretical limit is $f = 4098$ lines/mm, which means that the experiment was conducted at 97.6% of the theoretical limit of sensitivity.

The fringes were well resolved in the entire 51 x 76 mm (2 x 3 in.) area of the specimen grating. Enlargements of the very dense array of interference fringes around the hole are shown in Fig. 4.7. The patterns show remarkable clarity of the interference fringes and verify that moiré interferometry is effective even very near its theoretical limit.

The theoretical limit will be revisited in Sec. 5.2.1, where it is shown that greater sensitivity can be achieved if the virtual reference grating is formed inside a refractive medium.

Aside from the issue of theoretical limit, a comparison of Figs. 4.1 and 4.7 yields valuable insights. Notice the region near A in Fig. 4.7 and the corresponding region in Fig. 4.1a. The fringe orientations are dramatically different. The anisotropic composite specimen of Fig. 4.1 exhibits nearly vertical fringes in the region; the fringe gradient $\partial N_y / \partial x$ is very large, signifying a high shear strain (Eq. 4.3)

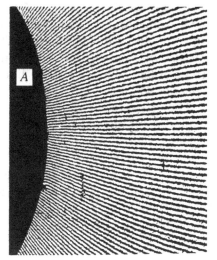

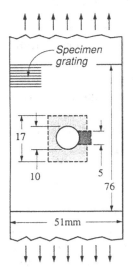

Fig. 4.7 Demonstration of moiré interferometry at 97.6% of the theoretical limit of sensitivity. f = 4000 lines/mm (101,600 lines/in.). The patterns correspond to the regions outlined by the dashed lines.

concentration. The gradient is much smaller in the isotropic specimen of Fig. 4.7, which exhibits much smaller shear strains.

4.6 Sensitivity and Resolution

The sensitivity of the displacement measurement is determined by the number of fringes generated per unit displacement, or N_x/U. By Eq. 4.1 ($f = N_x/U$), *displacement sensitivity* is equal to the frequency f of the reference grating. Sensitivity increases with f. In a region where the displacement accumulates from zero to U, the number of fringes that cross an x axis is $N_x = fU$. For the case of f = 2400 lines/mm (60,960 lines/in.), the sensitivity is 2.4 fringes/μm displacement (61 fringes/0.001 in. displacement). The inverse of sensitivity defines the *contour interval*, $1/f$. The contour interval is the displacement per fringe order. When f = 2400 lines/mm, the displacement per fringe order is 0.417 μm.

The *displacement resolution* reflects the reliability of a displacement measurement. It is the tolerance within which a measurement is assumed to be reliable. In the usual engineering analysis where the displacement varies smoothly between neighboring fringes, visual interpolation to 1/5 or 1/10 of a fringe order is reliable. Then the displacement at any point can be resolved to $1/(5f)$ or $1/(10f)$. When f = 2400 lines/mm, the displacement resolution is at least 0.1 μm (4 μin.).

4.7 Optical Systems for Moiré Interferometry

The apparatus usually comprises three subsystems:
* the illumination system, consisting of a coherent light source, beam expander and collimator
* the moiré interferometer, which divides the input beam into four separate beams and directs them onto the specimen grating as in Fig. 4.3, and
* the camera system, consisting of a collecting lens, imaging lens and photosensitive receiver in the image plane.

Figure 4.8 illustrates one version of such apparatus,[7] where the symmetry of optical paths in the interferometer makes it easy to understand and utilize. The illumination system consists of a laser, a spatial filter (beam expander and pinhole), a plane mirror and a parabolic mirror. For static analyses, the laser power requirement depends primarily on the diffraction efficiency of the specimen grating and the size of the magnified image. A 5 or 10 mW laser has generally been adequate. The pinhole (i.e., the filter) improves the cosmetic quality of the fringe pattern, but it is not essential for good results.

The camera system focuses a sharp image of the specimen surface onto the image plane, where it is recorded on film (or sometimes by a video camera). A rather small field of view is usually sufficient, e.g. 25 mm (1 in.) across, and in such a case the collecting and imaging functions can be performed by one lens as shown. Otherwise, a field lens should be positioned as close to the specimen as possible and an imaging lens should be positioned near the focal plane of the field lens (Fig. 2.25). To record the image, a high contrast, high resolution film is preferred for the usual case in which the fringe pattern exhibits a large number of fringes. A video camera can be used in special circumstances, especially when only a few fringes are to be recorded.

Several different moiré interferometers have been designed to provide two-beam or four-beam input to create virtual reference gratings. Any optical arrangement that provides coherent beams B_1, B_2 and/or B_3, B_4 of Fig. 4.3 is called a moiré interferometer, and many different arrangements are possible. Emphasis will be given to four-beam systems, which provide the complete state of in-plane displacements and surface strains; with two-beam systems, the accurate determination of shear strains is difficult (Sec. 4.12) since the rotation of the interferometer or specimen needed to obtain both fields must be precise.

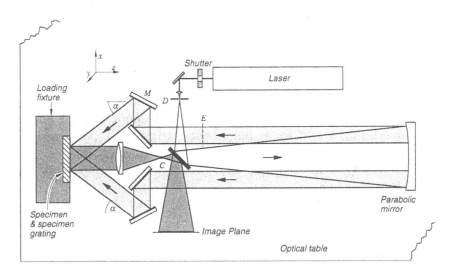

Fig. 4.8   Schematic view of a 4-beam moiré interferometer. Beams corresponding to B_3 and B_4 of Fig. 4.3 are not shown, but they are formed in the yz plane by a similar arrangement of mirrors.

The moiré interferometer of Fig. 4.8 is attractive because the optical elements do not lie very near the specimen, because the critical elements (the plane mirrors) can be mounted on a single fixture, and because the parabolic mirror can be used on-axis. This is a system of eight plane mirrors, with their centers in an xy plane; the four mirrors along x are illustrated and the four along y are not shown. Of these, the four inner mirrors can be fixed (non-adjustable) and the four outer mirrors should be mounted for universal angular adjustment through small angles. Another feature that makes the design attractive is insensitivity to the polarization angle of the incoming light. This results from the symmetry of the design and high contrast moiré patterns are obtained with any polarization.

Of the light from the parabolic mirror, the portion that strikes the inner group of four plane mirrors is useful and the remainder (perhaps 50%) is wasted. The sacrifice is relatively insignificant, however, with respect to other factors that influence light utilization. Instead, the diffraction efficiency of the specimen grating has the largest influence on light utilization.

4.7.1 Alignment

Alignment of the system of Fig. 4.8 proceeds as follows:

1. Temporarily insert a small plane mirror in the beam (near plane E). Adjust the axial position of the parabolic mirror to minimize the size of the bright spot reflected back to plane D, the plane of the pinhole (or the node of the diverging beam). This procedure is called *autocollimation* and it adjusts the parallelism of the beam. Adjust the angular position of the parabolic mirror to superimpose the bright spot (approximately) with the pinhole.

2. Adjust the four outer mirrors for a first approximation of the correct angle α of each input beam; angle α is calculated by Eq. 4.4.

3. Install an undeformed (no load) specimen and specimen grating of frequency $f/2$.

4. Insert white cards across the beams at C and D. Let light pass through a central hole in the card at D.

5. Block two opposite beams with cardboard obstructions, e.g., beams B_3 and B_4 in Fig. 4.3.

6. Light from beams B_1 and B_2 will be diffracted in ±1 diffraction orders of the specimen grating to emerge approximately normal to the specimen; they will be converged by the lens to two small dots in plane C. Light diffracted in the zeroth order of the specimen grating will trace paths through the optical system and converge to two bright dots in plane D. While observing these four dots, iteratively adjust the two outer mirrors until the two dots in C merge together and also the two dots in D merge together at a location near the source aperture. Superposition of the dots at C regulate angle α. Positioning the dots near the aperture at D controls the symmetry of the system; this adjustment is less critical than the superposition at C.

7. Remove the card from C and observe the moiré pattern in the

image plane of the camera. Fringes should be seen at this stage; use a magnifier to see them if necessary. If fringes are not seen, superimpose the dots at C more accurately. Then, adjust an outer mirror to decrease the number of fringes until a null field is obtained, i.e., a field with the minimum number of fringes.

8. Reinsert the card at C and remove the card that obstructs beam B_3. Adjust the corresponding outer mirror to superimpose its dot on the others at C. Repeat the operation for beam B_4.

9. Remove the card from C. Block beams B_1 and B_2 and adjust beams B_3 and B_4 to obtain a null field. This completes the alignment operation. Dots at C will be superimposed and dots at D will be separated, typically, but all close to the source aperture.

10. Focus the camera on the plane of the specimen grating by adjusting the position of the lens, or the image plane, or both. A fine cross scribed on the specimen grating aids focusing.

In general, a true null field—a field completely devoid of fringes—will not be obtained. This is because the beams forming the virtual grating do not have perfectly plane wave fronts and the specimen grating does not have perfectly straight and uniformly spaced lines. Typically one or two circular or saddle-shaped fringes will appear as the best null field when the field size is about 25 mm. This initial field is usually of no consequence when many fringes appear in the load-induced moiré pattern. In all cases, however, the fringe orders in the initial pattern can be subtracted from those in the final pattern to eliminate their influence (Sec. 4.17).

In operation, the N_x fringe pattern is recorded by using beams B_1, B_2, while blocking beams B_3, B_4. For the N_y pattern, beams B_1, B_2 are blocked. The procedure in this section applies generically to all the optical systems, whereby angle α is established by bringing together two first-order diffractions in the focal plane of the camera lens and symmetry is established by returning two zero-order diffractions to the source.

4.7.2 Size of Field

In principle, there are no limits to field size. In practice, however, the region of interest is usually rather small. Figure 4.1 is an example, where interest is centered on the region near a 3.2 mm (1/8 in.) radius hole. Figure 4.7 illustrates an exception, but the objective there was to investigate limits of the method rather than an engineering application. Accordingly, instrumentation of modest size is adequate. In special cases where a larger field must be investigated, the option of translating the specimen relative to the interferometer should be considered as an alternative to a large field instrument.

A field size of 30 to 40 mm diameter is recommended for general purposes. The modest size makes the apparatus compact, convenient and efficient. It reduces the investment in apparatus, reduces the problems of environmental control and appears to involve a shorter learning period and quicker success. The excellent fringe contrast and spatial resolution inherent in these techniques permits great enlargement of the moiré fringe patterns.

4.7.3 Environmental Concerns: Air Currents

The typical environmental concerns are air currents and vibrations. Air currents are disturbing because the index of refraction of air varies with pressure and temperature, and air currents are motivated by both. The interference patterns are formed by two beams of light, and wherever the two beams are separated in space the air currents might change the optical path length of one beam relative to the other. A relative change of *OPL* causes the virtual grating to shift and the regions of constructive and destructive interference in the fringe pattern to shift in harmony. Air currents cause movement of the interference fringes; the fringes *dance*.

Note that the susceptibility to air currents exists when the two beams are separated in space. The beams propagating from the specimen to the film plane of the camera occupy essentially the same space, and the pattern is (essentially) insensitive to currents in that space.

Under laboratory conditions, air currents seldom present a problem. Sometimes dancing fringes can be halted merely by switching off the room air conditioner. Alternatively, enclosures that shield the optical paths from air currents are effective. Temporary baffles assembled from blocks of foam rubber (rubber-like plastic foam) are excellent for the purpose.

4.7.4 Vibrations, Optical Tables

Interferometry measures tiny displacements and inadvertent vibrations can cause the fringes to dance at the vibration frequency. Vibrations of any element that causes a change of optical path length of more that about $\lambda/10$ should be avoided. In-plane motion of the specimen of more than about $g/10$ should be avoided, where g is the pitch of the virtual reference grating. Accordingly, the elements must be firmly mounted on essentially rigid fixtures.

Optical tables with pneumatic vibration dampers are becoming commonplace and they provide an excellent base for mounting the equipment. A massive slab supported on vibration damping material is almost as effective. Nearly complete freedom from vibration problems is achieved when the specimen is loaded (deformed) in a loading fixture that is attached firmly to the table.

Compact arrangements of the optical system can be mounted on a plate or frame that attaches to a mechanical testing machine.[8] Such systems have been employed successfully on large screw-driven testing machines, although dancing fringes are often encountered; they stop only after some element in the system is remounted, reinforced or damped. Optical systems have been proposed in which critical elements are attached directly on the specimen so that the specimen grating and reference grating vibrate together, without relative motion (Sec. 4.20).

Portable optical systems that rest directly on the structure under study have been proposed for measurements in the field, outside the laboratory.[9] Again, vibrations can be tolerated if the specimen and reference gratings vibrate in unison.

4.7.5 Loading Fixtures

A general purpose loading machine that can be mounted on the optical table might be suitable for some laboratories. It should be designed with the test section close to the base of the machine, such that the optical axis of the moiré apparatus is not elevated high off the

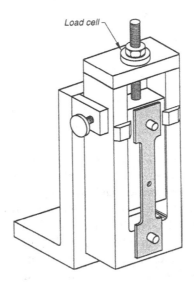

Fig. 4.9   Loading fixture for tension.

table surface. An alternative that is very effective is use of individual loading fixtures that can be customized to accommodate each application. In either case, fixed displacement devices are recommended, rather than fixed load systems. They should be sturdy and stiff. Their design and manufacture should minimize transverse motions when longitudinal loads are applied.

Figures 4.9 and 4.10 illustrate fixtures that proved effective. The tensile fixture utilizes a long plate that slides in a channel milled into the main body of the fixture. The load is measured by an electrical load cell. The fixture is mounted on a heavy angle plate in such a way that it can rotate about an axis that crosses the specimen near the specimen grating; a fine screw adjusts the angular orientation of the fixture. Although this design does not account for frictional forces along the slide, it is suitable for many purposes.

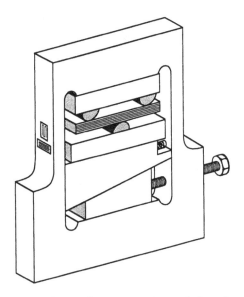

Fig. 4.10   Loading fixture for compression and short-beam flexure.

The fixture sketched in Fig. 4.10 is designed for compression and short-beam flexure tests, with loads up to 45 kN (10,000 lb.). The compressive displacements are achieved by driving the lower wedge to the left. The large contact area between the wedges makes the design compatible with large loads. The loads are measured by electrical strain gages mounted on both sides of the main frame; longitudinal and transverse gages are wired into adjacent arms of the strain gage bridge to increase sensitivity and provide temperature compensation. The load measurement is independent of frictional forces. The wedges are hardened steel and lubricated. The fixture can be attached to a heavy angle plate and allowed to rotate with fine angular adjustment.

4.7.6 Two-beam Systems

Numerous different optical schemes can be contrived to form the virtual reference grating. Any means that brings mutually coherent beams equivalent to B_1 and B_2 (Fig. 4.3) to the specimen grating would suffice. A particularly simple means is illustrated in Fig. 4.11. Here, half the incident beam impinges directly on the specimen surface while the other half impinges indirectly in a symmetrical direction after reflection from a plane mirror. The entire optical system is shown schematically. Light diffracted in the +1 and −1 diffraction orders of the specimen grating is admitted to the camera to form the moiré pattern. A parabolic mirror is shown as a collimator, but a

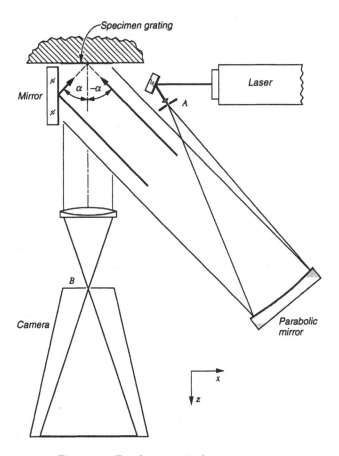

Fig. 4.11   Two-beam optical arrangement.

collimating lens is a feasible alternative. The collimator and plane mirror should have sufficient optical quality to produce wave fronts that are plane within a fraction of a wavelength. Typically, mirrors with profiles within $\lambda/8$ of the ideal surface are used.

Figures 4.12 and 4.13 illustrate assemblies of partial mirrors for creating virtual reference gratings. The complete optical systems are not shown, but systems analogous to that in Fig. 4.11 are applicable. The specimen with its specimen grating is to be located just to the left of the assembly in each case, in the zone of the virtual reference grating. Although mechanical mountings are not shown, adjusting screws control the frequency and angular orientation of the virtual gratings.

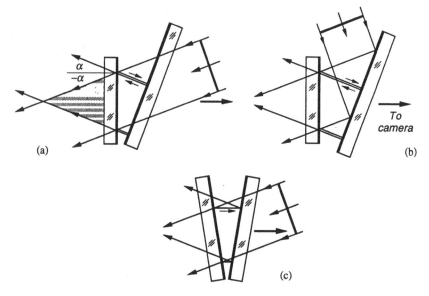

Fig. 4.12   Mirror assemblies for generating virtual reference gratings of 1200 lines/mm (30,480 lines/in.). Dielectric partial mirrors of ~60% reflectance are used. Refractions at the air/glass interfaces are not shown.

Multilayer dielectric mirror coatings of approximately 60% reflectance and 40% transmittance are appropriate. Other choices are satisfactory, too, noting that higher reflectance increases fringe contrast and reduces the intensity of light in the virtual grating. With 60% reflectance, the intensities of the two interfering beams are in the ratio of ~3:1. However, the fringe contrast remains a respectable 94% for this condition (Eqs. 2.21 and 2.26). In Fig. 4.13b, the lower element can be a full mirror, which increases the fringe contrast.

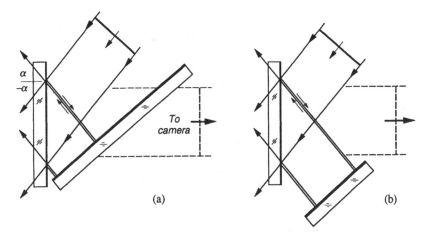

Fig. 4.13   Mirror assemblies for generating virtual reference gratings of 2400 lines/mm (60,960 lines/in.).

The diagrams of Fig. 4.12 are drawn with α = 22.3°, which corresponds to a virtual grating frequency of 1200 lines/mm (30,480 in.) for light of the red He-Ne laser (λ = 632.8 nm). The arrangement of Fig. 4.12b is particularly attractive when the moiré interference fringes are photographed with large magnification, since it allows the camera lens to be closer to the specimen.

The arrangements of Fig. 4.13 are effective for higher frequencies. They are drawn with α = 49.4° corresponding to f = 2400 lines/mm (60,960 in.) with λ = 632.8 nm. The system of Fig. 4.13b is superior inasmuch as the optical elements are smaller and light is utilized more efficiently, but the difference of optical path lengths of the two interfering beams is somewhat greater.

A useful variation of the system of Fig. 4.12a is the prism of Fig. 4.14. This scheme is well adapted to thermal analyses,[10] inasmuch as the prism is robust and can withstand uniform temperature changes without altering its angle Θ. If the gap between the wedge and specimen surface is very small, the influence of thermal currents in the air (or other gas) would be negligible. The frequency f of the virtual grating depends upon α, which in turn depends upon wedge angle Θ by

$$\sin \Theta = \frac{1}{n}\sin \alpha = \frac{\lambda f}{2n} \tag{4.17}$$

where n is the index of refraction of the prism material. For a frequency of 1200/mm (30,480 lines/in.), red light of the He-Ne laser, and n near 1.5, the wedge angle is $\Theta \approx 15°$. While angle Θ of the

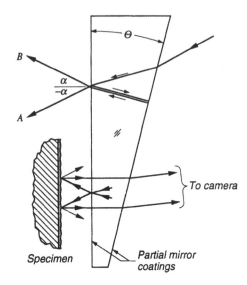

Fig. 4.14   Wedge prism for generating a virtual reference grating.

wedge does not change with temperature, its refractive index does change. Optical glasses that exhibit very low dn/dT, i.e., very low change of refractive index with temperature, may be selected. The wedge does not offer the advantage of adjustable reference grating frequency, so the initial frequency of the specimen grating must be

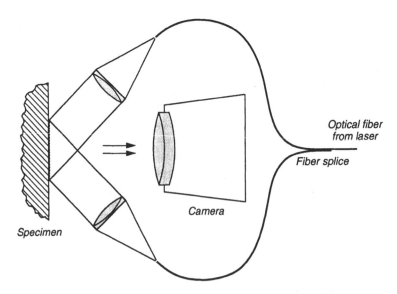

Fig. 4.15   Moiré interferometer system using optical fibers.

controlled to obtain a null field at room temperature (or at another temperature).

Figure 4.15 is a schematic illustration of a moiré interferometer constructed with two collimating lenses and coherent input by optical fibers.[9] Adjustments of the virtual reference grating can be made by altering the positions of the lenses or the fibers, or both. With appropriate mechanical design, two such systems could be coupled together to form a four-beam system.

The system illustrated in Fig. 4.16 is suitable for large fields. Note that a field lens is located as close as possible to the specimen, to converge light into the camera.

A virtual reference grating can be formed by a real grating of the same frequency, as illustrated in Fig. 4.17a. The zeroth and first diffraction orders emerge symmetrically when Eq. 2.57 is satisfied, and they interfere to form a virtual grating in their zone of

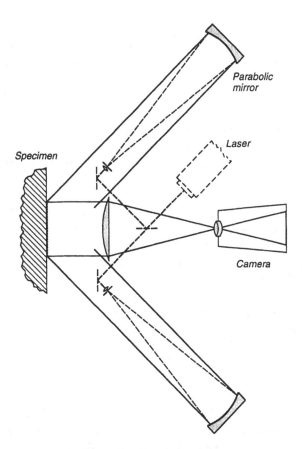

Fig. 4.16   Moiré interferometer for large fields. Elements and rays in dashed lines lie above or below those drawn in solid lines.

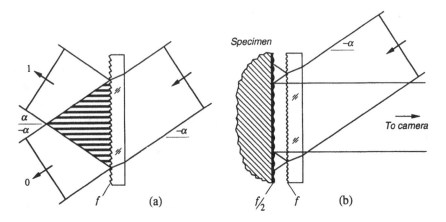

Fig. 4.17   (a) A real grating produces a virtual grating of the same frequency. (b) Optical arrangement utilizing a real reference grating.

intersection. When the grating frequency is high enough to satisfy Eq. 2.58, these are the only beams to emerge. Figure 4.17b illustrates such a grating used as a real reference grating. It forms a corresponding virtual grating which interacts with the specimen grating to form the moiré pattern. Light passes through the real reference grating in its zeroth diffraction order and propagates on to the camera. In its zeroth order the grating acts as a window and the moiré patterns produced by the arrangement have excellent quality. A cross-line real grating could be used to produce the U and V fields. It could be mounted directly on the specimen, in many cases, to cope with vibrations. Sensitivity to small variations of angle α is negligibly small. The coherence requirements are substantially relaxed when the gap between the gratings is small. The real reference grating can be a holographic grating with furrows of sinusoidal profile; its furrow depth could be made to give nearly equal intensities in the zeroth and first orders.

This arrangement is simple and potentially very useful. It has two shortcomings that make it less attractive for general use. One is that its frequency is not adjustable, so fine adjustments for a null field cannot be made, and also carrier fringes of extension (Sec. 4.13) cannot be applied. The second is that the perpendicularity error of the specimen and reference gratings might not be precisely matched, and initial fringes of rotation could not be nulled in both the U and V fields. For some applications, nevertheless, this could be an attractive optical arrangement. A variation that permits adjustability and either two-beam or four-beam capabilities is shown later in Fig. 4.48.

4.7.7 Four-beam Systems

Two systems have already been described, that of Fig. 4.8 and the four-beam version of Fig. 4.15. Many other arrangements are possible and some will be introduced here.

Figure 4.18 illustrates a four-beam optical system in which the two-beam arrangement of Fig. 4.11 is duplicated in the horizontal plane and the vertical plane.[11] A beam splitter, BS, directs a portion of the collimated beam downwards to mirror M_3, which in turn directs it to the specimen and to mirror M_4 at an angle α in the vertical plane. If an opaque card is placed in the beam on the reflection side of the beam splitter, the transmitted beam strikes the specimen and mirror M_2 and generates the N_x pattern (the U displacement field). If the card is placed on the transmission side of the beam splitter, the reflected beam generates the N_y pattern (or the V field).

Polarization of the light should be parallel to the y axis. The oblique reflections for the V field introduce some ellipticity into the

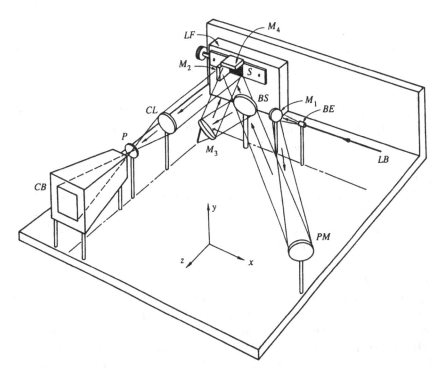

Fig. 4.18 Four-beam system that utilizes a beam-splitter to produce the N_x and N_y patterns. LB: laser beam. BE: beam expander. M_{1-4}: plane mirrors. PM: parabolic mirror. BS: beam splitter. S: specimen and spec. grating. LF: loading fixture. CL: camera lens. P: linear polarizer. CB: camera back.

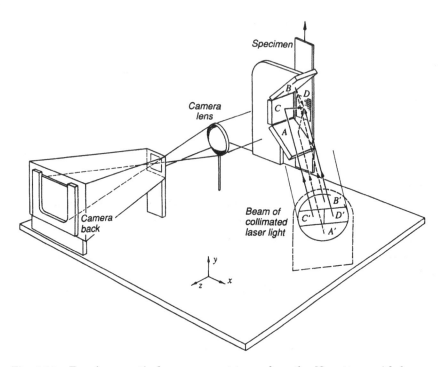

Fig. 4.19 Four-beam optical arrangement to produce the N_x pattern with beams C' and D' and the N_y pattern with beams A' and B'.

polarization, which diminishes the contrast of the N_y fringes. To cope with this, a polarizing filter P is introduced in the focal plane of the camera lens, with its polarizing axis parallel to the y axis. A polarizing filter laminated in glass, of the sort used on camera lenses, is suitable.

To avoid ghost fringes, the back side of the beam splitter should be antireflection coated. Another option is to order a beam splitter that has a strong wedge angle between its front and back faces, such that the unwanted light does not enter the camera.

Another four-beam system is illustrated in Fig. 4.19.[12] Two mirrors, A and B, are added to the basic system of Fig. 4.11. These 45° adjustable mirrors direct portions A' and B' of the collimated beam to angles $+\alpha$ and $-\alpha$ in the vertical plane. They form a virtual reference grating with its grating lines perpendicular to the y axis; it interacts with the corresponding specimen grating lines to form the N_y or V field. When light from A' and B' is blocked, beams from C' and D' form the N_x or U field. The system has the advantage that no polarizing filter is needed. It has the disadvantage relative to Fig.

4.18 that a larger collimated input beam is required, which means a larger space and larger optical table is needed.

A variation of the interferometer of Fig. 4.8—a variation that offers several attractive features—is illustrated in Fig. 4.20.[13,14] A collimated beam is reflected by the 45° mirror to strike a cross-line grating at normal incidence. The four beams of +1 and −1 diffraction

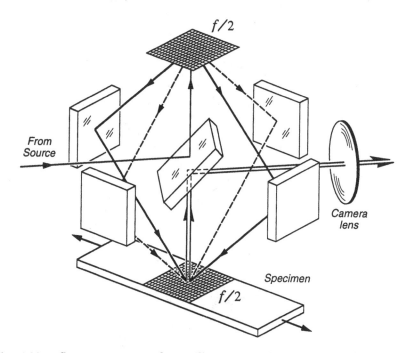

Fig. 4.20   Compact system that utilizes a cross-line grating and mirrors to produce the four incident beams.

orders are redirected by mirrors to strike the specimen at angles of incidence of α. Light of the zeroth order is wasted. Thus, x and y virtual gratings are formed on the specimen. Light diffracted by the specimen grating is collected by the camera lens, which focuses images of the moiré patterns onto the film plane of the camera. In practice, two opposite beams are blocked while the other two produce the N_x (or N_y) moiré pattern. Like the arrangement of Fig. 4.8, this interferometer is insensitive to polarization, and light of any polarization can be used. The virtual gratings can be controlled for null fields or carrier fringes by angular adjustments of the four vertical mirrors. The system is compact and its light utilization is efficient. When used with a loading fixture on an optical table, it could be arranged conveniently with the specimen in a horizontal plane as shown. An alternate arrangement suitable for a vertical specimen was constructed and employed for the studies reported in Chapter 9. Its compact mechanical design is illustrated in Fig. 9.8.

Another variation is illustrated in Fig. 4.21,[15] where transmission gratings direct the light instead of mirrors. Special properties of such systems are addressed in the following section.

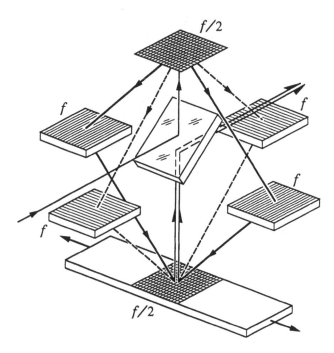

Fig. 4.21   Compact system utilizing gratings to direct the light.

4.7.8 Achromatic Systems

Consider Fig. 4.2 again. A virtual reference grating of frequency f is formed when the angle 2α between beams 1 and 2 satisfies Eq. 4.4, i.e.,

$$f = \frac{2\sin \alpha}{\lambda} \qquad (4.4)$$

The virtual grating is easily generated in monochromatic light, where λ is a fixed wavelength. It can be arranged for polychromatic light, too, by providing a system in which α and λ vary in precise harmony, such that $(\sin \alpha)/\lambda$ is a constant. This harmony is

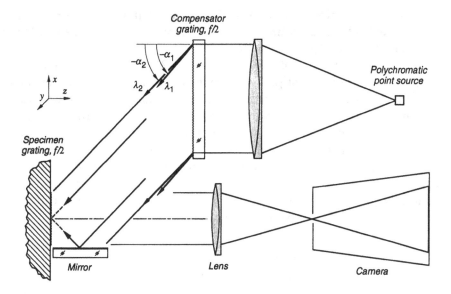

Fig. 4.22   A compensator grating permits use of temporally incoherent light.

provided by the compensator grating in Fig. 4.22. The grating equation, Eq. 2.54, gives precisely the relationship of Eq. 4.4 for the conditions illustrated, namely normal incidence on the compensator grating of frequency $f/2$ and emergence in diffraction order $m = -1$ at angle $-\alpha$. The compensator grating spreads the light of every wavelength into the correct direction, exactly as required for input to a moiré interferometer.[16]

Light from the compensator grating illuminates a moiré interferometry arrangement as shown in the same figure. In the specimen space, a virtual reference grating of unique frequency f is produced by each wavelength of the polychromatic light. These virtual reference gratings are in exact registration, with constructive and destructive interference occurring at identical locations for all wavelengths. The camera receives identical moiré patterns for every wavelength and records the scalar sum of the individual intensity distributions. The resulting pattern is the same as that produced with monochromatic light.

The width of the light source remains a critical parameter. In principle, the source must be a point or line, but in practice a laser diode is suitable. Its small size and convenient mechanical features make it especially attractive for instrumentation. Although it does not produce a broad spectrum, its spectral bandwidth is sufficiently wide to require compensation.

Note that requirements on the size of the light source are relaxed when a real reference grating (Fig. 4.17b) is used in close proximity to the specimen grating.

The interferometers of Figs. 4.20 and 4.21 also have achromatic properties, since the first grating provides angles of diffraction in the +1 and −1 orders in accord with Eq. 4.4. The system of Fig. 4.21 does not require any coherence, neither temporal nor spatial coherence.[15] With an extended source and broad spectrum, however, the depth of the virtual grating approaches zero; it occurs in a very narrow volume of space, approaching a plane, and to assure that the location of the virtual grating coincides with the specimen grating is generally impractical. Nevertheless, a finite source size and a finite spectral bandwidth can be used if neither is large.

4.7.9 Three-beam Interferometer

A cross-line holographic grating consists of an orthogonal array of hills, as pictured in Fig. 2.35; the hills are depicted in Fig. 4.23 by dark dots. The vertical rows of dots comprise a diffraction grating that diffracts light in the same directions as an array of vertical furrows. Similarly, the horizontal rows of dots act like a grating with horizontal furrows. The same concept applies to the rows in the + 45° and − 45° directions, and also in other directions like that shown by dashed lines in Fig. 4.23.

As illustrated in Fig. 4.24a, such a grating is normally illuminated by four beams in the directions AO, BO, CO, and DO, and light is diffracted out in the OQ direction. The beams are represented by rays. To achieve first order diffractions normal to the grating, the angle of intersection AOB, which is 2α, must satisfy Eq. 4.8,

$$\sin \alpha = \lambda f_s = \frac{\lambda f}{2} \qquad (4.8)$$

where f_s is the initial specimen grating frequency and f is the virtual reference grating frequency.

Figure 4.24b shows an alternative, where the grating is illuminated by three beams in the direction KO, LO, and MO, and light emerges again in the OQ direction. The planes of incidence (the planes of KOQ, LOQ, and MOQ) lie at 45° from the y axis. In order to emerge along OQ, incident beams diffract from the ± 45° gratings of Fig. 4.23. Their pitch is $g_s / \sqrt{2}$ and their frequency is $\sqrt{2}f_s$. The angle of incidence LOQ, or Θ, when the first-order diffraction emerges normal to the specimen, is determined by the diffraction equation (Eq. 2.54) as

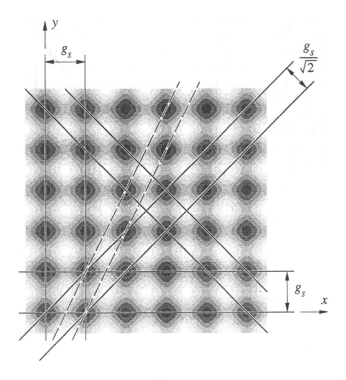

Fig. 4.23   A grating comprised of an orthogonal array of hills diffracts light in numerous directions: x, y, +45°, −45°, and other directions. The basic pitch is $g_s = 1/f_s$.

$$\sin \Theta = \sqrt{2}\lambda f_s \qquad (4.18)$$

From the geometry of the figure the angle of intersection LOM, or $2\alpha'$, must satisfy

$$\sin \alpha' = \lambda f_s \qquad (4.19)$$

How interesting! Angle α' equals angle α. The intersection of beams LO and MO creates the same virtual reference grating as the intersection of beams AO and BO. In both cases, the virtual grating has vertical walls of interference parallel to the yz plane with a frequency $f = 2f_s$. In both cases the virtual grating interacts with the specimen grating of initial frequency f_s to produce the moiré pattern. Consequently, the three-beam arrangement and the four-beam arrangement of Fig. 4.24 produce the same results, with the same sensitivity. Figure 4.24b represents the three-beam interferometer of Ref. 17.

Angle of incidence Θ is greater than angle α. Since $\Theta \to 90°$ is the theoretical upper limit, the maximum sensitivity of the three-beam

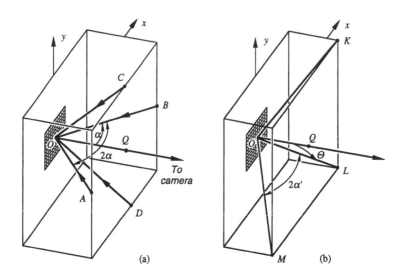

Fig. 4.24   (a) Representation of the normal four-beam optical system and (b) of the three-beam system.

system is less than that of the four-beam system. When f_s = 1200 lines/mm, which is the frequency used for most of the illustrations in this book, the angle of incidence Θ is rather large. The helium-neon laser (λ = 632.8 nm) cannot be used when f_s = 1200 lines/mm since it gives sin Θ > 1 (Eq. 4.18). With the 514.5 and 488.0 nm wavelengths of the argon ion laser, Θ = 60.8° and 55.9°, respectively.

4.7.10   ±45° Gratings

Let the grating of frequency f_s in Fig. 4.23 be applied to a specimen with the primary grating lines oriented in ±45° directions, as illustrated in Fig. 4.25. Now, the pitch of the grating with lines in the ±45° directions is g_s. The first order diffraction of incident beam LO emerges as OQ normal to the specimen when

$$\sin \alpha = \lambda f_s = \frac{\lambda f}{2} \tag{4.8}$$

The angle of intersection LOM is $2\alpha'$; the geometry of the figure relates α' to α, with the result

$$\sin \alpha' = \frac{\lambda f_s}{\sqrt{2}} \tag{4.20}$$

and the corresponding virtual grating frequency is f', where by Eq. 2.24,

$$f' = \sqrt{2}f_s \qquad (4.21)$$

With beams LO and MO, the walls of the virtual reference grating are vertical; their frequency is $f' = \sqrt{2}f_s$. They interact with the vertical lines of the specimen grating, which have a pitch $g_s / \sqrt{2}$ and frequency $\sqrt{2}f_s$. Thus, the initial frequency of the specimen grating is equal to that of the virtual reference grating in this case. The N_x pattern is formed when beams LO and MO are incident; the N_y pattern is formed with beams KO and LO. The sensitivity is $\sqrt{2}f_s$. When $f_s = 1200$ lines/mm, the contour interval is 0.59 μm/fringe order.

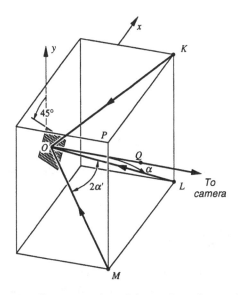

Fig. 4.25   Representation of the ±45° grating method.

This is the ± 45° gratings method of Ref. 18. An in-depth optical analysis is given in the reference. The displacement components in the ± 45° directions are *not* measured in the three-beam method or the ± 45° grating method. It is the N_x and N_y components that are measured, which give the U and V displacement fields.

A variation can be used to measure the 45° components. Referring again to Fig. 4.24a, let beams AO and DO be blocked. The beams BO and CO produce a virtual reference grating that interacts with the specimen grating to yield the U_{45} displacement field. Beams AO and CO are used to produce the U_{-45} field. When $f_s = 1200$

lines/mm, the contour interval is again 0.59 μm/fringe order. This technique was used to obtain Fig. 11.3.

Regardless of the scheme used to measure them, the four displacement components are not independent. As illustrated by the vector diagram of Fig. 4.26, the displacement of a point from position P to P' can be represented by either the sum of its 45° components or its U and V components. They are interrelated by

$$U = \frac{U_{-45}}{\sqrt{2}} - \frac{U_{45}}{\sqrt{2}}$$

$$V = \frac{U_{45}}{\sqrt{2}} + \frac{U_{-45}}{\sqrt{2}}$$

(4.22)

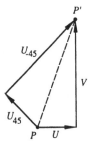

Fig. 4.26   The displacement of P to P' can be represented by either pair of vector components.

4.8 Transmission Systems

When the specimen is transparent, it is practical to employ moiré interferometry in transmission, as illustrated schematically in Fig. 4.27. With a transmission type cross-line grating applied to the specimen, light diffracted in its +1 and −1 diffraction orders is admitted into the camera to record the moiré pattern. The mechanism of fringe formation is the same in transmission and reflection systems. A feature of the transmission system is that there is no restriction on the location of the camera lens, and it can be very close to the specimen to obtain high magnification.

Under some circumstances, it is more practical to make replicas of the deformed specimen grating and view the replicas at a later time.[19] The replicas can be made on a transparent substrate and viewed in a transmission type moiré interferometer.

Every moiré interferometer designed for opaque specimens can also be used in transmission. In each case, the camera is merely

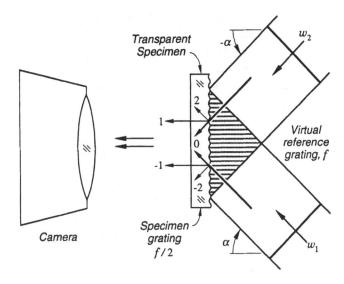

Fig. 4.27   Moiré interferometry in transmission with a transparent specimen.

moved to the opposite side of the specimen. The reader can visualize the transmission counterpart to each of the interferometer systems illustrated in Figs. 4.8, 4.11-4.22, 4.24, and 4.25.

4.9 Specimen Gratings

Specimen gratings are usually formed by a replication process, with the result illustrated by the cross-sectional view of Fig. 4.28. The process utilizes a special mold, which is a cross-line grating itself, and a liquid adhesive. The adhesive flows between the specimen surface and the mold to replicate every detail. When the liquid

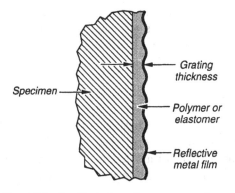

Fig. 4.28   Replicated grating on a specimen.

solidifies by polymerization and the mold is removed, the specimen is coated with a plastic that has the furrowed surface of a cross-line grating. An ultra thin reflective metal is applied, usually by evaporation (Sec. 2.2.3), to increase the diffraction efficiency in reflection. For transmission systems, the reflective film is deleted.

The features that enhance the performance of specimen gratings for moiré interferometry include the following:

- The grating must adhere to the specimen and deform together with the specimen, without delamination or slippage relative to the specimen surface. It should not reinforce the specimen, i.e., the specimen displacements should not be significantly altered.

- The grating should be thin to minimize shear lag (Sec. 4.24) in regions of large strain gradients and at geometric and material discontinuities. Replicated gratings are typically about 25 μm (0.001 in.) thick which is sufficiently small for most studies. For studies of micromechanics with microscopic fields of view, the thickness has been reduced to about 2 μm (80 μin.) in the small zones of interest (Sec. 5.4). So-called zero thickness gratings have been produced by other methods (Sec. 4.9.5), but not by replication.

- The grating surface should be specular, or mirror-like. The molds used for replicating these gratings have specular surfaces. After replication, the grating surface has the specular quality of the mold rather than the matte irregularities of the original specimen surface. This is the single most important reason for the excellent clarity of moiré interferometry fringes. The fringe patterns are relatively free of speckle noise, which occurs when matte surfaces are illuminated by coherent laser light.

- The grating should have a symmetrical furrow profile. Then, when incident beams 1 and 2 (Fig. 4.4) have equal intensities, the grating diffracts beams of equal strengths to produce the high contrast fringes of pure two-beam interference.

- The grating should have high diffraction efficiency in the diffraction orders that are utilized. Although this is not essential to produce fringe patterns of excellent quality, the associated efficiency of light utilization allows use of lower power lasers and/or shorter exposure times.

4.9.1 Replication

Figure 4.29 illustrates how a grating is replicated on the surface of a specimen. First, the mold is prepared by applying a reflective metal film, usually evaporated aluminum, on its grating surface. A double coating of aluminum is sometimes used, where the first film is

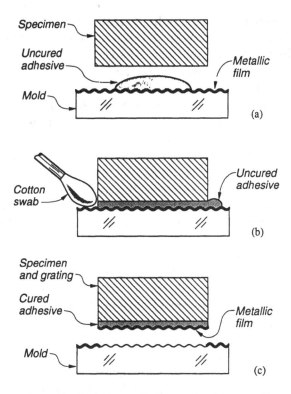

Fig. 4.29   Steps in producing the specimen grating by a casting or replication process; the reflective metallic film is transferred to the specimen grating.

oxidized by allowing air into the vacuum chamber and then a second coating is applied.

A pool of liquid adhesive is poured on the mold and squeezed into a thin film by pressing the specimen against the mold. As the excess liquid adhesive is squeezed out it is cleaned away from all specimen boundaries with cotton swabs; the swabs can be *slightly* dampened with alcohol. Cleaning is continued during the entire time that the liquid flows out. After polymerization, the mold is pried off—only a small prying force is required—leaving a reflective diffraction grating bonded to the surface of the specimen. The weakest interface occurs between the mold and the metal film, which accounts for the transfer of the reflective film to the specimen. The result is a reflective, high-frequency, phase-type diffraction grating formed on the specimen.

The reflective metal film acts as a parting agent to allow separation. Vacuum deposition techniques designed to enhance adhesion of the metal should *not* be used; in particular, the mold should be shielded from the radiant heat prior to deposition of the metal, since adhesion improves with substrate temperature.

As an alternative, a mold of silicone rubber is sometimes used. Silicone rubber is a non-stick material which does not require a parting agent. The reflective metal film is not applied to the mold, but instead it is applied to the specimen after replication.

Various room temperature curing adhesives have been successful, including epoxies, acrylics, urethanes and silicone rubbers. Unfilled polymers are generally used, since the fillers or pigments in the resin can be larger than the pitch of the grating and distort the groove profile. Various adhesives that cure by exposure to ultraviolet light can be convenient. They can be used with silicone rubber molds, since the opaque metallic film is not required for separation.

Minimum shrinkage of polymerization is desired. If the specimen surface is rough, and if the polymer shrinkage is 10%, then the surface of the grating will exhibit roughness, too; it will exhibit about 10% of the roughness of the original surface. The remaining roughness might be enough to create speckle noise in the moiré patterns. If the specimen cannot be smoothed by a fine abrasive paper, the problem can be alleviated by a double replication process. The first replication uses a smooth mold with a non-stick surface. For example, a smooth polymethylmethacrylate (pmma) plastic sheet can be used as a mold if the adhesive is epoxy, since epoxy does not bond to pmma very well. The excess adhesive should be squeezed out by heavy pressure on the mold to assure minimum thickness. If the shrinkage is 10% in each step, the roughness will be attenuated to 1% of the original by the two-step process.

The adhesive should polymerize uniformly. Otherwise, when the liquid solidifies in one zone and the remaining liquid polymerizes and shrinks later, the replica can exhibit severe defects; usually, a dendritic pattern of channels is developed. Adhesives tend to form channels when the cure is very rapid, but excellent results are obtained with slower cures. Various commercially available materials have been used successfully. At the time of publication, they included those listed in Appendix C.

4.9.2 The Mold

A photosensitive layer on a flat glass substrate is used for making the mold. In earlier work, the photosensitive material was a high resolution photographic emulsion—a holographic plate was used. As illustrated in Fig. 4.30, the plate was exposed to a virtual grating of frequency $f/2$, formed by two intersecting beams of coherent light. In this process, the emulsion is exposed in zones of constructive interference while it remains unexposed in zones of destructive interference. After the plate is developed, the photographic emulsion consists of clear gelatin in the unexposed zones and gelatin containing silver crystals in the exposed zones. Upon drying, the gelatin shrinks, but its shrinkage is partially restrained by the silver crystals, resulting in the undulating surface illustrated in Fig. 4.30c. It is this undulation that transforms the surface of the photographic plate into a phase-type diffraction grating. In practice, two exposures are made, one as shown and the second after the grating is rotated 90°. Thus, a cross-line grating is produced. The final step is to overcoat the grating with an ultra-thin aluminum film, or other reflective metal.

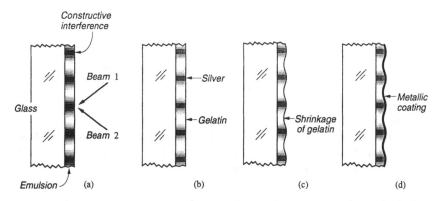

Fig. 4.30 A mold for producing phase-type specimen gratings can be generated optically on a high-resolution photographic plate: (a) expose photographic plate; (b) develop; (c) dry; (d) mirrorize.

Although specimen gratings replicated from these molds served nicely, they had low diffraction efficiency. The furrows were shallow and their depth could not be controlled. Subsequent use of photoresist as the photosensitive layer allowed optimization of the furrow depth and an order of magnitude improvement of diffraction efficiency was achieved.

Photoresists are photosensitive polymers. They are usually received in liquid form and a thin layer is spread on the glass substrate; spinning is used to spread a layer of uniform thickness by centrifugal force. After the resist dries, the plate is exposed to a virtual grating, as illustrated in Fig. 4.31, and given two exposures as

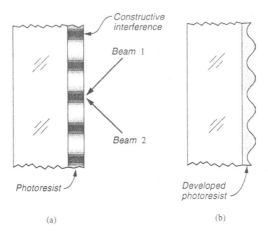

Fig. 4.31   With photoresist, the depth of the furrows can be optimized for maximum diffraction efficiency.

above. These resists are most sensitive to ultraviolet and blue light, so low wavelengths in the blue region are used with visible light lasers. The resist is developed in a solvent that preferentially dissolves the resist. The solubility of the resist polymer increases with exposure to light (for positive resists), and therefore the resist dissolves most rapidly in the zones of constructive interference. After development, the cross-sectional view is represented by Fig. 4.31b. A scanning electron micrograph is shown in Fig. 2.35.

With this photoresist process, the depth of the furrow can be optimized by control of the exposure and development. Cross-line gratings of 1200 lines/mm (30,480 line/in.) have been produced with 20% diffraction efficiencies in each of the four first-order diffractions.

In usual practice, the photoresist gratings are not used directly as molds. Instead, submasters are used. A two-step process is effective whereby a silicone rubber grating is first replicated by a process equivalent to that of Fig. 4.29, and then a more robust polymer (like epoxy) is replicated from it. The intermediate silicone rubber is used because of its non-stick nature, allowing separation without parting agents. The silicone rubber grating can be used repeatedly to produce many molds; there is no noticeable degradation of the molds as a result of successive replications.   No degradation is found in successive generations of submasters, each removed from the photoresist original by another replication step.   The final mold (submaster) can be considered as expendable and discarded after replication of the specimen grating. The production of subgratings is described in more detail in Appendix B.

A means of exposing a large photosensitive plate is illustrated in Fig. 4.32. An easy method to establish the required angle 2θ is to insert at P a commercial reflective diffraction grating of half the desired specimen grating frequency; for example, temporarily use a 600 lines/mm grating at P if a 1200 lines/mm specimen grating mold is to be produced. Then two beams of light reflect back from the grating in its +1 and −1 diffraction orders. They are converged by lens L to two bright spots on the screen. Coarse adjustment is achieved by varying angle 2θ until the spots are superimposed. If lens L is then removed, an interference pattern is seen on the screen. The angle is varied by fine adjustments until the pattern is reduced to a null field, whereupon the required angle and virtual grating frequency is established. Frequencies that are accurate to about 1 part in 10^5 can be achieved. Many of the moiré interferometers illustrated in previous figures could be used to produce smaller grating molds. For example, the system of Fig. 4.8 could be used by adjusting the outer mirrors for incident light at angle θ, where

$$\sin \theta = \frac{\lambda f_s}{2} \qquad (4.23)$$

Then, the angle would be refined as above using the commercial grating to null the field.

The method is called a holographic process, and the result a *holographic grating*, because the dual-beam illumination is commonly used in holography. The interference pattern captured by the plate has a simple harmonic intensity distribution (Eq. 2.20), assuring that the furrows have a symmetrical profile.

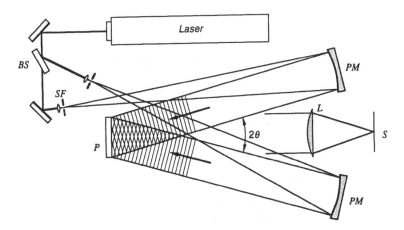

Fig. 4.32   Optical arrangement to expose a photosensitive plate to a virtual grating of frequency f_s. *P*: photosensitive plate. *BS*: beam splitter. *SF*: spatial filter. *PM*: parabolic mirror. *L*: lens. *S*: screen.

4.9.3 Alignment of Specimen Gratings

Before the grating is replicated on the specimen, some means should be provided to align the grating lines of the mold with respect to the specimen. This is usually done by means of an alignment bar as illustrated in Fig. 4.33. The bar, A, is first aligned with respect to the mold, B, and cemented to it. Then the bar rests along the edge of the specimen for grating-to-specimen alignment during the replication process of Fig. 4.29.

A convenient alignment fixture is illustrated in the figure. It consists of a flat base C, a bar D with a narrow groove parallel to the base and an adjustable fixture E. The mold is attached temporarily to the fixture with double-sided adhesive tape. An unexpanded low-power laser beam passes through a hole in the bar and intercepts the grating mold as shown. The fixture E is adjusted so that the +1 and −1 diffraction orders strike the parallel groove. This makes the grating lines accurately parallel to the base C. Next, a thin parallel bar A, e.g., 1 mm (0.04 in.) thick by 13 mm (1/2 in.) wide, is cemented to the mold, while the bar rests on the base as a reference surface; Fig. 4.33c illustrates this step. A quick-setting cyanoacrylate cement is effective. Care must be taken to avoid forming a filet of cement along the alignment edge of the bar, since that would effect the alignment accuracy.

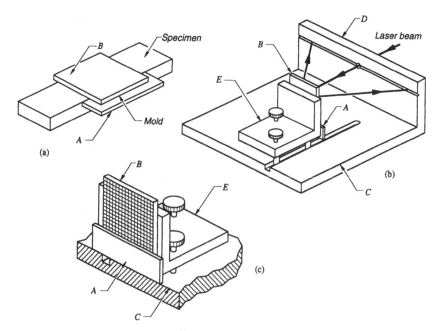

Fig. 4.33   Specimen grating alignment. (a) Use of the alignment bar. (b) Alignment fixture. (c) Alignment bar cemented to mold.

4.9.4 Curved Surfaces

The methods described here cannot be applied to the general analysis of curved surfaces. Local deformations can be determined, however, insofar as the local portion of the surface can be approximated by a flat surface. For a convex surface the specimen grating can be applied in the normal way, but with the flat mold positioned tangent to the local region of interest. Of course, the thickness of the specimen grating increases with distance from the tangent point and the analysis should be restricted to the zone where the thickness is sufficiently small. For concave surfaces, the flat mold must be cut to a narrow or small size to prevent excess thickness of the specimen grating.

While the above procedure is adequate for some problems, there is scope for other schemes. Examples might include the replication of curved or flexible grating molds, or the direct generation of gratings on the specimen (Sec. 4.21).

4.9.5 Zero-thickness Gratings and High-temperature Gratings

Zero-thickness gratings are cross-line amplitude gratings comprised of an orthogonal array of metallic dots separated by bars of zero thickness. Since the dots are tiny islands, they move freely to follow the specimen distortions; such gratings are attractive candidates for micromechanics studies involving severe strain gradients or material discontinuities, as in fracture studies. They circumvent the shear lag problem (Sec. 4.24) of finite thickness gratings, whereby strains are attenuated and redistributed as they are transmitted through the grating thickness. They also feature temperature stability, making them good candidates for high-temperature applications.

The steps followed to produce these gratings are illustrated in Fig. 4.34. The steps proceed as follows:[20]

- The specimen surface is polished to a smooth mirror-like finish.
- A highly adherent film of gold is formed on the specimen by high-vacuum evaporation or by sputtering. The film is ultra-thin, e.g., 30-50 nm (1-2 μin.), just enough for good reflectivity.
- A uniform layer of photoresist is applied over the gold, dried and prebaked.
- The photoresist is exposed to a virtual grating formed by interference of two intersecting beams of coherent light. The specimen is rotated 90° and exposed again. After development, the

photoresist has a cross-line furrowed surface with 1200 peaks per mm (30,480 per in.).

- Dry ion etching is used to uniformly erode the photo-resist. Subsequently, the reflective metal film is eroded below the valleys of the photoresist to expose the substrate.
- The remaining photoresist is stripped chemically, leaving a grating of ultra-thin gold dots on the specimen.

Gratings produced by this process are amplitude gratings. They exhibit excellent moiré interferometry fringe patterns, approximately equal to those from replicated phase gratings. Their diffraction efficiency is not as high as that obtained by replication from photoresist molds, but it is adequate. Whereas amplitude gratings of low frequency exhibit low diffraction efficiencies, the efficiency of high-frequency amplitude gratings is much greater. This is because the light is divided into only a few orders—only three orders for 1200 lines/mm gratings—and the first-order diffraction efficiency is reasonably high.

Such gratings can also serve for high-temperature applications. The development of survivable specimen gratings is an important step in the development of moiré interferometry for very high temperature applications. The choice of refractory metal depends upon the operating temperature for the moiré investigation. Gold would suffice for many high-temperature applications. Even higher temperatures can be sustained by other refractory metals, and tungsten can be considered for very high-temperature work. Gold might tolerate a working environment of oxygen or air, but more reactive metals would require an inert or vacuum environment.

Kang et al.[21] reported survivability at 1100°C for two types of zero-thickness gratings. One used the process outlined above, whereby gratings of 600 lines/mm were ion-etched in gold on low-expansion substrates. The other used chemical etching, whereby cross-line gratings of 300 lines/mm were etched directly into the surfaces of metallic (Inconel 718) and quartz specimens. Instrumentation was reported for successful moiré tests to 1100°C in an air environment.

A somewhat simpler procedure is under development at the National Physical Laboratory (England) where the refractory coating is applied over a photoresist grating, and the photoresist is subsequently decomposed and vaporized at elevated temperature. Survivability at 1100°C has been reported for 1200 lines/mm gratings by Kearney and Forno.[22]

Further advances for zero-thickness gratings and high-temperature gratings are imminent. The technology is mature for producing gratings in the specimen material itself by reactive ion etching. With this technique, a photoresist grating is first applied to

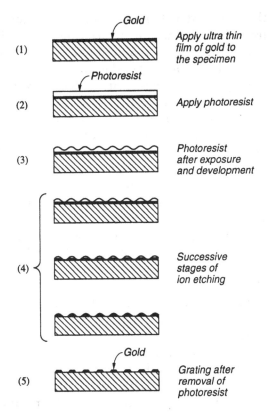

Fig. 4.34   Process to form a 'zero thickness' specimen grating.

the polished specimen. Then it is ion etched in a gas atmosphere that reacts chemically with the specimen material; the rate of attack on the specimen material is much greater than on the photoresist and gratings can be etched deeply into the specimen surface. The method can be used on a wide variety of metals and ceramics.

Additional techniques are becoming available to 'write' high-frequency gratings by focused electron beams and ion beams. The beams can be deflected systematically to draw gratings in special resists. In addition, focused ion beams can have enough power to scribe a grating directly into the specimen material. As this technology becomes more widespread, it will be practical to utilize it for moiré interferometry.

4.10 Fringe Counting

The location of the zero-order fringe in a moiré pattern can be selected arbitrarily. Any fringe—black, white or gray—can be assigned as the zero-order fringe. This is because rigid-body translations are not important in deformation analysis. Absolute displacement information is not required and relative displacements can be determined using an arbitrary datum.

The rules of topography of continuous surfaces govern the order of fringes.[3,23] Adjacent fringes differ by plus or minus one fringe order, except in zones of local maximum or minima where they can have equal fringe orders. Local maxima and minima are usually distinguished by closed loops or intersecting fringes; in topography, such contours represent hills and saddle-like features, respectively. Fringes of unequal orders cannot intersect. The fringe order at any point is unique, independent of the path of the fringe count used to reach the point.

Figure 4.35 is an example. The zero-orders are assigned to the centers of the dark fringes marked 0 in the N_x and N_y fields. The fringe $N_x = 4$ that passes through point A is a saddle-shaped contour. The fringe orders increase to the left and right of A, and they decrease above and below A. Elsewhere, the fringe orders increase or decrease monotonically in the U field. Another saddle appears near point B in the V field.

These fringe patterns are taken from a series of tests of the compact shear specimen,[24] which is a specimen geometry proposed for determining the shear-stress/shear-strain properties of composite materials. The jagged and wavy nature of fringes results from small local variations of material properties; the fringe irregularities are characteristic of many composites when investigated with high sensitivity. The two displacement fields are shown for the case of a small load. A small region near each notch is obscured in the U field, and the top and bottom is obscured in the V field. The reason is that light was obstructed by the loading fixture and these portions of the specimen were in the shadow region.

Once the zero-order fringe is selected, the next question is whether the fringe order of a neighboring fringe is greater or less—i.e., what is the sign of the fringe gradient? First, a sign convention is required.

Sign Convention: As the N_x pattern is scanned in the x direction, the sign of $\partial N_x / \partial x$ is positive if the fringe order is increasing. Similarly, $\partial N_y / \partial y$ is positive if N_y increases with y, i.e., along a path of increasing y values. The gradient $\partial N_y / \partial x$ is positive when N_y increases in the positive x direction. The gradient $\partial N_x / \partial y$ is positive when N_x increases with $+y$.

This convention assures that tensile strains correspond to positive fringe gradients and compressive strains to negative gradients. The

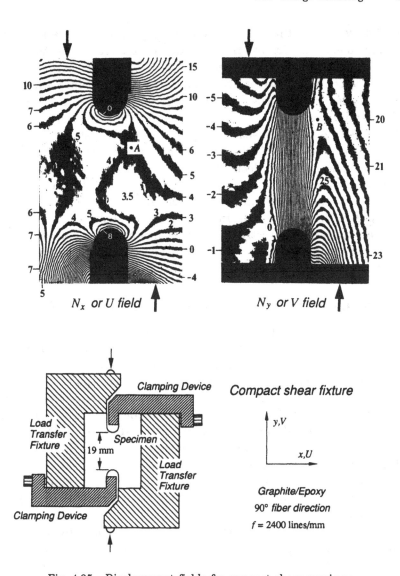

Fig. 4.35   Displacement fields for compact shear specimen.

question of fringe ordering reduces to a determination of the sign of the fringe gradient.

Frequently the sign can be inferred from knowledge of the geometry and loading conditions. For example, in Fig. 4.35 it is known that the right side of the specimen moves upwards (positive V or N_y) with respect to the left side. Therefore $\partial N_y / \partial x$ must be positive in the central region of the N_y pattern, and the fringe orders increase with x.

It is easy to infer that the gradient $\partial N_x / \partial y$ must be negative near the notch roots. This follows from the fact that there can be no shear stress τ_{xy} at the free boundary, and therefore γ_{xy} must be zero. By Eq. 4.3, the two cross-derivatives are equal and opposite. Since $\partial N_y / \partial x$ is positive (as determined above), then $\partial N_x / \partial y$ must be negative near the notch boundaries.

There are more clues. Near the applied forces, the strains in the y direction must be compressive. This means $\partial N_y / \partial y$ is negative in these zones and N_y must decrease in the $+y$ direction. Additionally, the fixture clamps induce compressive ε_x in the two clamped legs of the specimen, the upper-right and lower-left portions. Thus, N_x is decreasing in the $+x$ direction in these regions. Only some of these clues are required to correctly order the fringe numbers, but all of the decisions are self-consistent, as they must be. Accordingly a unique fringe order is assigned to each fringe.

4.10.1 Sign of Fringe Gradient: Experimental Method

Circumstances could arise in which there are not enough clues, but the sign of the fringe gradient can always be determined experimentally. If, during the experiment, the specimen is moved gently in the $+x$ direction, the fringe order N_x at every point increases. This means that the fringes all move toward the direction of *lower-order fringes*. If the fringe movement is in the $+x$ direction in any region, the gradient $\partial N_x / \partial x$ is negative in that region. If the fringe moves in the $-x$ direction, the gradient is positive.

The argument is the same for the y direction. While the specimen is moved in the $+y$ direction by pressing a pencil gently against it, the fringes that move in the opposite direction signify positive $\partial N_y / \partial y$ in their region.

In the case of Fig. 4.35, when the specimen is moved gently upwards ($+y$ direction), the N_y fringes in the left and right portions of the specimen move upwards, too, indicating negative $\partial N_y / \partial y$. The vertical fringes in the central portion move to the left, indicating that the fringes on the left side have lower fringe orders. When the specimen is moved gently to the right ($+x$ direction), the fringes in the left and right portions of the N_x field move downwards toward lower fringe orders, while the fringes near the notches move upwards toward lower fringe orders. The question of whether the order of a fringe is greater or less than that of its neighbor can always be determined by observation during the experiment.

4.11 Strain Analysis

The basic techniques of data analysis for strain determinations will be illustrated for the specimen of Fig. 4.35. Here, the primary interest is shear strains, determined by Eq. 4.3. Where the moiré fringes are closely and regularly spaced, it is practical to approximate the derivative by its finite increment, e.g., $\partial N_y / \partial x \approx \Delta N_y / \Delta x$. Note that distances measured on the fringe pattern must be divided by the image magnification M to obtain the corresponding distance on the specimen. Thus,

$$\frac{\partial N_y}{\partial x} \approx \frac{\Delta N_y}{\Delta x / M} \tag{4.24}$$

and equivalent relationships apply for the other three derivatives. In the case of Fig. 4.35, the cross derivative of N_y can be determined along the vertical centerline by measuring the distances between the central 5 fringes; then ΔN_y is +5 and Δx is the measured distance between the fringes. This method was followed by measuring Δx at several locations along the vertical centerline in order to determine $\partial N_y / \partial x$ plotted in Fig. 4.36.

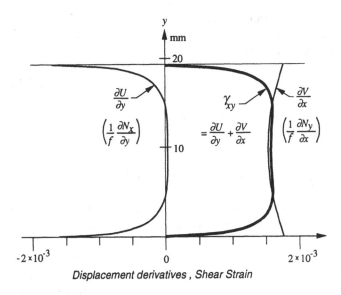

Fig. 4.36   The distribution of shear strain along the vertical centerline of the compact shear specimen of Fig. 4.35.

Where the moiré fringes are spaced irregularly, the curve of fringe order versus position should be plotted along a line parallel to x (or y). Then the slope of that curve at any point represents the derivative with respect to x (or y). This method was used along the vertical centerline of the N_x field to obtain $\partial N_x / \partial y$ in Fig. 4.36. The technique is demonstrated in Sec. 7.2 and Fig. 7.4.

The shear strain γ_{xy} in Fig. 4.36 is plotted from the sum of these two derivatives. The result is the highly desirable condition of nearly uniform shear strain between the notches.

In this specimen, the normal strains are extremely small in the region between the notches, i.e., in the test zone. By Eq. 4.2, strain ε_x is proportional to $\partial N_x / \partial x$, which is essentially zero near the notches where the fringe slope is essentially zero; it is very small in the central region, where the N_x fringes are widely spaced. Strain ε_y is essentially zero along the vertical centerline, since $\partial N_y / \partial y$ is essentially zero. A condition of nearly pure shear is achieved in this specimen.

At higher load levels the fringes are closely spaced but nevertheless the whole field is well resolved on photographic film. Normally a highly enlarged photographic print is made for the strain analysis. A portion of such a print is shown in Fig. 4.37a for the same specimen. For this figure, the average shear strain in the test zone is 0.35%, or about 2.3 times that in Fig. 4.35.

Strain distributions are not always so orderly. Figure 4.37b shows the N_y or V displacements in a compact shear specimen, but the specimen material was a three-dimensional weave of glass fibers in an epoxy matrix. A systematic distribution of zones of higher shear and zones of lower shear is evident. The importance of a whole-field method of observation and measurement is exemplified by this complex deformation field.

4.12 In-plane Rigid-body Rotations

In practice, it is sometimes difficult to control the in-plane rigid-body rotation of the specimen while applying external loads. Rigid-body rotations relative to the virtual reference grating introduce fringes that are (essentially) perpendicular to the grating lines. The rotation changes the fringe order of every point in the N_x field by a linear function of y. However, the gradient $\partial N_x / \partial x$ is not changed.[*] Similarly $\partial N_y / \partial y$ is unaffected by rotations. Thus, accidental rigid-

---

[*] There is a second-order influence because the fringes of rotation are inclined by half the angle of rotation (Sec. 3.1.4). With high-sensitivity moiré interferometry, the influence is negligible.

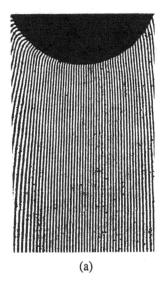 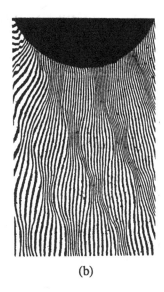

(a)   (b)

Fig. 4.37   (a) N_y or V field in the notch region of the specimen of Fig. 4.35, but for a greater applied load.   (b) N_y or V field for a composite specimen with a 3-dimensional weave of fibers. $f = 2400$ lines/mm (60,960 lines/in.).

body rotations can be tolerated, without consequence, for determination of normal strains ε_x and ε_y.

A rigid body rotation of the specimen generates strong cross-derivatives that are mutually dependent; their relationship is [25]

$$\frac{\partial N_x}{\partial y} = -\frac{\partial N_y}{\partial x}$$

When these contributions of pure rotation are introduced into the shear-strain equation, Eq. 4.3, their effect is nil. The four-beam and three-beam moiré interferometers capitalize on this relationship. With them, the x and y virtual reference gratings are fixed in space and the magnitude of the rigid-body specimen rotation, relative to each of them, is identical. Although accidental rigid-body rotation of the specimen alters the fringe patterns, the rotation has no influence on the calculated shear strains. Accordingly, none of the strains (ε_x, ε_y, γ_{xy}) are affected adversely by rigid-body rotations.

In fact, this insensitivity to rotations can be an asset. Figure 4.38 shows fringe patterns for the same specimen and load as Fig. 4.35. The only difference is rigid-body rotation, which was adjusted to minimize $\partial N_y / \partial x$ in the test section of the V field. The cross-derivative that was canceled at each point in the V field is added automatically at the corresponding point in the U field. Thus, their sum is unchanged. Why is this transformation useful? Now the gradient $\partial N_y / \partial y$ is clearly depicted and readily evaluated. The normal strain ε_y is very small in the test zone, and zero along the centerline between notches.

4.13 Carrier Fringes

Any moiré pattern can be altered by changing the virtual reference grating. Visualize first, the effect of changing the reference grating before the specimen is deformed. When the frequency of the virtual reference grating is changed (by a tiny adjustment of the angle of incidence α of one beam), the moiré pattern changes from a null field to a series of uniformly spaced fringes parallel to the grating lines. This uniform array of fringes is called a *carrier pattern*. It is equivalent to a uniform apparent strain, ε_x = constant, across the field, and the fringes are called *carrier fringes of extension*.[26] In this context, the term *extension* is used for positive and negative fringe gradients, i.e., it includes carrier fringes of apparent compressive strain.

When the virtual reference grating is rotated with respect to the specimen, the moiré pattern changes from a null field to a series of uniformly spaced fringes perpendicular to the grating lines; rigorously, the fringes are perpendicular to the bisector of the reference grating and specimen grating lines, but the difference is negligible for small rotations. The fringes correspond to those of rigid body rotation of the specimen, and they are called *carrier fringes of rotation*.

Subsequently, when the specimen is loaded and deformed, the change of fringe order at each x,y point caused by the deformation is added to the fringe order of the carrier fringes of the initial field. The initial fringes are changed by the deformation and they are said to *carry* the information.

Carrier fringes in moiré interferometry are the same as *mismatch* fringes of geometric moiré. With moiré interferometry, however, the carrier fringes are produced very easily, merely by turning a screw that controls a mirror or other element of the apparatus; they can be sparse or very dense; they can be positive or negative, i.e., they can correspond to tensile or compressive apparent strains, or clockwise or counterclockwise rotations.

Referring again to Fig. 4.8, carrier fringes of extension can be introduced by rotating mirror M about an axis normal to the page. When α is increased, the virtual reference grating frequency

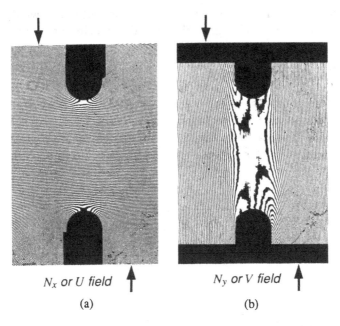

N_x or U field   ↑ N_y or V field   ↑

(a)   (b)

Fig. 4.38   Patterns after rigid-body rotation of the specimen of Fig. 4.35. The loading conditions and deformations are identical.

increases. This adds a positive fringe gradient to the pattern; it can reduce or nullify a negative fringe gradient caused by a compressive deformation of the specimen. Decreasing α has the opposite effect.

Carrier fringes of rotation can be introduced by tilting mirror M about an axis that is both parallel to the mirror face and parallel to the plane of the page. Alternatively, they can be introduced by rotating the interferometer about the z axis. As in the case of Fig. 4.38b, the carrier fringes can nullify cross-derivatives. The use and virtues of carrier fringes are demonstrated in subsequent sections and chapters, where specific experimental analyses are addressed.

The frequency of the virtual reference grating is altered by the carrier of extension, and a question arises. What value of f should be used in Eqs. 4.1, 4.2 and 4.3? Should it be the reference grating frequency before or after the introduction of the carrier fringes? The question was addressed in Sec. 3.2.3 in connection with geometric moiré and the answer is the same. It depends in the same way upon whether specimen coordinates or space coordinates are used for the x,y coordinate system. However, the question is less important in moiré interferometry. If the original frequency is 2400 lines/mm and a rather dense carrier of extension of 10 fringes/mm (254 fringes/in.)

is introduced, the change of f is only 0.4%. Its influence would rarely be significant in contributing to the overall experimental error.

4.14 Fringe Vectors

Fringe vectors represent local fringe gradients. They were introduced in connection with geometric moiré (Sec. 3.1.6) and they apply here, too. As illustrated in Fig. 3.11, the fringe vector represents the local gradient of fringe orders: its magnitude, direction and sense.

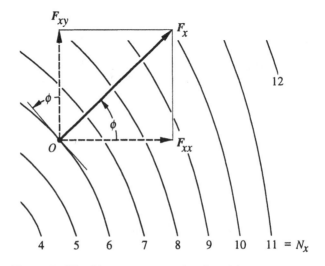

Fig. 4.39   The fringe vector at point O and its components.

Accordingly, Fig. 4.39 illustrates the fringe vector and its components for the local region surrounding point O. The N_x moiré fringe pattern is represented by the numbered lines, which depict the centerlines of fringes. The x and y components of the fringe gradient are tied to the local angle of the fringes by

$$F_{xy} = F_{xx} \tan \phi \qquad (4.25)$$

For the N_y field, the components are related by

$$F_{yy} = F_{yx} \tan \phi \qquad (4.25a)$$

In this notation F_{xy} represents $\partial N_x / \partial y$. The first subscript indicates the x or y field and the second indicates the x or y vector component. Angle ϕ is the acute angle from the x axis to the fringe vector; counterclockwise ϕ is positive. The implication of this relationship is important: instead of measuring the two fringe gradients, the fringe angle and one of the gradients can be measured to provide the same information.

Figure 4.40 illustrates how carrier fringes transform the pattern. When carrier fringes are added to the deformation fringes, the resulting fringe pattern is that labeled $N_x^{\sigma,c}$. Note that the principle of parametric curves (Sec. 3.1.5) is operational here, whereby $N_x^{\sigma,c} = N_x^{\sigma} + N_x^{c}$ at every point in the field. The superscripts σ and c refer to the load-induced fringes of Fig. 4.39 and the carrier fringes, respectively.

At point O, the fringe vector is $F_x^{\sigma,c}$. Since it is the vector sum of the load-induced fringe vector of Fig. 4.39, F_x^{σ}, and the carrier fringe vector, F_x^{c}, the pattern yields the load-induced vector by

$$F_x^{\sigma} = F_x^{\sigma,c} - F_x^{c} \tag{4.26}$$

The vector F_x^{c} points downward in this case, and its negative, $-F_x^{c}$, is added to $F_x^{\sigma,c}$. The resultant is the vector representing the load-induced fringes, F_x^{σ}, which is the same as the fringe vector in Fig. 4.39, of course. It is the load-induced information that we require, or its x and y components, and these are defined at point O by the vector diagram.

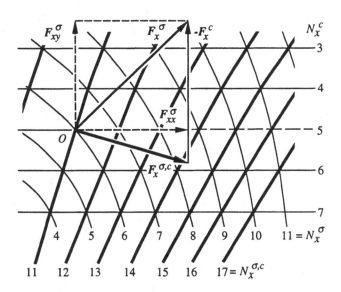

Fig. 4.40 The $N_x^{\sigma,c}$ fringe pattern is formed when carrier fringes N_x^{c} are added to the moiré pattern of Fig. 4.39. The load-induced fringe vector is determined by $F_x^{\sigma} = F_x^{\sigma,c} - F_x^{c}$. It is the same as the fringe vector in Fig. 4.39.

4.15 Anomalies and Subtraction of Uniform Gradients

When a specimen is loaded in tension, it exhibits a uniform array of moiré fringes—closely spaced, straight fringes—if the material behavior is uniform. Anomalous, or nonuniform behavior shows up as slight irregularities of the otherwise straight fringes.

Moiré interferometry offers a unique capability to analyze the nonuniformities. After the specimen is loaded in tension, the frequency of the virtual reference grating can be adjusted to cancel the uniform part of the fringe pattern. In effect, a carrier pattern is added, where the gradient of carrier fringes has the same magnitude as the uniform part of the moiré pattern, but the opposite sign. The remaining pattern shows the irregularities as bold variations of displacements across the sample.

This method is an exception to the general rule of metrology—that sensitivity diminishes as range is increased. The high sensitivity of moiré interferometry is retained and the anomalies can be analyzed in detail, even when the canceled uniform field represents a large fringe gradient and very high fringe orders.

An example is shown in Fig. 4.41. The specimen is a tension coupon of annealed, virtually pure copper. A 1200 lines/mm grating was applied to the specimen in its virgin condition. Then the specimen was loaded into its small plasticity range and unloaded. When viewed in a moiré interferometer with a 2400 lines/mm (60,960 lines/in.) virtual reference grating, the moiré pattern exhibited essentially horizontal fringes. They were very densely packed, averaging about 20.4 fringes/mm (520 fringes/in.), which corresponds to an average strain $\varepsilon_y = 0.85\%$. The fringes exhibited local irregularities or waviness that indicated a nonuniform strain field.

An opposite carrier pattern of –20.4 fringes/mm was applied (by adjusting the angle of incidence α) to produce the moiré pattern of Fig. 4.41; this was the most sparse pattern obtainable. The uniform part of the permanent strain field was canceled and the nonuniform or anomalous part remained. The remaining fringe gradient $\partial N_y^{\sigma,c} / \partial y$ is zero in some regions, like region A. Elsewhere, it is positive in some regions and negative in others. Strong fringe gradients appear in bands along 45° directions, corresponding to plasticity or yielding on planes of maximum shear stress.

The largest normal strain occurs at B, where the fringe gradient corresponds to $\varepsilon_y^{\sigma,c} = 0.53\%$. This is the anomalous strain at B. The total strain at B is 1.38%, the anomalous part plus the canceled uniform part.

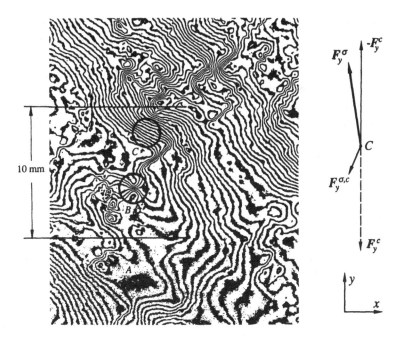

Fig. 4.41 Nonuniform part of the N_y displacement field after the uniform part was subtracted by an opposite carrier pattern. The specimen was annealed OFHC copper. It was plastically deformed in tension, and unloaded. The permanent strains in plastically deformed copper are highly heterogeneous.

At B, the fringe orders increase in the $+y$ direction, whereas they decrease at C. The fringe gradient at C is $F_y^{\sigma,c}$ It is directed downwards as shown in the vector diagram. When the carrier fringe vector F_y^c is subtracted, the load-induced fringe gradient is determined as F_y^σ; it represents the fringes at C prior to introducing the carrier. The anomalous normal strain at C is determined from the vertical component of $F_y^{\sigma,c}$ as $\varepsilon_y^{\sigma,c} = -0.19\%$.

The technique can be used for shear strains, too. Anomalies in otherwise uniform shear strain fields can be detected by canceling the uniform part by rotation of the specimen or the virtual reference grating. Figure 4.38b is an example, where the jagged fringes in the central region indicate slight anomalies, or small irregularities of the material behavior.

4.16 Two-body Problems, Bridges

Problems arise in which relative displacements of two separate elements of an assembly must be measured. Figure 4.42 is an example. The specimen is an adhesively bonded lap joint with aluminum adherends and a rubber-modified epoxy adhesive. The figure shows the U displacement fields when the applied loads are about (a) 25% of the failure load and (b) 70%. The fringes in the narrow adhesive zone are well defined in (a) and there is no difficulty in establishing fringe orders on both sides of the joint. In (b), the fringe gradients in the adhesive were very much larger as a result of the increased load and also because of nonlinearity of adhesive properties, and the fringes were not resolved in the adhesive zone.

It remains possible, however, to determine the displacements across the adhesive by establishing the fringe orders in the aluminum adherends. This was done with the help of the bridges bonded to the aluminum adherends along the dashed lines drawn in (a). The bridges allow a unique count of fringe orders on one side of the joint with respect to the other. While fringes in the bridges are not clearly reproduced in Fig. 4.42b, they were clearly visible on the photographic film and they were counted with the aid of magnification. In Fig. 4.42b, the change of fringe order across the adhesive is 59 fringes at the center, and the change increases to 63 fringes at the ends. Comprehensive results appear in Ref. 27. The adhesive was 0.22 mm thick, which means that the average fringe gradient across the adhesive thickness was about 270 fringes/mm (6800 fringes/in.). These fringes were not resolved in the photograph.

The bridges were made from a low modulus epoxy, which had a modulus of elasticity of about 210 MPa (30,000 psi). With their low modulus and low stiffness geometry, the bridges did not contribute in any significant amount to the load-bearing characteristics of the specimen. In retrospect, a less complicated bridge geometry would have been effective, namely, a straight bridge without the omega-shaped path. One bridge would have been sufficient for this specimen, but the second provided a reassuring check on the fringe count.

Artificial bridges are proposed as an effective scheme to quantify displacements in two-body problems, where the displacement of one body with respect to the other would otherwise remain unknown.

4.17 Coping with the Initial Field

In general, the specimen grating will not have lines that are perfectly straight and uniformly spaced. Also, the optical elements used to form the reference grating will not be so accurate that a perfect reference grating is formed. The result is that fringes will usually appear in the field of view before the specimen is loaded. These initial

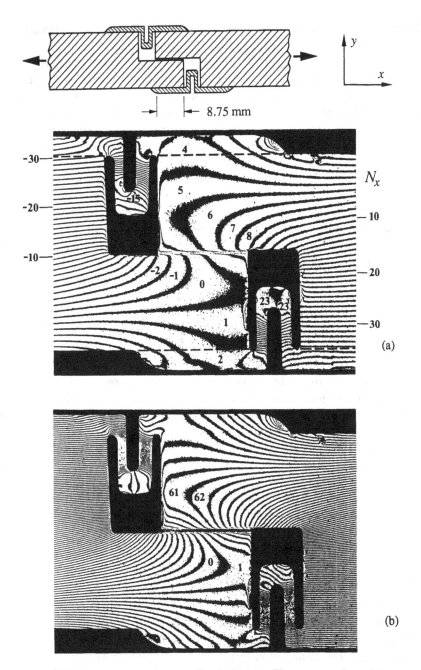

Fig. 4.42 N_x fringe patterns for an adhesive joint. The omega-shaped features are bridges of low stiffness used to ensure a continuous path between the adherends for fringe counting. f = 2400 lines/mm (60,960 lines/in.).

fringes must be subtracted from the pattern obtained after loading in order to determine the load-induced displacements.

There are at least two conditions, however, in which no subtraction is required. (1) When the initial pattern is sparse, but the full-load pattern exhibits a great number of fringes, the subtraction may be unnecessary to achieve the desired accuracy. (2) When the field of view is small, especially when it is so small that magnification is used in photographing the fringe pattern, the initial pattern in the field of view would usually exhibit a small fraction of a fringe order and it could be neglected in most analyses.

When subtraction is required, it can be done manually by subtracting the fringe orders at corresponding points in the load and no-load patterns. Alternatively, the subtraction can be performed simultaneously for the entire field of view by geometric moiré. A practical technique is to superimpose photographic transparencies (or negatives) of the moiré interferometry patterns for the load and no-load conditions. Both patterns should be modified by identical carrier fringes to increase the number of parametric curves (Sec. 3.15) and thus obtain a subtractive moiré pattern of good visibility. Use of an office copying machine is usually practical for making the transparencies.

Another technique is to make a double-exposure on the same film, first of the no-load moiré interferometry pattern and then the as-loaded pattern. Again, carrier fringes should be used to densify the patterns. After the film is developed, the pattern is one of additive intensities (Sec. 3.1.3), which is characterized by poor visibility of the moiré fringes. However, excellent fringe visibility can be obtained by optical filtering, which is described in the following section. Although optical filtering is usually employed to enhance double-exposure moiré patterns, it can be used with superimposed patterns (multiplicative intensity), too.

4.18 Optical Filtering

Optical filtering makes use of the diffraction properties of gratings, and therefore, patterns should be modified by a dense array of carrier fringes. Carriers of extension or rotation, or both, can be used. A rather high frequency of carrier fringes should be applied for optical filtering, e.g., 10 fringes/mm (250 fringes/in.) or more.

Figure 4.43 is an enlarged view of a double exposure of two moiré interferometry patterns, a no-load (or initial) pattern and a full-load pattern. A carrier pattern of rotation of about 12 fringes/mm (300 fringes/in.) was identical in both patterns. As a result of deformation

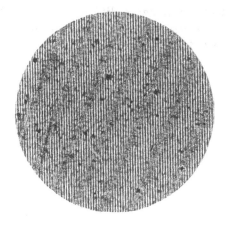

Fig. 4.43   Enlarged view of a double-exposure negative of two moiré interferometry patterns: a no-load and a full-load pattern. Identical carrier patterns of 12 fringes/mm (300 fringes/in.) were added to the patterns.

of the specimen, the full-load and initial patterns are different from each other, and therefore fringes cross each other in a systematic way. They form a geometric moiré pattern that depicts the difference of fringe orders present in the two constituent moiré interferometry patterns. This geometric moiré pattern is the desired contour map of the load-induced displacements, independent of any distortions present in the no-load pattern.

The double-exposure pattern is a case of additive moiré (Sec. 3.1.3). Each geometric moiré fringe of Fig. 4.43 is comprised of an area of continuous tone (an area where the constructive interference fringes of the full-load pattern fall midway between the constructive interference fringes of the initial pattern) and an area of black and clear stripes (where constructive interference fringes of both patterns fall upon each other in registration). The filtering process is possible because the striped areas diffract light while the continuous-tone areas do not.

The filtering apparatus is illustrated in Fig. 4.44 with the double-exposure negative (or two superimposed negatives) placed in the system. Light that passes through the striped areas is diffracted into several diffraction orders; light in each order converges to a separate point in the plane of the aperture plate. Light that passes through the gray continuous-tone areas does not experience diffraction and converges to point 0. Since light from only one (nonzero) diffraction order is admitted into the camera, the image on the camera screen is bright at all points corresponding to striped areas, and it is black at all other points. With optical filtering, an enhanced contrast image is made, using light of any nonzero diffraction order. In fact, only the zero order must be obstructed and light from any and all the other orders can be admitted into the camera.

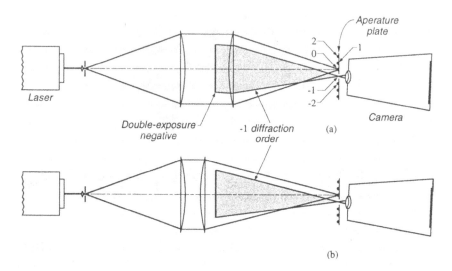

Fig. 4.44   Optical filtering to enhance the contrast of geometric moiré fringes in the double-exposure negative.   Light diffracted by the negative is converged to discrete points in the plane of the aperture plate, where the numbers correspond to the diffraction orders.   Either arrangement (a) or (b) can be used.

The result is seen in Fig. 4.45, in which the circled zone corresponds to Fig. 4.43. The enhancement of contrast is dramatic. The patterns depict the U and V displacement fields for the upper half of a composite tensile specimen with 15° off-axis fibers. The loading condition was axial translation of the grips with (almost) no rotation of the grips. The initial field is canceled and load-induced fringes are displayed uniquely.

The 15° orientation of the fibers makes the specimen asymmetrical. It deforms into a shallow S-shaped member and the central region becomes a zone of uniform shear strain. This example is taken from an early experiment with moiré interferometry,[28] in which the reference grating frequency was 1200 lines/mm (30,480 lines/in.).

4.19 Sensitivity to Out-of-plane Motion

It was shown in Sec. 4.3.1 that the mathematical analysis of moiré interferometry accounts fully for *load-induced* out-of-plane

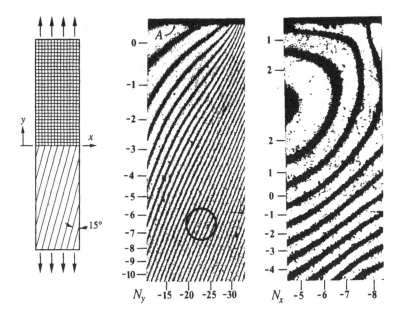

Fig. 4.45   Result of optical filtering, where the circled region corresponds to Fig. 4.43. The initial or no-load fringes are canceled and the patterns reveal load-induced displacements uniquely.   The specimen is a graphite-polyimide composite with 15° off-axis fibers. f = 1200 lines/mm (30,480 lines/in.).

displacements; the $U_{(x,y)}$ and $V_{(x,y)}$ displacements are determined correctly by the moiré fringe patterns, regardless of the presence of $W_{(x,y)}$ displacements and out-of-plane slopes of the specimen surface.

One caution should be noted.   Under some circumstances, *accidental rigid-body rotations* can occur during application of the loads. The out-of-plane slope resulting from an accidental rigid rotation about an axis parallel to the grating lines is seen as a fore-shortening of the specimen grating. The foreshortening produces a uniform apparent compressive strain throughout the specimen.

Figure 4.46 shows that a rigid-body rotation about the y axis, Ψ_y, introduces an apparent U displacement. The pitch g_s of the specimen grating changes by $\Delta g_s = g_s \Psi_y^2 / 2$. The grating experiences an apparent compressive strain

$$\varepsilon_x = -\frac{\Delta g_s}{g_s} = -\frac{\Psi_y^2}{2} \tag{4.27}$$

provided the rotation Ψ is not large. By Eq. 4.6, an extraneous fringe gradient F_{xxe} is produced, where

$$F_{xxe} = -f\frac{\Psi_y^2}{2} \tag{4.28}$$

The extraneous fringe gradient is a second-order effect and it can be neglected in the usual case where it is small compared to the strain-induced fringe gradient F_{xx}. Otherwise, a correction can be applied.

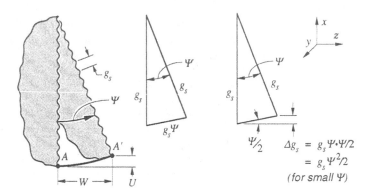

Fig. 4.46   Accidental rigid-body rotation about the y-axis causes foreshortening of the specimen grating.

In practice, the angle Ψ_y is determined by observing the spot of light reflected back to the source. Referring to Fig. 4.8, light reflected in the zeroth order of the specimen grating travels through the optical system to converge on plane D. The spot moves horizontally through a distance δ when the specimen rotates by Ψ_y. The angle of accidental rigid-body rotation is determined by

$$\Psi = \frac{\delta}{FL} \tag{4.29}$$

where FL is the focal length of the collimating mirror (or collimating lens). Although the accidental rotation is almost always negligible, its magnitude can be determined and the extraneous fringe gradient can be calculated by Eq. 4.28. Thus, a correction can be made in these special problems where the correction is warranted.

Other out-of-plane motions could occur. However, moiré interferometry is insensitive to out-of-plane translations and to out-of-plane rotations about an axis perpendicular to the grating lines.

4.19.1 Black Holes and Out-of-Plane Slopes

It is reassuring that load-induced out-of-plane slopes do not influence the positions of fringes in the U and V moiré patterns. This feature stems from a property of the camera lens, whereby the optical path length is the same for all rays between an object point and the corresponding image point. Accordingly, the emergent rays can have any slope. When the slope of a ray is large enough, however, it can miss the camera lens and never arrive at the image point. Regions of the specimen where the load-induced deformation causes large out-of-plane slopes might appear black in the image plane because light from such regions does not enter the camera. Such regions will be called *black holes*. In the event that one of the two interfering beams passes through the lens while the other misses the lens aperture, no interference fringes are formed. Such regions appear as *white holes* in high-contrast prints of the moiré pattern.

It is rare that a homogeneous body would exhibit large slopes. A composite body, fabricated by joining two dissimilar materials, can exhibit large slopes near the interface as a result of a Poisson's ratio mismatch. Experience with laminated fiber-reinforced composites indicates that the slopes are seldom severe enough to cause black regions until very high strain levels are reached and failure is imminent.

The camera lens can be translated laterally to capture the light and the moiré pattern that is otherwise lost. In doing so, it is possible that other regions will become black. A caution should be noted. When the slope is large, $\partial W / \partial x$ or $\partial W / \partial y$ might not be small compared to the in-plane derivatives ($\partial U / \partial x$, etc.) and finite strain relationships[3] might be appropriate for such regions.

4.20 Coping with Vibrations

Whenever it is practical to do so, the specimen should be loaded in a special fixture that is mounted on the optical table (Sec. 4.7.4, and 4.7.5). That arrangement provides optimum control for viewing the deformation under live loading, i.e., simultaneously with load application.

When very large loads are required, or when the specimen is very large, it becomes necessary to use a large mechanical testing machine. These machines usually cause tiny extraneous motions of the specimen, either by vibrations caused by the machine itself, or vibrations transmitted through the floor from other machines in the mechanical testing laboratory. Various schemes can cope with the vibrations. Compact optical systems that are mounted on the testing machine are practical; see Sec. 4.7.4 and Sec. 9.5 (Fig. 9.9).

Perhaps the best scheme is to load the specimen in the testing machine and make a replica of the deformed specimen grating, as outlined in Sec. 4.9.1. Replication is discussed in Sec. 4.21, too. The replica can be analyzed in the optical laboratory with a table-mounted moiré interferometer. This approach allows complete flexibility, including application of carrier fringes. It provides a permanent record of the deformation field, which can be investigated repeatedly if additional questions arise. Replicas can be made for several load levels. Regarding disadvantages, real-time observations cannot be made and data collection can be more time consuming.

Another solution lies in the use of real reference gratings, as in Fig. 4.17, mounted directly on the specimen. The objective is to assure that the reference grating vibrates in harmony with the specimen grating, without relative motion between them. Then, the moiré pattern is fixed, or unchanging, and its tiny motion caused by specimen vibration does not cause a blur of its image in the camera. Cross-line reference gratings and cross-line specimen gratings are used in Fig. 4.47. The system is relatively simple and highly effective in coping with vibrations. It lacks flexibility, inasmuch as carrier fringes of extension cannot be applied; in addition, null fields cannot be established unless the specimen and reference grating frequencies are in an exact 1:2 ratio and the cross-gratings have identical angles near 90°. Nevertheless, the scheme can be very effective when used in

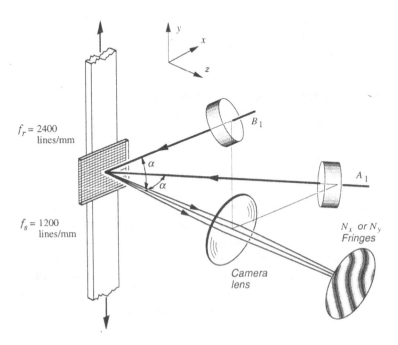

Fig. 4.47   A real reference grating, suitably attached to the specimen, vibrates in harmony with the specimen.

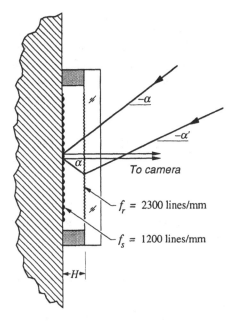

Fig. 4.48   Separate incoming beams permit control of the null field and carrier fringes.

conjunction with the methods of Sec. 4.17 and 4.18, whereby carrier patterns are applied and double exposure (or superposition) and optical filtering are employed.

A variation of this approach permits control of null fields and carrier patterns.[29] It is illustrated in Fig. 4.48, where the frequencies of the specimen grating and the real reference grating are 1200 and 2300 lines/mm (30,480 and 58,420 lines/in.), respectively. Two separate beams of mutually coherent light are used to form the virtual reference grating. One is incident at angle $-\alpha'$, such that its first-order diffraction is α. Thus, the two beams intersect at an angle 2α and form a virtual reference grating of 2400 lines/mm (60,960 lines/in.). The null field and carrier fringes of extension and rotation are controlled by tiny adjustments of the angle of one of the beams. Fringes of extension vary with tiny adjustments of α in the plane of the diagram; fringes of rotation are controlled by tiny adjustments of angles perpendicular to the plane of the diagram. These adjustments are made remote from the specimen by elements that control the angles of the incoming beams.

With the grating frequencies shown in Fig. 4.48 and the 632.8 nm wavelength of the helium-neon laser, the angles are $\alpha = 49.4°$ and $\alpha' = 44.1°$. The beams collected by the camera emerge essentially normal to the specimen. All other diffractions emerge at other angles, at angles $\alpha - \alpha' = 5.3°$ different from these, and they are not admitted into the camera lens. In practice the reference grating separation H can be very small, e.g., less than 1 mm. Cross-line gratings can be used, in conjunction with two pairs of incoming beams.

4.21 Variations of Moiré Interferometry

Transmission systems were addressed in Sec. 4.8. They are convenient for real-time observation with two-dimensional transparent specimens. They are convenient, too, for analysis of replicas of deformed gratings, when a glass plate is used as the substrate for the replica.

4.21.1 Replication of Deformed Specimen Gratings

Techniques were introduced in the late 1970s whereby a deformed specimen grating on a loaded structure was replicated and the replica was brought to the laboratory for analysis.[19] There it was installed in an optical system to record the moiré pattern. The method was originally used with low-frequency specimen gratings and fringe multiplication (Sec. 4.21.3) by high factors. It can be used, of course, with high-frequency specimen gratings and either transmission or reflection optical systems. It is an attractive option for specimens loaded in a mechanical testing machine and for studies of structures outside the laboratory. In addition, it is an attractive option for analysis of thermal strains. Figure 10.10b-d is an impressive example where the technique was applied at –40°C for macroscopic and microscopic analyses.

A replica of the specimen grating can be made while the specimen is under load by essentially the same technique as that illustrated in Fig. 4.29. A liquid adhesive is squeezed into a thin layer between the deformed specimen grating and a glass plate. When the adhesive solidifies by polymerization, the replica is pried off and held for subsequent analysis. A suitable scheme is to make the specimen grating of silicone rubber. Then, the replica can be made with a rapid curing epoxy, acrylic, or other polymer, which does not bond to silicone rubber. Different combinations of materials that do not bond well to each other can be used, too. Adhesives that are polymerized by ultraviolet light are good options. Ultraviolet light was used for Fig. 10.10.

4.21.2 Replication for Thermal Strains

A technique for steady-state thermal strain analyses is presented in Sec. 8.2.3. It utilizes a procedure to replicate the deformed specimen grating at elevated temperature onto a zero-expansion substrate. Then, the deformation field recorded in the replica is analyzed at room temperature.

Many important problems involve steady-state elevated temperatures. Deformations around cracks, inclusions, interfaces, etc., in specimens at elevated temperature are examples that can be analyzed by the replica methods. Transient thermal analyses might be approached by replication, too, using fast-setting replication materials.

Analysis in the very high temperature domain requires speculation. In many cases, specimen gratings can be formed in the specimen material itself, using reactive ion etching or focused ion beam etching (Sec. 4.9.5). Suitable replicating materials must be found. A possibility appears to be a substrate coated with a film of glass. At elevated temperatures, the softened glass is pressed against the specimen to form an embossed replica of the grating in the glass.[30] Materials other than glass could be applicable, too. In many cases the substrate can be a plate of the same material as the specimen. Research based upon this approach might be highly productive.

4.21.3 Moiré Fringe Multiplication

The sensitivity of moiré measurements depends on the frequency of the reference grating. At one stage, high-frequency real reference gratings were practical, but specimen gratings were restricted to low frequencies. They could be used together to provide a sensitivity corresponding to the reference grating. The ratio of reference grating frequency to specimen grating frequency was called the fringe multiplication factor. The next step was to use a high frequency virtual reference grating instead of a real reference grating.

The typical arrangements described in this chapter use a fringe multiplication factor of 2. As an alternative, consider Fig. 4.4b again, but now let the initial frequency of the specimen grating be one-half that illustrated in the figure, or $f_s = f/4$. The new condition is illustrated in Fig. 4.49; in this case, incoming beams A and B have the same directions as before, where α is determined by Eq. 4.4. Twice as

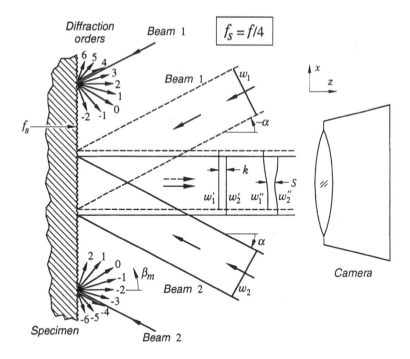

Fig. 4.49 Moiré fringe multiplication by a factor of four. Light from the +2 and −2 diffraction orders enters the camera.

many diffraction orders emerge from the specimen grating, since its frequency was halved. From beam 1, light of the +2 diffraction order enters the camera; from beam 2, light of the −2 order enters the camera. They combine and produce the same result as that obtained in Fig. 4.4. The wave fronts w_1'' and w_2'' have the same warpages in both cases.

Alternatively, the specimen grating frequency could be 1/6 that of the virtual reference grating; light of the +3 and −3 diffraction orders would enter the camera and produce the same moiré pattern again. The same scheme applies for any even fringe multiplication factor, β, where

$$f = \beta f_s = \frac{2 \sin \alpha}{\lambda}$$

$$\beta = 2m$$

(4.30)

where f_s is the initial frequency of the specimen grating and $\pm m$ are the diffraction orders used to produce the moiré pattern. Moiré fringe multiplication by a factor of 20 has been reported for strain analysis[31] and higher factors have been demonstrated in the classroom. High quality specimen gratings are required for high multiplication factors, since the signal-to-noise ratio decreases as diffraction order increases. The lowest optical noise is achieved in the +1 and −1 diffraction orders, and these orders were chosen for current methods of moiré interferometry, where $\beta = 2$. Only the transmission mode of moiré fringe multiplication is reported in the early literature, but the reflection mode is now an obvious counterpart.

4.21.4 Multiplication by One

Several schemes have been put forward that do not involve multiplication. Figure 4.50 illustrates an example where dashed lines represent a real reference grating and a specimen grating of the same initial frequency. When illuminated at angle α determined by Eq. 4.4, symmetrical first and zeroth-order diffraction beams emerge from the reference grating. These are diffracted again by the specimen grating in its zeroth and first orders, respectively, to emerge in the direction of the camera. The camera collects these two beams, which interfere to create the moiré pattern of specimen displacements. Guild[1] calls this optical arrangement a configuration of *minimum deviation*. Oblique viewing of the specimen is a disadvantage for strain analysis.

In another scheme, the specimen is coated with a photosensitive material—a photographic emulsion or a photosensitive resist.[32,33] It is exposed by two beams of coherent light at angles ±α, (i.e., by a virtual grating) which imprints a real grating in the photosensitive surface. The specimen is deformed and it is illuminated again by the same virtual grating to form a moiré with the real grating. If the

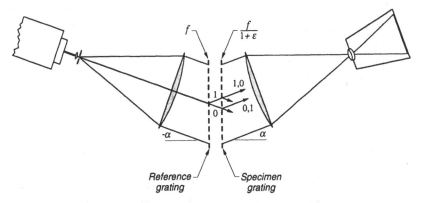

Fig. 4.50   Multiplication by one using a real reference grating. The numbers are diffraction order sequences.

specimen surface is partially matte, the moiré pattern can be observed by viewing the specimen from a convenient direction. However, speckle noise from the matte surface reduces fringe visibility. The method has potential for analysis of curved surfaces as well as flat surfaces.[34]

4.22 Mechanical Differentiation

Derivatives of displacements are required to calculate strains ε_x, ε_y and γ_{xy} by Eqs. 4.2 and 4.3. They can be obtained on a whole-field basis by *mechanical differentiation*, by shearing interferometry, and by computerized data analysis techniques. In Fig. 4.51, geometric moiré is used to perform mechanical differentiation.

When two identical transparencies of a fringe pattern are superimposed in registration with each other, they appear as the original. However, if the top transparency is shifted, the two patterns interweave to form a pattern of geometric moiré fringes. The effect is demonstrated in Fig. 4.51, where a moiré interferometry pattern of N_y is shifted by increments Δy and Δx. The moiré formed by the superposition is called a super-moiré pattern.

Why does this shifting of patterns produce the derivatives? Remember that

$$\frac{\partial N_y}{\partial y} \sim \frac{\Delta N_y}{\Delta y} = \frac{N'_y - N_y}{\Delta y} = \frac{N^D_{yy}}{\Delta y} \tag{4.31}$$

The prime signifies the shifted pattern and N^D_{yy} is the fringe order in the super-moiré pattern generated by the shift Δy. Using point A on Fig. 4.51a as an example, the increment Δy depicted by the vertical arrow crosses 4 fringes, from $N_y = 27$ to $N'_y = 31$. Thus $N^D_{yy} = \Delta N_y = 4$, as revealed by the super-moiré pattern of Fig. 4.51b. Similarly, the super-moiré fringes reveal the number of N_y fringes that are crossed at every point in the field when a shift of Δy is performed. The phenomenon is another realization of moiré as a whole-field subtraction process. From Eq. 4.31,

$$N^D_{yy} \approx \Delta y \frac{\partial N_y}{\partial y} \tag{4.32}$$

Accordingly, the super-moiré fringe pattern is a contour map of the derivative field.

In the same way, the arrow depicting Δx at point A moves from $N_y = 27$ to $N'_y = 25$ and $N^D_{yx} = \Delta N_y / \Delta x = -2$, as seen in Fig. 4.51c. The N^D_{yx} pattern reveals the cross-derivative field

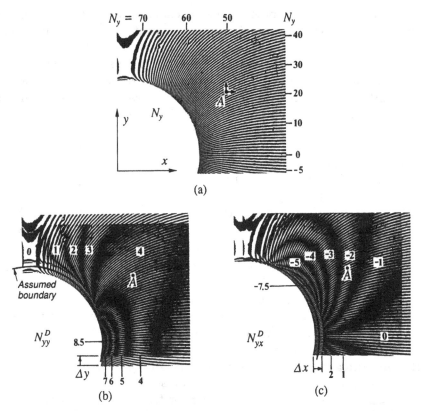

Fig. 4.51 Demonstration of the mechanical differentiation technique: (a) fringe pattern of N_y; (b) super-moiré pattern obtained by superimposing two transparencies of the fringe pattern with a shift Δy; (c) super-moiré pattern with a shift Δx. The numbers are fringe orders N_{yy}^D and N_{yx}^D, which depict the $\partial N_y / \partial y$ and $\partial N_y / \partial x$ fields, respectively.

$$N_{yx}^D \approx \Delta x \, \frac{\partial N_y}{\partial x} \tag{4.32a}$$

Similarly, N_{xx}^D and N_{xy}^D represent the super-moiré fringes of the N_x displacement field, and depict the $\partial N_x / \partial x$ and $\partial N_x / \partial y$ contours. The symbol N^D represents the fringe orders of the super-moiré pattern, i.e., the derivative pattern. The double subscript refers first to the displacement component of the original fringe pattern (Fig. 4.51a) and second to the direction of the shift.

In practice, the number of geometric moiré fringes thus formed increases as the shifts Δx and Δy increase The shifts must be large enough to produce several fringes, but they must be small enough that the finite increments reasonably represent the true derivatives.

After the shift, a point A on the specimen is seen at two separate points in the geometric moiré pattern, separated by Δx. In practice, the point A on the specimen should be assumed to lie midway between the two corresponding points in the pattern. Thus, the boundary of the specimen should be assumed to lie midway between the shifted boundaries, as drawn in the figure. With the specimen defined by this boundary, points in the pattern represent corresponding points in the specimen.

In Sec. 4.1 it was noted that every method of extracting derivatives from experimental data involves a substantial loss of accuracy. In this case, the displacement field is represented by 70 fringes, whereas the derivative fields are represented by fewer fringes, fewer by an order of magnitude. Thus, the data in the derivative field are decreased by an order of magnitude. This example illustrates that it is important to have an abundance of displacement information to extract reasonably reliable strain information.

One might be tempted to add carrier fringes to the displacement field prior to mechanical differentiation. A caution is required. The effect of $\Delta N^c / \Delta y$ (where N^c represents the carrier fringes) is a constant, but the constant might not be an integer. Then, the bright fringes of the derivative field would not represent fringe orders that are integers. The practical answer is to use carrier fringes that are parallel to the direction of the shifts, so that $\Delta N^c / \Delta y = 0$. A more practical answer is to avoid using carrier fringes in this case.

Mechanical differentiation is used later in Fig. 8.7 in connection with an important thermal stress problem. The geometric moiré fringes reveal the two direct-derivative fields and the two cross-derivative fields. The normal strains are proportional to the fringe orders at each point of the direct-derivative patterns, and the shear strains are proportional to the sum of the fringe orders at corresponding points of the cross-derivative patterns.

Mechanical differentiation has its virtues and its shortcomings. Note that it is an approximate method, but the finite increment is *exactly* equal to the derivative at one or more points in the interval Δx or Δy. This is true even for regions of high strain gradients. In addition, the finite increment approximation is very effective at all points in regions of small strain gradients or linear strain gradients. On the other hand, the data vanishes near specimen boundaries, where it is often most important. This characteristic is evident in Figs. 4.51 and 8.7.

In Chapter 8, mechanical differentiation was used effectively to provide a qualitative view of whole-field strain and stress distributions. Quantitative measurements were taken from the displacement fields themselves, the N_x and N_y patterns. This follows the general philosophy of the authors, which advocates use of mechanical differentiation for qualitative interpretation—sometimes very important interpretation—and displacement fields for quantitative data analysis.

While the method of Fig. 4.51 is very easy, numerous other techniques appear in the literature. In Refs. 35 and 36, double exposure of moiré interferometry fields is used, with a shift Δx (or Δy) of the photographic film between exposures; subsequent optical filtering is employed, since double-exposure moiré fringes have low visibility. Several optical techniques have been proposed to *shear* the moiré interferometry pattern and thus project the mutually shifted images of it simultaneously, or else to shear the wave fronts emerging from the specimen grating. These *shearing interferometry* techniques include work by Weissman,[37] Fang,[38] Patorski,[39] and others.

4.23 Comments on Data Analysis

As whole-field displacement maps, moiré interferometry patterns usually provide vast quantities of information. Automated analysis of these patterns is a great advantage when whole-field strains must be extracted. Numerous schemes are under development and it is fair to predict that many more investigators will undertake the challenge of developing an optimum computerized strain analysis technique for moiré interferometry. That is good. We suggest that those who undertake the challenge should pursue a serious program of manual analyses first. A particular reason lies in the versatility of the data gathering process, especially the opportunity to select carrier patterns that best elucidate the deformation.

Whole-field strains are not always needed. Evaluation of theoretical or numerical solutions can be made on the basis of whole-field displacements in most cases. Where strains are needed, it is frequent that interest lies in local zones. In Fig. 4.1, for example, the eye is quickly drawn to the point of maximum shear strain and quantitative measurement at that point is likely to be sufficient. In Fig. 4.35, strain analysis along the vertical centerline is sufficient. As the reader surveys the examples throughout the book, it becomes clear that whole-field strain analysis would be useful only in special cases.

For most of the applications shown here, strains were extracted in local zones, usually along a line of interest. Measurements were made on an enlarged image of the fringe pattern using a digitizing tablet. Lines were drawn on the fringe pattern and the locations of fringes that crossed these lines were recorded by the digitizer. To do this, the operator aligned cross-hairs over each data point and pressed a button to send the coordinates of the point to a personal computer. Then the fringe spacings and local strains were calculated by simple algorithms.

These comments advance two points. First, important and excellent experimental analysis can be accomplished without elaborate automatic data analysis systems. Second, when whole-field automation is justified, versatile and mature systems should be sought.

4.24 Shear Lag and the Thickness of Specimen Gratings

Although we desire to measure the deformation of the surface of the specimen, we *actually* measure the deformation of the surface of the specimen grating. The difference can be important when the specimen grating is not sufficiently thin. The reason is *shear lag*, whereby shear stresses in the grating attenuate as they propagate through the grating thickness.

Shear lag is illustrated in Fig. 4.52, where the specimen is an elastic tension member comprised of a stiff bar on the right bonded to another bar of lower stiffness. For this hypothetical specimen, let $E_2 = 2E_1$ and $v_2 = 2v_1$. Let the cross-sectional area, A, of the member be constant. Under these conditions, the transverse strains ε_y and ε_z are equal in the two members and constant along the entire bar. The three-dimension stress system that would otherwise occur near the butt joint is avoided and the stress remains one-dimensional; $\sigma_x = P/A$ throughout the specimen, while $\sigma_y = \sigma_z = 0$. The tensile strain, $\varepsilon_x = \sigma_x/E$, is discontinuous since the stiffnesses, E, of the two materials are unequal. The strain distribution is illustrated in Fig. 4.52a.

Let a specimen grating of finite thickness be applied as illustrated in Fig. 4.52b (where the grating thickness is exaggerated). Let the grating material be elastic and let it be compliant enough that its presence does not significantly alter the deformation of the specimen, such that the interface strain is still represented correctly in (a). The forces acting on the grating are represented in (c). Away from the discontinuity, the specimen and grating stretch equally and therefore no shear tractions are developed at the interface. Instead, tensile stresses and corresponding forces P' and P'' act in the grating. Force P' is larger because the tensile strain is greater on the left side. Near the discontinuity, shear forces act at the interface to balance the horizontal force system and allow the left side of the grating to stretch

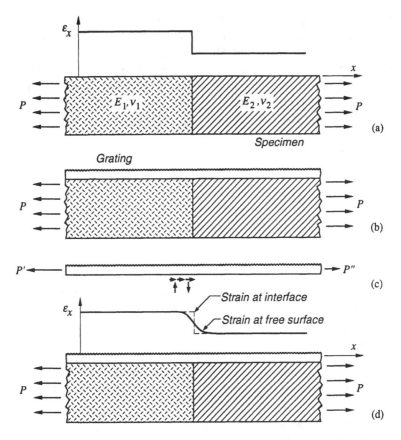

Fig. 4.52   Shear lag in a tensile bar that has a discontinuity of stiffness.

more than the right side. The vertical arrows represent distributed forces required to keep the grating interface flat and in contact with the specimen.

The grating material deforms under the force system illustrated in Fig. 4.52c. Near the discontinuity, shear tractions act on its bottom surface, but none act on its top surface. The corresponding shear stresses moderate or attenuate through the grating thickness, and this effect is known as shear lag. The result is a redistribution of the axial strain ε_x through the grating thickness, whereby the strain at the grating surface is changed to that illustrated schematically in Fig. 4.52d. The areas under the strain curves in (a) and (d) are equal (since the overall change of length is the same), but the grating causes a redistribution of strains in a region adjacent to the discontinuity.

The shear lag always attenuates the largest strains near a discontinuity or strain gradient. Its region of influence is reduced by minimizing the thickness of the grating. As a general rule, the strain on the outer surface of the grating differs significantly from the strain at the interface in a region that extends away from the discontinuity by about one grating thickness. In the usual case of 25 μm (0.001 in.) grating thickness, reliable results can be expected everywhere except in a zone of about 25 μm width bordering a severe strain concentration.

4.25 References

1. J. Guild, *The Interference Systems of Crossed Diffraction Gratings*, Clarendon Press, Oxford (1956).

2. R. Czarnek, D. Post and Y. Guo, "Strain Concentration Factors in Composite Tensile Members with Central Holes," *Proc. 1987 SEM Spring Conference on Experimental Mechanics*, pp. 657-663, Society for Experimental Mechanics, Bethel, Connecticut (1987).

3. A. J. Durelli and V. J. Parks, *Moiré Analysis of Strain*, Prentice Hall, Englewood Cliffs, New Jersey (1970).

4. A. Livnat and D. Post, "The Governing Equations for Moiré Interferometry and Their Identity to Equations of Geometrical Moiré," *Experimental Mechanics*, Vol. 25, No. 4, pp. 360-366 (1985).

5. F.-L. Dai, J. McKelvie and D. Post, "An Interpretation of Moiré Interferometry from Wavefront Interference Theory," *Optics and Lasers in Engineering*, Vol. 12, No. 2, pp. 101-118 (1990).

6. E. M. Weissman and D. Post, "Moiré Interferometry Near the Theoretical Limit," *Applied Optics*, Vol. 21, No. 9, pp. 1621-1623 (1982).

7. M. E. Tuttle and R. J. Klein, "Compressive Creep Strain Measurements using Moiré Interferometry," *Optical Engineering*, Vol. 27, No. 8, pp. 630-635 (1988).

8. Y. Guo and P. Ifju, "A Practical Moiré Interferometry System for Testing Machine Applications," *Experimental Techniques*, Vol. 15, No. 1, pp. 29-31 (1991).

9. V. A. Deason and M. B. Ward, "A Compact Portable Diffraction Moiré Interferometer," *Laser Interferometry: Quantitative Analysis of Interferograms*, Vol. 1162, pp. 26-35, SPIE, Bellingham, Washington (1989).

10. W. M. Hyer, C. T. Herakovich and D. Post, "Thermal Expansion of Graphite Epoxy," *1982 Advances in Aerospace Structures and Materials*, pp. 107-114, ASME, New York (1982).

11. J. D. Wood, "Detection of Delamination Onset in a Composite Laminate Using Moiré Interferometry," *Composite Technology Review*, Vol. 7, No. 4, pp. 121-128 (1985).

12. R. Czarnek and D. Post, "Moiré Interferometry with ±45° Gratings Applied to the Dovetail Joint," *Proc. 1985 SEM Spring Conference on Experimental Mechanics*, pp. 553-559, Society for Experimental Mechanics, Bethel, Connecticut (1985). Also R. Czarnek, "Three-Mirror, Four-Beam Moiré Interferometer and its Capabilities," *Optics and Lasers in Engineering*, Vol. 15, No. 2 pp. 93-101 (1991).

13. B. J. Chang, R. Alferness and E. N. Leith, "Space-invariant Achromatic Grating Interferometers: Theory," *Applied Optics*, Vol. 14, No. 7, pp. 1592-1600 (1975).

14. R. Czarnek, "High Sensitivity Moiré Interferometry with Compact Achromatic Interferometer," *Optics and Lasers in Engineering*, Vol. 13, pp. 99-115 (1990).

15. G. J. Swanson, "Broad-source Fringes in Grating and Conventional Interferometers," *J. Optical Society of America A*, Vol. 1, No. 12, pp. 1147-1153 (1984).

16. D. Post, "Moiré Interferometry in White Light," *Applied Optics*, Vol. 18, No. 24, pp. 4163-4167 (1979).

17. A. McDonach, J. McKelvie and C. A. Walker, "Stress Analysis of Fibrous Composites using Moiré Interferometry," *Optics and Lasers in Engineering*, Vol. 1, pp. 85-105 (1980).

18. R. Czarnek and D. Post, "Moiré Interferometry with ±45° Gratings," *Experimental Mechanics*, Vol. 24, No. 1, pp. 68-74 (1984).

19. J. McKelvie and C. A. Walker, "A Practical Multiplied Moiré-Fringe Technique," *Experimental Mechanics*, Vol. 18, No. 8, pp. 316-320 (1978).

20. P. Ifju and D. Post, "Zero Thickness Specimen Gratings for Moiré Interferometry," *Experimental Techniques*, Vol. 15, No. 2, pp. 45-47 (1991).

21. B. S.-J. Kang, G. Z.. Zhang, M. G. Jenkins, M. Ferber and P. Ifju, "Development of Moiré Interferometry for In-situ Material Surface Deformation Measurement at High Temperature," *Proc. 1993 SEM Spring Conference on Experimental Mechanics*, pp. 964-976, Society for Experimental Mechanics, Bethel, Connecticut (1993).

22. A. Kearney and C. Forno, "High Temperature Resistant Gratings for Moiré Interferometry," *Experimental Techniques*, Vol. 17, No. 6 (1993).

23. A. J. Durelli, C. A. Sciammarella and V. J. Parks, "Interpretation of Moiré Patterns," *J. Engineering Mechanics Division*, Proc. ASCE, EM2, No. 3487, pp. 71-88 (1963).

24. P. G. Ifju and D. Post, "The Shear Gage and Compact Shear Fixture for Shear Property Measurements of Composite Materials," Report CCMS-92-19, Center for Composite Materials and Structures, Virginia Polytechnic Institute and State University, Blacksburg, Virginia (1992).

25. D. Post, "The Moiré Grid-Analyzer Method for Strain Analysis," *Experimental Mechanics*, Vol. 5, No. 11, pp. 368-377 (1965).

26. Y. Guo, D. Post and R. Czarnek, "The Magic of Carrier Fringes in Moiré Interferometry," *Experimental Mechanics*, Vol. 29, No. 2, pp. 169-173 (1989).

27. D. Post, R. Czarnek, J. D. Wood and D. Joh, "Deformations and Strains in a Thick Adherend Lap Joint," *Adhesively Bonded Joints: Testing, Analysis and Design*, ASTM STP 981, W. S. Johnson, Editor, American Society for Testing and Materials, Philadelphia, pp. 107-118 (1988). Also D. Post, R. Czarnek, J. Wood, D. Joh, and S. Lubowinski, "Deformations and Strains in Adhesive Joints by Moiré Interferometry," *NASA Contractor Report* 172474 (1984).

28. D. Post and W. A. Baracat, "High-Sensitivity Moiré Interferometry—A Simplified Approach," *Experimental Mechanics*, Vol. 21, No. 3, pp. 100-104 (1981).

29. Y. Guo, "Vibration Insensitive Moiré Interferometry System for Off-table Applications," *Proc. 2nd International Conference on Photomechanics and Speckle Metrology*, Vol. 1554B, pp. 411-419, SPIE, Bellingham, Washington (1991).

30. R. Lloyd, Idaho National Engineering Laboratory: suggestion by private communication (1990).

31. D. Post and T. F. MacLaughlin, "Strain Analysis by Moiré Fringe Multiplication," *Experimental Mechanics*, Vol. 11, No. 9, pp. 408-413 (1971).

32. M. Marchant and S. M. Bishop, "An Interference Technique for the Measurement of In-plane Displacements of Opaque Surfaces," *J. Strain Analysis*, Vol. 9, No. 1, pp. 36-43 (1974).

33. L. Pirodda, "Improvements in the Experimental Techniques of Moiré Interferometry," *Proc. International Conference on Photomechanics and Speckle Metrology*, Vol. 814, pp. 456-463, SPIE, Bellingham, Washington (1987).

34. F. P. Chiang and C. C. Kin, "Three-beam Interferometric Technique for Determination of Strain of Curved Surfaces," *Optical Engineering*, Vol.. 23, No. 6, pp. 766-768 (1984).

35. E. M. Weissman and D. Post, "Full-field Displacement and Strain Rosettes by Moiré Interferometry," *Experimental Mechanics*, Vol. 22, No. 9, pp. 324-328 (1982).

36. D. Post, R. Czarnek and D. Joh, "Shear-Strain Contours from Moiré Interferometry," *Experimental Mechanics*, Vol. 25, No. 3, pp. 282-287 (1985).

37. E. M. Weissman, D. Post and A. Asundi, "Whole-field Strain Determinations by Moiré Shearing Interferometry," *Jl. Strain Analysis*, Vol. 19, No. 2, pp. 77-80 (1984).

38. J. Fang and F.-L. Dai, "Holographic Moiré for Strain Fringe Patterns," *Experimental Mechanics*, Vol. 31, No. 2, pp. 163-167 (1991).

39. K. Patorski, D. Post, R. Czarnek and Y. Guo, "Real-Time Optical Differentiation for Moiré Interferometry," *Applied Optics*, Vol. 26, pp. 1977-1982 (1987).

4.26 Key Relationships

displacements

$$U(x,y) = \frac{1}{f}N_x(x,y)$$
$$V(x,y) = \frac{1}{f}N_y(x,y)$$

(4.1)

strains

$$\varepsilon_x = \frac{\partial U}{\partial x} = \frac{1}{f}\left[\frac{\partial N_x}{\partial x}\right]$$
$$\varepsilon_y = \frac{\partial V}{\partial y} = \frac{1}{f}\left[\frac{\partial N_y}{\partial y}\right]$$

(4.2)

$$\gamma_{xy} = \frac{\partial U}{\partial y} + \frac{\partial V}{\partial x} = \frac{1}{f}\left[\frac{\partial N_x}{\partial y} + \frac{\partial N_y}{\partial x}\right]$$

(4.3)

virtual reference grating

$$f = \frac{2}{\lambda}\sin\alpha$$

(4.4)

specimen grating

$$f_s = \frac{f}{2}$$

(4.5)

fringe frequency, normal strain (fringe gradient)

$$F_{xx} = f\varepsilon_x$$

(4.6)

fringe frequency, rotation (fringe gradient)

$$F = f\theta$$

(3.4)

notation: fringe frequency (fringe gradient)

$$F_{xx} = \frac{\partial N_x}{\partial x} \approx \frac{\Delta N_x}{\Delta x} \qquad F_{xy} = \frac{\partial N_x}{\partial y} \approx \frac{\Delta N_x}{\Delta y}$$
$$F_{yy} = \frac{\partial N_y}{\partial y} \approx \frac{\Delta N_y}{\Delta y} \qquad F_{yx} = \frac{\partial N_y}{\partial x} \approx \frac{\Delta N_y}{\Delta x}$$

sign convention for fringe frequency (fringe gradient)   (Sec. 4.10)

Sign Convention: As the N_x pattern is scanned in the x direction, the sign of $\partial N_x/\partial x$ is positive if the fringe order is increasing.

Similarly, $\partial N_y / \partial y$ is positive if N_y increases with y, i.e., along a path of increasing y values. The gradient $\partial N_y / \partial x$ is positive when N_y increases in the positive x direction. The gradient $\partial N_x / \partial y$ is positive when N_x increases with $+y$.

in-plane rigid-body rotation

$$\frac{\partial N_x}{\partial y} = -\frac{\partial N_y}{\partial x} \qquad \text{(Sec 4.12)}$$

fringe vector angles, fringe angles

$$F_{xy} = F_{xx} \tan \phi \qquad (4.25)$$

$$F_{yy} = F_{yx} \tan \phi \qquad (4.25a)$$

carrier fringes, fringe vectors

$$F_x^\sigma = F_x^{\sigma,c} - F_x^c \qquad (4.26)$$

out-of-plane accidental rigid body rotation

$$F_{xxe} = -f\,\frac{\Psi_y^2}{2} \qquad (4.28)$$

4.27 Exercises

Theory

4.1 For the moiré interferometer of Fig. 4.11, the plane of polarization of the incident light should be either perpendicular or parallel to the plane of incidence. Explain why.

Illustrative solution

The plane mirror in Fig. 4.11 causes a left to right reversal of the plane of polarization, but no top to bottom reversal. With perpendicular or parallel polarization, the reversal does not change the plane of polarization. Therefore, the two beams that create the interference pattern in the camera remain mutually coherent. Note that other angles of polarization reduce the coherence. With 45° polarization, the two beams that propagate to the camera are fully incoherent inasmuch as they are orthogonally polarized, as illustrated in the figure, and no interference fringes are formed.

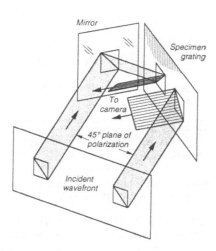

4.2 The moiré interferometer of Fig. 4.8 can be used with any angle of polarization. Explain why.

Illustrative solution

In the system of Fig. 4.8, the two beams that create the interference pattern in the camera experience the same number of mirror reflections, and therefore the same number of left to right reversals of their polarization. Thus, they always remain mutually coherent.

4.3 Under what circumstances can an unpolarized laser be used for moiré interferometry?

4.4. Derive Eq. 4.8 from the diffraction equations.

4.5 In Fig. 4.1b, the pitch of the virtual reference grating is $1/f$. What changes in the fringe pattern would occur if the specimen is translated by $3/f$ in a direction 60° from the x axis.

4.6 Given a moiré interferometry system with f_s = 1200 lines/mm, λ = 632.8 nm, and α = 49.50°, find, for the no-load condition (a) the frequency of the virtual reference grating; (b) the angles β_1 and β_{-1} of the ±1 diffraction orders; (c) the frequency of the moiré pattern in the camera screen, assuming magnification M = 1. (d) What is the relationship between the frequencies in answers (a) and (c)?

4.7 A tensile specimen with a 25 mm x 3 mm cross-section is loaded to produce a tensile strain of 0.1% in the y (loading) direction. The specimen material is isotropic and its Poisson's ratio is 0.30. With f_s = 1200 lines/mm, λ = 632.8 nm and starting with null U and V fields, determine (a) the fringe frequencies in the U and V fields and (b) the included angle $(\beta_1 - \beta_{-1})$ between the ±1 diffraction orders. (c) What change of incident angle α is needed to subtract off the fringes of the V field?

4.8 Referring to Fig. 4.6, what initial frequency of the specimen grating would produce a null field when α = 45°? Let (a) λ = 632.8 nm, (b) λ = 514.5 nm.

Analysis

4.9 In Fig. 4.1, mark the locations of the maximum strains ε_x, ε_y, γ_{xy}.

4.10 For Fig. 4.7, determine the maximum strain ε_y. If E = 3.5 GPa and ν = 0.35, determine the maximum stress σ_y.

4.11 Reproduce the pattern of Fig. 9.7 (using an office-type copying machine). (a) Assign fringe orders N_x and N_y by marking numbers on the copy. (b) Determine the maximum tensile strain ε_y.

4.12 Reproduce the pattern of Fig. 11.3 (using an office-type copying machine). Assign fringe orders N_x by marking numbers on the copy.

4.13 In the fringe pattern of Fig. 11.3, the shaded zone lies between two neighboring fringes of fringe orders Q and Q + 1. The paths of these fringes seem ambiguous. How can we prove that each of these irregular fringes is a contour of constant fringe order?

4.14 Make an enlarged image of the V field of Fig. 11.7. (a) Mark fringe orders on the pattern. (b) Plot graphs of displacements V and strains ε_y along the vertical centerline. (c) List any assumptions you made.

4.15 For the N_x field in Fig. 4.35, draw the vertical centerline between the notches and plot N_x along that line. Use that graph to determine $\partial N_x / \partial y$ at several points. Plot $\partial U / \partial y$ versus y and compare the curve to that in Fig. 4.36.

4.16 Make an enlarged image of Fig. 4.38 and use it to determine γ_{xy} along the vertical centerline. Compare the result to that in Fig. 4.36.

4.17 Determine the magnitude and direction of rigid-body rotation of the specimen to transform the patterns in Fig. 4.35 to those in Fig. 4.38.

4.18 For Fig. 4.45, determine the shear strain along the x axis (y = 0).

4.19 Determine the normal strain ε_y from Fig. 7.2. Describe the measurements and procedure.

4.20 Determine the coefficient of thermal expansion of the brass from Fig. 8.4. $\Delta T = -133°C$. Describe the measurements and procedure.

4.21 Referring to Fig. 9.2, determine ε_x at the center of AA'. Describe the measurements and procedure.

4.22 Referring to Fig. 9.5, explain why ε_y is considered essentially independent of x.

4.23 Find the location and magnitude of the maximum shear strain γ_{xy} in the patterns of Fig. 10.10.

Instrumentation

4.24 To align the system of Fig. 4.11 for a null field and symmetry, two bright dots of first diffraction orders are superimposed in plane B while two bright dots of zeroth diffraction order are superimposed upon the source in plane A. Devise a method to utilize the second diffraction orders instead of the zeroth and first orders.

Illustrative solution
Block the left half of the incident beam. Adjust the parabolic mirror to direct the +2 diffraction order from the specimen grating back to the source at A. This adjustment provides the correct angle of incidence $(-\alpha)$. Next, cover the right half of the incident beam. Adjust the plane mirror to direct the −2 diffraction order back to the source. Now, the interference pattern should be visible in the camera screen. Make the tiny adjustments of the plane mirror needed to establish a null field.
Note 1. An equivalent procedure can be used with any 2-beam and 4-beam moiré interferometer.
Note 2. The second diffraction order is usually dim and the procedure would be carried out in a darkened room.

4.25 (a) For the system shown in Fig. 4.48, sketch the rays corresponding to all the diffraction orders emerging from the specimen grating. (b) Under what conditions will the two emergent rays shown in the figure represent the only light that enters the camera lens?

4.26 In the transmission system of Fig. 4.27, let the specimen be turned around so that the specimen grating faces the camera. The index of refraction of the specimen material is n. If the directions of the external beams remain unchanged, i.e., such that the frequency of the virtual reference grating in air remains $f = (2/\lambda) \sin \alpha$, determine (a) the directions of the intersecting beams inside the specimen, and (b) the frequency of the virtual reference grating formed inside the specimen material. Hint: For a refractive medium, the wavelength in Fig. 2.21 changes to λ/n (Eq. 2.8).

4.27 The copper rod on the left side is heated by a torch. (a) Design an experiment using moiré interferometry to determine its change of length. Show the optical arrangement and specify the parts. (b) Sketch the fringe pattern. (c) Derive the relationship between the fringe pattern and the change of length of the rod. (d) Estimate the uncertainty of the change of length measurements. (e) List the assumptions that were made.

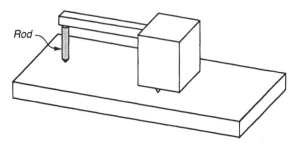

4.28 Small errors can occur in the alignment of specimen gratings (Fig. 4.33). (a) Sketch the fringe pattern for a uniaxial strain field, $\varepsilon_y = k$ and $\varepsilon_x = 0$, for the case of a large alignment error of $\phi = 10°$. Assume the pattern was a null field before the strain was applied. (b) Derive the percent error of V displacement measurements when $\phi = 10°$. Hint: The reference grating is also inclined by $\phi = 10°$. Answer: 1.5%

4.29 In exercise 4.28, derive the % error of V displacements for $\phi = 10°$ and a biaxial strain field caused by thermal expansion, where $\varepsilon_y = \varepsilon_x = k$. Hint: How many lines of the reference grating are crossed when P moves to P'? Answer: 18.9%.

4.30 In exercise 4.29, derive the % error of V displacements for $\phi = 10°$ and a biaxial strain field in which $\varepsilon_y = k$ and $\varepsilon_x = -0.25k$. Answer: −2.8%.

5
Microscopic Moiré Interferometry: Very High Sensitivity

5.1 Introduction

Many fields of study require deformation measurements of tiny specimens or tiny regions of larger specimens. The mechanics of microelectronic assemblies is an example, where the ever increasing demand for closer packing exacerbates the problems of thermal stresses. Other fields include crack-tip analyses in fracture mechanics; grain and intragranular deformations of metals and ceramics; interface problems; etc. Moiré interferometry adapted for such micromechanics studies is the subject of this chapter.

The small size and the need for high spatial resolution requires microscopic viewing of the specimen. The optical microscope with visible light will be used. The choice of visible light reduces the problems associated with the specimen size, loading systems, environmental limitations, and coherence of the light source, which are otherwise entailed in the use of shorter wavelengths, e.g., with the electron microscope.

Enhanced sensitivity of displacement measurements is demanded. Within a small field of view, the relative displacements are small even when the strains are not small. Two techniques will be introduced to further enhance sensitivity, relative to the high-sensitivity method already discussed. They are the *immersion interferometer* and *optical/digital fringe multiplication* (O/DFM).

5.1.1 Scope and Summary

A specific system will be described.[1,2] The approach is based on the premise that the moiré pattern will contain only a few fringes. Accordingly, it is practical to record the pattern by a video camera. Good fringe resolution is preserved because the pattern is recorded with numerous pixels per fringe.

The apparatus is illustrated in Fig. 5.1. The specimen is mounted in a loading frame of 45 kN (10,000 lb.) capacity in compression or flexure (Fig. 4.10), which is fixed to a rotary table for angular adjustments. The imaging system is comprised of a long-working distance microscope lens and a CCD video camera (without lens). The interferometer is a compact four-beam unit described later; it provides a virtual reference grating of 4800 lines/mm (122,000 lines/in.), which doubles the displacement sensitivity relative to that in Chapter 4. Its illumination is provided by optical fibers to two collimators. The microscope, interferometer and collimators are

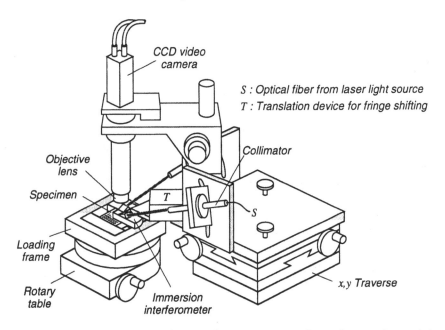

Fig. 5.1   Mechanical and optical arrangement for microscopic moiré interferometry.

mounted on an x,y traverse that can move the system to observe any portion of the specimen. Additional equipment includes a personal computer with a frame-grabber board and TV monitors. The translation device will be discussed in connection with fringe shifting.

An abundance of data is provided by fringe shifting, as illustrated in Fig. 5.2. A series of β shifted moiré patterns is recorded by the video camera. Each pattern is different from its neighbor, but they all represent the same displacement field. All of this data is manip-

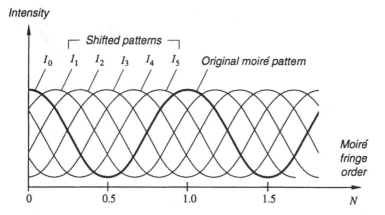

Fig. 5.2   Fringe shifting provides an abundance of data, all representing the same displacement field. Here $\beta = 6$.

ulated digitally to generate a contour map of the displacement field, where the map contains β times as many fringe contours as the original moiré pattern; β is the fringe multiplication factor. Sensitivity, measured in fringe contours per unit displacement, is increased by a factor of β; a multiplication of $\beta = 12$ has been achieved in practical applications.

The fringe multiplication process involves (a) data collection, wherein β relatively sparse moiré patterns are recorded; (b) fringe sharpening by digital techniques; and (c) superposition of the sharpened contours. The result is a contour map in which β correctly positioned contours are formed for each fringe of the original moiré pattern.

An example of the method is shown in Fig. 5.3. It is applied to a microelectronics application in which the solder ball interconnection between an active silicon chip and ceramic circuit board is studied. When such assemblies are subjected to thermal cycles in their normal operation, the thermal expansion mismatch between the chip and circuit board causes high stresses at the solder joints, notably shear stresses. These have led to electrical and mechanical fatigue failures. An epoxy cement can be used to fill the open gap between the chip and circuit board, thus restraining their relative expansion and reducing shear strains in the solder balls. This type of assembly is illustrated.

A special experimental technique to reveal steady-state thermal strains was used, wherein the specimen grating was applied at elevated temperature and the deformation incurred during cooling was observed at room temperature. Details are given in Sec. 8.2.3.

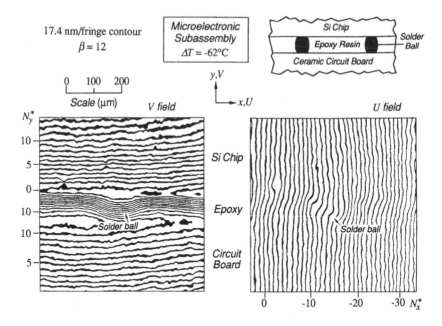

Fig. 5.3   Enhanced sensitivity displacement maps for thermal deformations of a microelectronic subassembly. Carrier fringes of extension were added to both patterns.

The displacement fields induced by a temperature change of −62°C are shown in Fig. 5.3. Carrier fringes of extension were used in the V field to reduce the fringe density in the solder and to clarify the gradients near solder interfaces. Carrier fringes of extension were used in the U field to increase the density of fringes. The zero-order contours were selected arbitrarily, since deformations and not absolute displacements were studied. The sensitivity was 57.6 fringe contours per μm displacement; the contour interval was 17.4 nm/fringe contour. This enhanced sensitivity was achieved with an immersion interferometer that produced moiré patterns with 4.8 fringes/μm displacement and O/D fringe multiplication by a factor of 12.   Relative displacements, which include the apparent displacements introduced by the carrier fringes, are determined by

$$U = \frac{1}{\beta f} N_x^*, \qquad V = \frac{1}{\beta f} N_y^* \qquad (5.1)$$

where N^* is the contour number in the pattern and f is 4800 lines/mm.

While the location of the solder ball is evident in both fields, the presence of the epoxy eliminates the serious shear strains that would otherwise occur at the solder joint. The derivatives $\partial N_y^* / \partial x$ and

$\partial N_x^* / \partial y$ are both small. These derivatives are independent of extensional strains and therefore they are not influenced by the carrier fringes of extension or the free thermal contractions of the component parts. By Eq. 4.3, the shear strains are directly proportional to the sum of these derivatives which means that the shear strains and stresses are small. The epoxy layer was effective.

Since a CCD video camera was used, the normal specification of magnification is not meaningful. Instead, object size per TV frame is meaningful. The object field was about 600 μm (0.024 in.) wide. With the 480 x 480 pixel format employed, this corresponds to 1.2 μm (50 μin.) per pixel.

In the following sections, a specific optical, mechanical and electronic system is described. Certain variations are indicated and other variations are likely to evolve as the technology advances.

5.2 The Immersion Interferometer

The immersion interferometer was developed for two goals, to increase the basic sensitivity of moiré interferometry and to devise an especially stable system compatible with the requirements of fringe shifting. As background, remember that the sensitivity of moiré interferometry increases with the frequency, f, of the virtual reference grating. Since

$$f = \frac{2}{\lambda} \sin \alpha \qquad (4.4)$$

the sensitivity increases with α (the angle of incidence of the illuminating beams) and it increases as the wavelength, λ, decreases. In air, the theoretical upper limit of sensitivity is approached as α approaches 90°. However, when $\alpha = 64°$, $\sin \alpha$ is already 90% of its maximum value and not much can be gained by greater angles. For practical reasons of instrument design, angles approaching 90° were not implemented, but instead, angles between 45° and 65° were considered practical.

5.2.1 Sensitivity of Moiré Interferometry in a Refractive Medium

The theoretical upper limit of sensitivity can be increased by forming the virtual reference grating inside a refractive material. Note that the wavelength of light can be manipulated without changing its visibility. In vacuum, the velocity of light is $C = \omega\lambda$ (Eq. 2.3). In a refractive medium, the velocity is reduced to $C' = C/n$ (Eq. 2.7), and therefore $C' = \omega\lambda/n$. Since ω is independent of the material medium, the wavelength must decrease by

$$\lambda_m = \frac{\lambda}{n} \tag{2.8}$$

where λ_m is the wavelength of light in the refractive material.

What is the frequency, f_m, of a virtual reference grating in a refractive medium? The conditions of Fig. 2.21 prevail, but in the refractive medium λ becomes λ_m and G becomes G_m or $1/f_m$. Accordingly, Eq. 4.4 becomes

$$f_m = \frac{2\sin\alpha}{\lambda_m} \tag{5.2}$$

and by Eq. 2.8,

$$f_m = \frac{2n\sin\alpha}{\lambda} \tag{5.3}$$

For a given light source of wavelength λ in vacuum, the frequency of the virtual reference grating in a refractive medium is increased by a factor of n. For example, the wavelength of green argon laser light is 514 nm in air, but it is reduced to 338 nm in optical glass (BK7) of refractive index 1.52. Thus, the maximum frequency of the virtual reference grating is increased from 3890 lines/mm (in air) to 5910 lines/mm (in glass). Since sensitivity increases with f, the theoretical upper limit of sensitivity is also increased by the same factor, n.

5.2.2 Optical Configurations

Various optical configurations that generate a virtual reference grating in a refractive medium can be devised. Several configurations are described below.

Configuration 1

Figure 5.4 is a schematic illustration of configuration 1, where the shaded element represents a prism fabricated from optical glass. A coherent collimated beam enters through an inclined entrance plane. The lower portion of the beam impinges directly on the specimen grating while the upper portion impinges symmetrically after reflection from the mirrorized surface; they create a virtual reference grating on the specimen.

The gap between the interferometer and the specimen grating is filled with an immersion fluid. When the refractive index of the immersion fluid is not the same as that of the interferometer, angle α

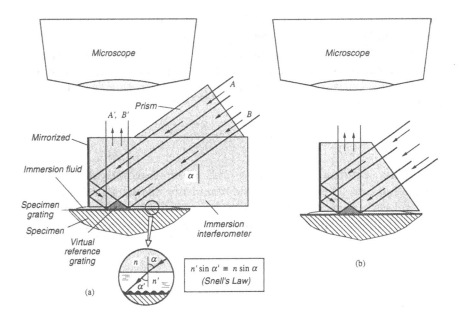

Fig. 5.4 Optical paths in the immersion interferometer; configuration 1.

changes to α' as illustrated in the insert. However, this does not alter the frequency of the virtual reference grating, since $n' \sin \alpha' \equiv n \sin \alpha$ by Snell's law and Eq. 5.3 remains uneffected. The same result is obtained from the equations of diffraction and refraction without the conceptualization of a virtual reference grating. A reasonable match of refractive indices of the immersion fluid and the immersion interferometer is desired, but mismatch can be tolerated until it is so large that the critical angle of complete internal reflection occurs. The refractive index of the immersion fluid should be considered a limiting factor in Eq. 5.3, as well as the refractive index of the interferometer.

In Fig. 5.4a, the inclined entrance plane is provided by a separate prism bonded with optical cement to the main interferometer element. In (b) the interferometer is fabricated as a single element, with an integral entrance plane.

Configuration 2

Configuration 2 is basically the same as configuration 1, but as illustrated in Fig. 5.5, a collimated beam experiences multiple internal reflections before it strikes the specimen grating. This configuration would be more suitable for systems in which the interferometer must be very thin, for example when the working distance of the microscope lens is not large.

It is interesting to remember that total internal reflection is obtained when the angles of incidence are greater than the critical angle. The critical angle θ_c is determined from Snell's law as Eq. 2.15, which reduces to

$$\theta_c = \sin^{-1}\left(\frac{1}{n}\right) \tag{5.4}$$

when the external medium is air. For the optical materials considered here, the critical angle would not exceed 42°. Thus, total internal reflections will occur when $\alpha > 42°$ and it is not necessary to mirrorize the top and bottom surfaces of the interferometer.

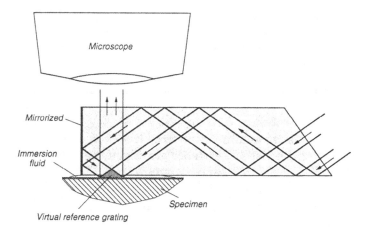

Fig. 5.5   Configuration 2.

Configuration 3

Configuration 3 is different from the other two in that a real grating is utilized to achieve the desired angle α. As illustrated in Fig. 5.6, light diffracted by the first grating—called a compensator grating— illuminates the specimen grating in the same way as in configuration 2. If the frequency of the compensator grating is $f_m/2$, its first order diffraction angle α is exactly right to generate a virtual reference grating of frequency f_m.

Configuration 3 is an achromatic interferometer (Sec. 4.7.8) and permits use of temporally incoherent light. A laser diode light source is an attractive option, which could improve the compactness and reduce the cost of the physical system.

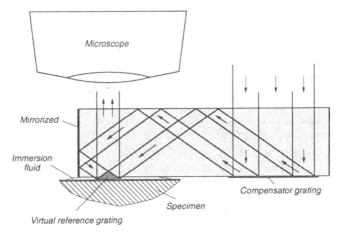

Fig. 5.6   Configuration 3.

Configuration 4

This immersion interferometer is illustrated in Fig. 5.7. It is fabricated by attaching prisms to a flat plate, using an optical cement. The plate can be relatively thin, and that is an advantage for microscopic viewing. Unlike the other configurations, there is no sharp corner of the interferometer that lies close to the specimen, and damage of the interferometer or specimen caused by accidental contact is greatly reduced.

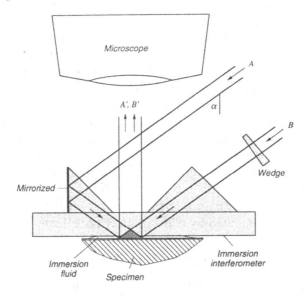

Fig. 5.7   Configuration 4.

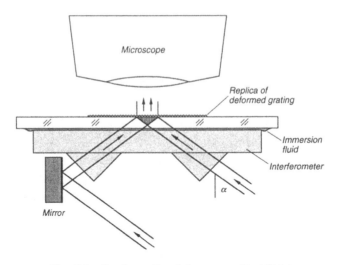

Fig. 5.8   Configuration 5 for transmitted light.

Configuration 5 for Transmission.

A transmission system is extremely attractive for microscopic moiré interferometry. As illustrated in Fig. 5.8, the microscope can be brought as close to the specimen grating as desired; long working distance lenses are not required and magnification is not restricted. In addition, there is no refractive material between the grating and microscope to affect the image.

A powerful technique lies in the ability to make replicas of a deformed specimen grating from the loaded workpiece (Sec. 4.21) and view these replicas in a transmission system. The replica is made on a glass plate, which is brought into optical contact with the interferometer by immersion fluid. The advantage of forming the virtual reference grating in a refractive material is retained, while the grating is not contacted by immersion fluid. The opportunity to load the specimen separately with any testing machine improves the versatility of microscopic moiré interferometry. In addition, the method provides a permanent record of the deformation field.

Numerous different designs can be implemented for transmission systems. In the schematic design of Fig. 5.8, the virtual grating can be controlled by adjustments of the mirror, and fringe shifting (Sec. 5.3.1) can be implemented by translation of the mirror.

5.2.3 Four-beam Immersion Interferometer

The immersion interferometer concept was implemented in the form of a very compact four-beam moiré interferometer for microscopic

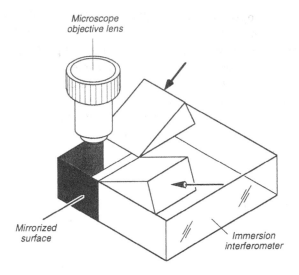

Fig. 5.9   Immersion interferometer for U and V fields.

viewing. It is illustrated in Fig. 5.9, which is a four-beam version of configuration 1. Two collimated beams enter the interferometer through the cemented prisms and they create the virtual reference gratings for the U and V fields.

Since the interferometer requires accurately perpendicular surfaces of high optical quality, it was cut out of a corner cube reflector. Figure 5.10 illustrates this approach. Corner cube reflectors are manufactured in high volume and thus they are inexpensive, which makes this an economical scheme. To complete the interferometer, planes 1 and 2 were mirrorized by vacuum deposition of aluminum and two prisms were bonded to the main

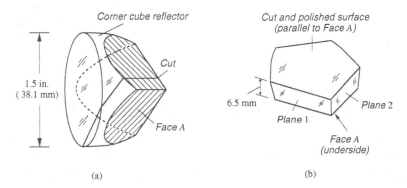

Fig. 5.10   The main element of the immersion interferometer was cut from a corner cube reflector.

element with an optical cement. They are illuminated by separate collimators, as shown in Fig. 5.1.

The interferometer was designed for a virtual grating frequency of 4800 lines/mm. (122,000 lines/in.). The parameters in Eq. 5.3 are n = 1.52, λ = 514.5 nm and α = 54.3°. Angle α could be adjusted between 47° and 63° to provide a ±10% variation of frequency f_m; thus, carrier patterns could be used to subtract off uniform strain fields to extend the dynamic range of measurement to ±10% strain. Figure 5.3 and other figures that show the results of microscopic moiré interferometry were made with the interferometer of Fig. 5.9.

The frequency could be increased in a practical design by using optical glass and immersion fluid of $n \approx 2$, laser light of λ = 488 nm and incidence at α = 64°. With these parameters, f_m = 7200 lines/mm, providing a 50% increase.

An issue associated with these four-beam designs relates to the perpendicularity of the cross-line specimen grating. In general, the x and y gratings on the specimen are not *exactly* perpendicular; when the system is adjusted for a null field of U, some fringes of rotation will typically appear in the V field. An adjustment is needed to modify the fringes of rotation of one field.

In Fig. 5.7, the shallow wedge inserted in beam B provides the means for adjustment. When the wedge is rotated about an axis parallel to beam B, the beam tilts out of the plane of the page. This rotates the virtual reference grating, thus adjusting the fringes of rotation. The adjustment is needed for only one channel, to rotate either the x or y fringes. Of course, this method can be applied to the other configurations, too. Note that the adjustable mirror in Fig. 5.8 serves the same purpose.

Both null fields can be obtained if the interferometer of Fig. 5.7 is combined with the physical arrangement of Fig. 5.1. First, adjust one field by the inclination α of the collimator and the orientation of the rotary table. Then adjust the second field by the inclination of the second collimator and the orientation of the wedge. Subsequently, carrier fringes of rotation can be introduced by adjustments of the rotary table; carrier fringes of extension can be introduced by adjustments of the angle α of the collimators.

5.3 Optical/Digital Fringe Multiplication

The moiré pattern produced by the specimen grating and the immersion interferometer is recorded by the video camera. The intensity at each x,y point in the pattern varies with the fringe order at the point as indicated by the bold curve in Fig. 5.2. Fringe shifting is needed to amass the additional data in Fig. 5.2.

5.3.1 Fringe Shifting

Figure 5.4 can be used to illustrate fringe shifting. When the interferometer is translated horizontally by a fraction of the pitch g_m of the virtual reference grating, the relative phase of rays A' and B' (from each point in the field) changes by the same fraction of 2π. Accordingly, when the interferometer is translated by $d = g_m / \beta$, the relative phase changes by $\Delta\phi = 2\pi / \beta$ everywhere in the field of view. This proposition is proved in the next paragraph. A new intensity is present at every x,y point. Figure 5.2 depicts the intensity distributions for a series of β equal phase shifts, where $\beta = 6$ in this case. Six intensities represent the fringe order at each x,y point. These will be manipulated subsequently by the fringe sharpening and multiplication algorithm.

The relationship between the interferometer translation, d, and the phase change $\Delta\phi$ between A' and B' is developed with the help of Fig. 5.11. Before the translation, incoming rays A and B travel unique paths and meet at point P on the specimen grating. There, they are diffracted as rays A' and B'. After the translation, ray B^* strikes point P; its path length is greater than that of ray B by distance $a + b$. Ray A is not affected by the translation. Thus, the change of optical path length at P is $S = (a + b)n$. This change determines the change of phase, i.e., the state of constructive or destructive interference of light emerging from P. Since the path length of A' and B' between point P and its conjugate point in the recording plane of the video camera is invariant (Sec. 2.3.6), the phase change at the camera is

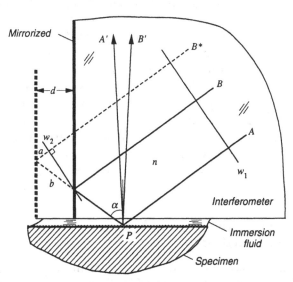

Fig. 5.11   A relative phase shift between A' and B' occurs when the interferometer is shifted by d; w_1 and w_2 are wave fronts.

$$\Delta\phi = 2\pi\frac{S}{\lambda} = 2\pi\frac{(a+b)n}{\lambda} \qquad (5.5)$$

The change of path length can be determined by trigonometry, with the result

$$a + b = 2d\sin\alpha \qquad (5.6)$$

By substituting Eqs. 5.6 and 5.3 into Eq. 5.5,

$$\Delta\phi = 2\pi d f_m \qquad (5.7)$$

When the interferometer is shifted by

$$d = \frac{g_m}{\beta} = \frac{1}{\beta f_m} \qquad (5.8)$$

the phase change in Eq. 5.7 is

$$\Delta\phi = \frac{2\pi}{\beta} \qquad (5.9)$$

and the proposition is proved.

Figure 5.12 illustrates the physical shape of the translation device and the immersion interferometer. The displacement generated by the piezoelectric actuator deflects the flexible arms of the actuator holder. This deflection translates the interferometer with respect to the specimen grating, as illustrated in Fig. 5.12b where the deflection is much exaggerated. Since the direction of motion is 45° from the x and y axes of the specimen, a displacement of $\sqrt{2}g_m/\beta$ is required for each fringe shift. The calibrations for fringe shifting of the U and V fields are identical. Translation errors are less than 1 nm.

The piezoelectric translator is actuated by a series of analog voltages that are controlled by a personal computer. For each step, the moiré pattern received by the video camera is digitized and stored in the personal computer, i.e., the intensity at each pixel is measured and assigned a digital intensity level on a 256 step gray scale. This sequence is repeated for β steps, and β moiré patterns are stored in digital form. Figure 5.2 depicts the stored data. Since it takes only 1/30 second to digitize one frame, shifting and grabbing a series of shifted patterns is accomplished within a fraction of a second.

5.3.2 Fringe Sharpening and Multiplication: Physical Analysis

Digital image processing is used to sharpen the series of fringe-shifted moiré patterns. The process is illustrated graphically in Fig. 5.13. In (a), the curves $I_{(x)}$ and $I^\pi_{(x)}$ represent a moiré pattern and its

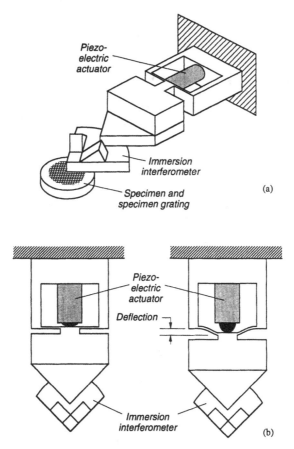

Fig. 5.12   Piezoelectric translation device for fringe shifting. It moves the
interferometer relative to the specimen in a series of tiny steps of g_m / β.

complementary pattern. The complementary pattern is the moiré
pattern obtained when the interferometer is shifted by $g_m/2$, which
causes a phase shift of π and a corresponding intensity shift
everywhere in the field. If β is an even integer, the series of fringe-
shifted moiré patterns always contain complementary pairs, i.e., for
each pattern in the series there is a complementary pattern.

The example in Fig. 5.13a illustrates the influence of nonuniform
illumination and noise. The uppermost curve represents the
nonuniform, noisy intensity distribution when there is constructive
interference at every point. The moiré fringes modulate that curve
and exhibit attenuated levels of the noise at every location. At points
where the intensities of the idealized patterns are equal, the
attenuation is identical in each pattern. Consequently, the noise does
not alter the locations where $I(x) = I^{\pi}(x)$.

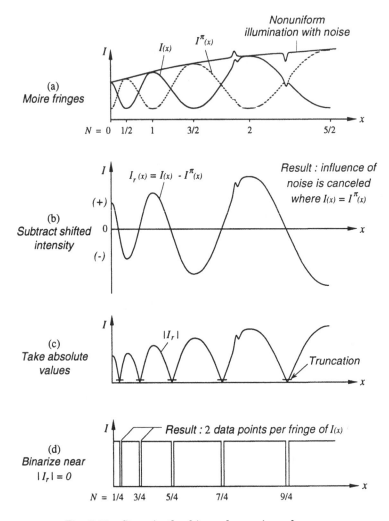

Fig. 5.13   Steps in the fringe sharpening scheme.

With these digitized intensities in storage, the algorithm proceeds by subtracting the intensities at each pixel.[3-5] The result is illustrated in (b), where the crossing points of the complementary patterns lie at $I_r(x) = 0$. These locations are not altered by the nonuniform illumination and noise. Note, too, that negative intensities are unrealistic, but the difference between two digitized intensity levels can be a negative number. The next step is to invert the negative portions, as illustrated in (c), to obtain highly sharpened intensity minima near $|I_r| = 0$. Then the digital intensities are binarized by truncation near $|I_r| = 0$; to produce the distribution in (d), the intensities below the truncation level are assigned zero, and those above are assigned one. The result, on a whole field basis, is narrow black fringe contours on a white background, with two contours generated for each fringe order of the original moiré pattern.

A special feature of the subtraction process is that the fringe contours lie at the quarter fringe order points (N = 1/4, 3/4, 5/4,---), called *quarter points*. At these points the intensities in the complementary moiré patterns vary most rapidly with fringe order, so the fringe order represented by a video gray level is defined most accurately. Near intensity maxima and minima, on the other hand, the gray levels each represent the broadest possible increment of fringe orders, so they define the fringe orders least accurately. The quarter points are located with the highest fidelity (see Sec. 5.3.4: *Data Sites*).

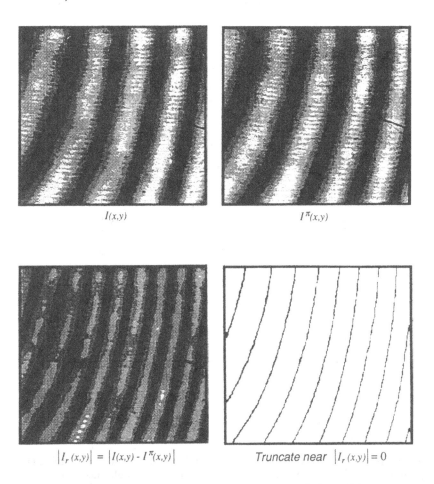

$I(x,y)$

$I^{\pi}(x,y)$

$\left|I_r(x,y)\right| = \left|I(x,y) - I^{\pi}(x,y)\right|$

Truncate near $\left|I_r(x,y)\right| = 0$

Fig. 5.14   Illustration of fringe sharpening on a whole-field basis.

The fringe sharpening technique is demonstrated in Fig. 5.14 on a whole field basis. The original and complementary patterns are shown as $I(x,y)$ and $I^\pi(x,y)$. Optical noise is evident in both patterns. The sequence is continued with the double-frequency plot corresponding to Fig. 5.13c. Its true sharpness is not illustrated; although the patterns were recorded and manipulated with a 256 gray level capacity, they are printed out here with only a few gray

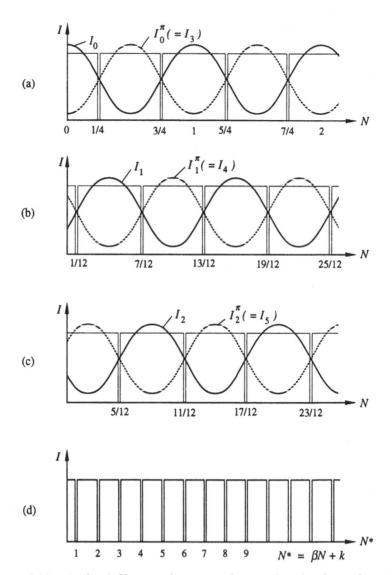

Figure 5.15   (a, b, c) Sharpened contours from each pair of complementary patterns. (d) The contours are combined to produce the multiplied contour map of I versus N^*.

levels. The next step, truncation and binarization, does show the true sharpness on a whole field basis.

The final step is to combine the results from the several pairs of complementary patterns. It is illustrated in Fig. 5.15 for the case of $\beta = 6$. The six fringe-shifted moiré patterns are grouped in the figure as complementary pairs. The sharpened fringe contours generated from each pair are combined in the fringe processing algorithm to produce the multiplied contour map represented by Fig. 5.15d.

The contour maps are interpreted in the normal way, where the contour order N^* is proportional to the displacement by Eq. 5.1. Since rigid body displacements are unimportant, $N^* = 0$ can be assigned arbitrarily to any contour, as before. The technique is illustrated by various applications in subsequent chapters.

5.3.3 Mathematical Analysis

The moiré pattern resulting from interference of A' and B' in Fig. 5.4 is, by Eq. 2.21,

$$I_{(x,y)} = I_A(x,y) + I_B(x,y) + 2\sqrt{I_A(x,y)I_B(x,y)} \, \cos\left[\phi(x,y)\right] \qquad (5.10)$$

where the phase ϕ represents the fringe order N at each point by

$$\phi(x,y) = 2\pi N(x,y) \qquad (5.11)$$

The terms I_A and I_B are functions of x,y because the intensities of these incoming beams are not constant across the field. In addition, the diffraction efficiency of the specimen grating might vary across the field, and that variation is imbedded in I_A and I_B.

For a series of β shifts of the immersion interferometer, the phase of each moiré pattern is shifted by $2\pi/\beta$ relative to its neighbors. An additional phase term is added to $\phi(x,y)$ by each shift. Since β is an even integer, the patterns can be divided into two groups: the patterns of the first half and their complements. These intensity distributions are

$$I_i(x,y) = I_A(x,y) + I_B(x,y) + 2\sqrt{I_A(x,y)I_B(x,y)} \, \cos\left[\phi(x,y) + \frac{2i\pi}{\beta}\right]$$

$$I_i^\pi(x,y) = I_A(x,y) + I_B(x,y) - 2\sqrt{I_A(x,y)I_B(x,y)} \, \cos\left[\phi(x,y) + \frac{2i\pi}{\beta}\right]$$

$$i = 0, 1, 2, \cdots, \frac{\beta}{2} - 1 \qquad (5.12)$$

where I_i is the intensity distribution of the ith shifted pattern which is shifted by $2i\pi/\beta$ with respect to the original pattern. I^π is the intensity distribution of the corresponding complementary pattern which is shifted by π with respect to the ith shifted pattern.

When these intensity distributions are captured by a video camera and the intensities at every pixel are converted to digital form, they can be subtracted to give the resulting intensity distribution as

$$I_r(x,y) = I_i(x,y) - I_i^\pi(x,y)$$

$$= 4\sqrt{I_A(x,y)I_B(x,y)} \cos\left[\phi(x,y) + \frac{2i\pi}{\beta}\right] \tag{5.13}$$

which shows that the resulting intensity is a cosine curve modulated by a term containing two input beams. Equation 5.13 indicates that at the points where the phase $\phi(x,y)$ satisfies the condition $\cos(\phi(x,y) + 2i\pi/\beta) = 0$, the *phase* is independent of both I_A and I_B and therefore independent of systematic noise in I_A and I_B. These phase values are

$$\phi(x,y) = \pi\left(\frac{2m+1}{2} - \frac{2i}{\beta}\right) \qquad m = 0, \pm1, \pm2, \cdots \tag{5.14}$$

Thus, $I_r = 0$ at every point where, from Eq. 5.11,

$$N(x,y) = \frac{2m+1}{4} - \frac{i}{\beta} \tag{5.15}$$

These are the quarter-points of Figs. 5.13 and 5.15.

The computer algorithm continues by inverting the negative intensity values and truncating them near $|I_r| = 0$, as illustrated in Fig. 5.13c. The intensities are binarized by assigning 0 and 1 to intensities below and above the truncation value, respectively, as illustrated in (d). The process is repeated for each of the complementary pairs, as in Fig. 5.15. Lastly, all the pixels of zero intensity are printed out to produce a composite map of all the fringe contours. The result is a pattern consisting of β narrow contours for each fringe of the original moiré pattern.

If rigid-body displacements are dismissed and contour numbers N^* are assigned to the combined pattern in the normal way, the patterns represent U and V displacements by Eq. 5.1, repeated here,

$$U = \frac{1}{\beta f}N_x^*, \qquad V = \frac{1}{\beta f}N_y^* \tag{5.1}$$

Thus, $N^* = \beta N + k$ where k is a constant associated with the choice of an arbitrary zero-order datum. With this algorithm, the fringe multiplication factor β is equal to the number of moiré patterns (original and shifted) utilized to create the combined contour map.

General Periodic Intensity Distributions

This section takes the mathematical analysis a step further. It proves that the O/DFM method is more general—that it applies for a broad range of periodic intensity distributions. Equation 5.10 and the succeeding analysis assumes that the intensity distribution in the

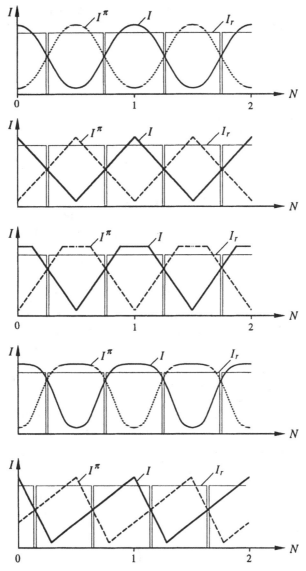

Fig. 5.16 Diverse intensity distributions. Sharpened fringe contours are formed wherever $I = I^\pi$.

physical description of Fig. 5.13 does not restrict the algorithm to simple functions. Actually, any periodic intensity distribution is applicable, provided the complementary curves of I versus N cross in a zone of high intensity gradient.[6] Figure 5.16 illustrates a variety of such curves.

The intensity distribution of a general interferogram, including cases that are not simple harmonic distributions, can be expressed as a periodic function by the summation of a Fourier cosine series as

$$I(x,y) = I_m(x,y) + \sum_{n=1}^{\infty} C_n(x,y) \cos[n\phi(x,y)] \qquad n = 1, 2, 3, \cdots \qquad (5.16)$$

where I is the intensity distribution of the interferogram or moiré pattern, I_m is the mean intensity, C_n are the coefficients of the harmonic series, and ϕ is the angular phase information of the interferogram; ϕ represents the fringe order N at each point of the pattern by Eq. 5.11. Here n is an integer unrelated to index of refraction.

Equation 5.16 can express any periodic intensity distribution by prescribing C_n. For example, the familiar simple harmonic function is represented when $n = 1$ is the only non-zero term. A triangular distribution is represented when

$$C_n = \frac{I_m}{(n\pi)^2} (1 - \cos n\pi) \qquad (5.17)$$

The terms C_n and I_m are functions of x,y in Eq. 5.16 when the intensity of the initial illumination is nonuniform across the field. Equations 5.16 and 5.17 combine to produce a triangular distribution of I versus ϕ or I versus N.

A series of β fringe-shifted patterns can be recorded with sequential changes of phase by a constant increment $2\pi/\beta$, where β is an even integer. As before, the patterns can be divided into two groups, the patterns of the first half and their complements, with intensity distributions

$$I_i(x,y) = I_m(x,y) + \sum_{n=1}^{\infty} C_n(x,y) \cos\left\{n\left[\phi(x,y) + \frac{2i\pi}{\beta}\right]\right\}$$

$$I_i^*(x,y) = I_m(x,y) + \sum_{n=1}^{\infty} (-1)^n C_n(x,y) \cos\left\{n\left[\phi(x,y) + \frac{2i\pi}{\beta}\right]\right\} \qquad (5.18)$$

$$i = 0, 1, 2, \cdots, \frac{\beta}{2} - 1$$

By subtracting these intensities, one obtains the intensity distribution I_r as

$$I_r(x,y) = I_i(x,y) - I_i^\pi(x,y)$$

$$= 2 \sum_{n=1}^{\infty} C_{2n-1} \cos\left\{(2n-1)\left[\phi(x,y) + \frac{2i\pi}{\beta}\right]\right\} \qquad (5.19)$$

Since $C_{2n-1} \neq 0$, the phase values $\phi(x,y)$ of the points where the condition $I_r(x,y) = 0$ is satisfied can be expressed as

$$\phi(x,y) = \pi\left(\frac{2m+1}{2} - \frac{2i}{\beta}\right) \qquad m = 0, \pm 1, \pm 2, \cdots \qquad (5.14)$$

which is identical to Eq. 5.14. By Eq. 5.11, the fringe orders where $I_r(x,y)$ is zero are

$$N(x,y) = \frac{2m+1}{4} - \frac{i}{\beta} \qquad (5.15)$$

Again, the x,y points where $I_r = 0$ are determined by taking the absolute values of $I_r(x,y)$, and then truncating and binarizing the intensity near $|I_r(x,y)| = 0$. The results are the same as those for simple harmonic intensity distributions, as might be expected from inspection of Fig. 5.16.

5.3.4 Implications of the Analyses

The *quasi-heterodyne (or phase stepping) method* is well known and commonly utilized for computerized fringe analysis of interferograms.[7-9] It was developed prior to the O/DFM method. The purpose of both is to enhance the resolution and accuracy of data extraction from interferograms, and they should be compared. There are advantages and disadvantages of both. The quasi-heterodyne method computes the fringe order at every pixel in the field, using three or more fringe-shifted interferograms. The O/DFM method provides a contour map with discrete contour intervals.

Linearity

The quasi-heterodyne method is dependent upon a simple harmonic intensity distribution. While many interferometers produce two-beam interference and faithful simple harmonic distributions, the requirement precludes certain applications involving multiple beam and mechanical interference. Examples include Fizeau interferometry (when reflectivity is not small) and geometric moiré systems.*

Perhaps more importantly, the requirement for simple harmonic intensities prescribes stringent conditions on linearity of the image

---

* Filtering techniques to suppress higher harmonics have been used for such applications.

recording system. A nonlinear response would produce systematic errors of the data that are subsequently manipulated by an algorithm developed for simple harmonic intensities. On the other hand, O/D fringe multiplication is compatible with a great range of periodic intensity distributions, and therefore the method tolerates the systematic intensity variations introduced by nonlinearity of the camera system.

Data Sites

With the quasi-heterodyne method, some of the data will usually lie in the low-gradient region (near intensity minima or maxima) where the intensity changes slowly with changes of the phase angle. With the O/D multiplication method, however, all the data that is utilized lies at the points where the gradient of intensity versus fringe order (or phase) is largest. This provides superior reliability of the data used in the O/D multiplication method.

The details are illustrated in Fig. 5.17a for a sinusoidal intensity distribution and a 256 gray level intensity scale. The phase that corresponds to one intensity increment at the maxima (and minima) can be any value within a range $\Delta\phi = 14.4°$. At the data sites of the O/D multiplication method, the range is $\Delta\phi = 0.45°$. The ratio is 32:1.

Excess intensity and saturation can be used with the O/DFM method, with the result illustrated in Fig. 5.17b. Whereas 256 gray levels are recorded, the phase increment near the crossing point corresponds to the increment that would be present for 400 gray levels. The phase increment per gray level is reduced to $\Delta\phi = 0.29°$, which compares to 14.4° as 1:50. Furthermore, the quasi-heterodyne method uses fewer than 256 gray levels in actual practice for fear of saturation and nonlinearity, which means that the ratio favoring the O/DFM data sites exceeds 50:1. Although the argument is based on a 256 step gray scale, the conclusion remains the same for all gray scales, regardless of the number of steps.

Computations

The quasi-heterodyne method computes the fringe order (or the phase value) at every pixel in the field, whereas the O/D multiplication method utilizes data from selected sites to produce contour maps with discrete contour intervals. As a consequence, the quasi-heterodyne method is much more computer-intensive. Nevertheless, the results of the extensive computations are frequently displayed in the form of contour maps, thus reducing the importance of the information available at pixels between the contours. In addition, the O/D multiplication method requires only simple arithmetic operations such as subtraction, addition and binarization. It is not computer-intensive, and use of a modest personal computer is practical.

Noise, Ghosts and Filtering

The O/D multiplication method cancels the influence of optical noise caused by nonuniform intensities in the two incoming beams A and B (Fig. 5.4) and caused by point-to-point variations of diffraction efficiency of the specimen grating. The quasi-heterodyne method cancels noise from these sources, too.

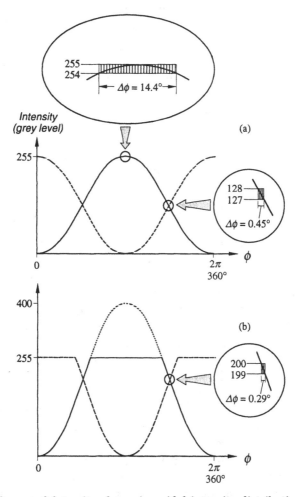

Fig. 5.17   Influence of data sites for a sinusoidal intensity distribution. (a) The full 256 gray level intensity scale is used. (b) Overexposure and truncation by the camera. Observe that the phase increment per gray level is smallest at the crossing points.

There are other sources of noise. One is random electrical noise introduced by the video camera and image grabber, which is not nullified or attenuated by either method. Another is optical noise introduced in the observation path between the specimen and video camera. Light from extraneous sources within this path, e.g., multiple reflections from the elements within the microscope and reflection from the top surface of the interferometer create ghost patterns of two types.

One type is a fixed ghost pattern, which does not change with fringe shifting. This extraneous light does not influence the O/D multiplication method since it contributes equally to the two complementary patterns and therefore it does not change the locations where $I_{(x,y)} = I^{\pi}_{(x,y)}$. Similarly, the quasi-heterodyne method is not affected by fixed ghost patterns.

The second type is a changing ghost pattern that is created when an extraneous wave front combines with a phase-shifted wave front diffracted from the specimen grating. These ghost fringes change as the main patterns are phase-stepped. In the O/D multiplication method, the ghosts add unequal contributions to the complementary patterns and affect the results. Similarly, in the quasi-heterodyne method, the changing contributions of ghosts to the phase stepped intensities upset the intensity differences needed to calculate phase.

Fortunately for both methods, the intensities of ghosts are low. They are minimized by antireflection coatings on the optical elements of the microscope and on the upper surface of the immersion interferometer. Investigators might consider developing an achromatic system such as Fig. 5.6 for use with a laser diode with modest temporal coherence. Then extraneous wave fronts could not interfere with a main wave front diffracted by the specimen grating because the difference of path lengths would be too great. Instead, the extraneous light would appear as equal background intensities in the complementary patterns and it would have no deleterious effects when the O/D multiplication method is used.

Another source of noise is the specimen grating itself. While the algorithms cope nicely with variations of diffraction efficiency, they cannot assist for local points where the diffraction efficiency is zero, i.e., where no light is diffracted into the microscope. The condition arises from tiny defects in the specimen grating—defects that are insignificant and unnoticed in macroscopic moiré interferometry. The intensity is zero at pixels corresponding to such points and it remains zero throughout the fringe analysis process. The result is random black dots in the white areas of the final fringe-contour map. The obvious remedy is defect-free specimen gratings, and there is great scope for improvement relative to current practices (Sec. 4.9.5).

A minute change of focus can change the black dots to white dots; this is a diffraction effect. The white dots become visible at pixels within the black fringe contours. They introduce a so-called salt-and-pepper distribution of dots in the fringe-contour maps. Digital filtering can be used to minimize their influence. Median filtering after the binarization step is effective for dealing with the salt-and-pepper, whereby the median intensity of each 5x5 pixel array (or 7x7 array) is assigned to the central pixel. Another means of dealing with the dots is by misfocus of the image, but it is not recommended for cases of large strain gradients because the warped wave fronts that create the pattern change with changes of focus (Fig. 2.2c).

5.4 Shear Lag and Grating Thickness

Shear lag was discussed in Sec. 4.23. For measurements in micromechanics, where strain gradients can be very severe, the ideal specimen grating is one of zero thickness (Sec. 4.9.5). When that is impractical, gratings of very small thickness should be sought. Fortunately, very small thickness can be achieved by replication if the size of the specimen grating is small. Gratings of a few square millimeters in extent and 2 μm thickness have been produced routinely by the following procedure: (1) the specimen is polished to produce a flat smooth surface; (2) a tiny volume of catalyzed, quick-setting epoxy is applied to the surface with a sharp-pointed tool; (3) a silicone-rubber mold is pressed against the epoxy to spread it into a thin film; (4) when the epoxy is cured, the mold is removed and the specimen grating is coated by vacuum-deposited aluminum. Such gratings, with 2 μm thickness, have been replicated at room temperature for tests with mechanical loading and at elevated temperatures for thermal strain tests.

5.5 References

1. B. Han and D. Post, "Immersion Interferometer for Microscopic Moiré Interferometry," *Experimental Mechanics*, Vol. 32, No. 1, pp. 38-41 (1992).

2. B. Han, "Higher Sensitivity Moiré Interferometry for Micromechanics Studies," *Optical Engineering*, Vol. 31, No. 7, pp. 1517-1526 (1992).

3. V. T. Chitnis, Y. Uchida, K. Hane and S. Hattori, "Moiré Signals in Reflection," *Optics Communications*, Vol. 54, No. 4, pp. 207-211 (1985).

4. S. Hattori, Y. Uchida and V. T. Chitnis, "An Automatic Super-accurate Positioning Technique Using Moiré Interference, *Bulletin Japan Society of Precision Engineering*, Vol. 20, No. 2, pp. 73-78 (1986).

5. T. W. Shield and K.-S. Kim, "Diffraction Theory of Optical Interference Moiré and a Device for Production of Variable Virtual Reference Gratings: A Moiré Microscope," *Experimental Mechanics*, Vol. 31, No. 23, pp. 126-134 (1991).

6. B. Han, "Interferometric Methods with Enhanced Sensitivity by Optical/Digital Fringe Multiplication," *Applied Optics*, Vol. 32, No. 25, pp. 4713-4718 (1993).

7. P. Hariharan, "Quasi-heterodyne Hologram Interferometry," *Optical Engineering*, Vol. 24, No. 4, pp. 632-638 (1985).

8. R. Dandliker and R. Thalmann, "Heterodyne and Quasi-heterodyne Holographic Interferometry," *Optical Engineering*, Vol. 24, No. 5, pp. 824-831 (1985).

9. J. B. Brownell and R. J. Parker, "Automated Fringe Analysis for Moiré Interferometry," *Proc. 2nd International Conference on Photomechanics and Speckle Metrology*, Vol. 1554B, pp. 481-492, SPIE, Bellingham, Washington (1991).

5.6 Exercises

5.1   Conceive and sketch an interferometer system, different from those shown in Figs. 5.4–5.7, for practical application with mechanically loaded specimens.

5.2   Conceive and sketch two interferometer systems, different from that of Fig. 5.8, for practical application with transparent replicas of deformed bodies.

5.3   With regard to problem 1 or 2, discuss advantages and disadvantages of the proposed systems in relation to the schemes illustrated in the chapter.

5.4   Explain why a mismatch of the refractive index of the immersion fluid does not affect the fringe formation. Discuss the advantages of having a good match of the refractive indices.

5.5   What other characteristics of the immersion fluid are important? Explain why they are important. Consider viscosity, chemical reactivity, volatility, temperature stability, flammability, purity, transparency, other factors.

5.6   Discuss the coherence length requirements of light sources employed for the systems shown in Figs. 5.4–5.8.

5.7   The interferometer of Figs. 5.4 and 5.9 was designed for f_m = 4800 lines/mm. Sketch the directions of diffracted rays when it is used in conjunction with specimen gratings of frequencies (a) f_s = 2400 lines/mm, (b) f_s = 1200 lines/mm, (c) f_s = 800 lines/mm. (d) Explain why the moiré patterns are the same for the three cases, assuming the specimen is subjected to the same strain field. (e) Discuss advantages and disadvantages of the three cases.

5.8   The basic sensitivity of the immersion interferometer increases with the frequency, f_m, of the virtual reference grating.   Design a realistic interferometer that provides f_m = 8000 lines/mm.

5.9   What specimen grating frequency would you select for use with the above interferometer?   Discuss the advantages and disadvantages of this choice. Propose a method for producing the grating mold.

5.10   Prove Eq. 5.6.

5.11   Conceive and sketch two fringe shifting schemes, different from that of Fig. 5.12. Discuss advantages and disadvantages of the proposals relative to the device of Fig. 5.12.

5.12   What are the major parameters that govern the accuracy of fringe shifting? Propose a calibration procedure for fringe shifting.

5.13   Draw a flow chart (block diagram) for computer execution of the fringe sharpening algorithm.

5.14   Draw a flow chart (block diagram) for computer execution of fringe multiplication by superposition of sharpened fringes.

5.15   Conceive a practical technique to extract whole-field strain maps from the contour maps of microscopic moiré interferometry. Discuss its advantages and disadvantages.

5.16   Referring to Exercise 5.15, conceive another practical technique that circumvents or minimizes the main disadvantages.

6

On the Limits of Moiré Interferometry

6.1 Introduction

The main focus of this chapter is potential limitations imposed by the aperture of the camera lens. Numerical and experimental investigations were performed to validate moiré interferometry for extremely large strain gradients and discontinuities.

Additional topics relating to the legitimate scope of moiré interferometry include resolution, gage length, dynamic range, and absolute accuracy. These topics are discussed.

6.2 Warped Wave Front Model

The warped wave front model of optical interference is depicted in Fig. 2.23. It is proposed for all types of two-beam interferometry, including moiré interferometry. It proclaims that the interference fringe pattern is a contour map of the separation $S_{(x,y)}$ between two warped wave fronts of equal phase. The model further stipulates that the x,y points in the interference pattern correspond exactly to the x,y points in the specimen. This condition is achieved when the camera is focused on the specimen surface, such that the two wave trains have the same wave front separations when they cross the image plane as they had when they emerged from the specimen.

The statement is imprecise because the two warped wave fronts of equal phase do not emerge from the specimen at the same time. Instead, let $S_{(x,y)}$ be expressed as $C\Delta t_{(x,y)}$, i.e., the velocity of light multiplied by the increment of time between the emergence of wave fronts of equal phase from each x,y point. Of course, $\Delta t_{(x,y)}$ is the same at the corresponding image points. The model will use this interpretation of $S_{(x,y)}$.

What determines $S_{(x,y)}$ or $\Delta t_{(x,y)}$? A remarkable answer emerged from the mathematical analysis (Sect. 4.3). The relative phase and

$S_{(x,y)}$ is determined by the changes of optical path lengths *to each point* on the deformed specimen grating. No other information was used. Accordingly, the information that maps the deformation is fully established before the light leaves the specimen grating. The model assumes that the information is conveyed to the film plane by two beams with warped wave fronts; one of the two beams is illustrated in Fig. 6.1a. The deformation of the specimen grating is conveyed with a marvelous code, namely, the diffraction equations (Eqs. 2.54-2.56). The direction of each ray, and the wave front element perpendicular to the ray, contributes to the whole field wave fronts in such a way that their separation is $S_{(x,y)}$.

The information is conveyed to the moiré pattern if the rays of the two diffractions from each point pass through the camera lens. Otherwise, *a black hole* or *white hole* appears in the moiré pattern (Sec. 4.19.1).

6.3 Fourier Wave Front Model

The warped wave front model is an elementary treatment. A more comprehensive treatment would use the Fourier wave front model. With the Fourier model, the analysis proceeds through the mathematical derivation of Sec. 4.3 without modification. As before, the optical path lengths define two warped wave fronts that emerge from the specimen grating. These are the two warped wave fronts that propagate toward the camera. When $f = 2f_s$, they are the diffractions of orders $m = 1$ and $m = -1$ in Eqs. 2.54-2.56. Upon emergence, however, the models become different.

The two models are illustrated in Fig. 6.1 for one of the two input beams. In (a), the light propagates toward the camera with a single warped wave front. In (b), the same warped wave front is decomposed into a Fourier series of component waves.[1] The components are beams with plane wave fronts, each propagating at a different angle $\theta_{m,i}$. The decomposition can produce a large number of component beams, called *mini-orders*, but only a few are shown in the figure. The mini-orders of local amplitude maxima are represented by the subscript i. These mini-orders have various amplitudes which are determined by the coefficients of the Fourier series. Generally the amplitudes become smaller for higher mini-orders (larger i). A multitude of additional mini-orders surround each local maxima, but these have smaller amplitudes.

As an alternative to the decomposition of the emergent wave front, one can visualize that the deformed grating itself is decomposed into a series of Fourier components, called subgratings. Each subgrating has a uniform and unique frequency f_i and light emerges from each subgrating with the unique directions $\theta_{m,i}$, which are determined (for 2-D diffraction) by the grating equation

$$\sin \theta_{m,i} = \sin \alpha + m\lambda f_i \qquad i = 0, \pm 1, \pm 2, \cdots \qquad (6.1)$$

Usually we are interested in only one diffraction order, e.g., $m = -1$, for each input beam. Then one beam directed along $\theta_{-1,i}$ emerges for each subgrating frequency f_i, to generate the mini-order beams in Fig. 6.1b. The series of plane wave fronts that results from the Fourier decomposition is the same, regardless of whether it is the deformed grating or the warped wave front that is decomposed.

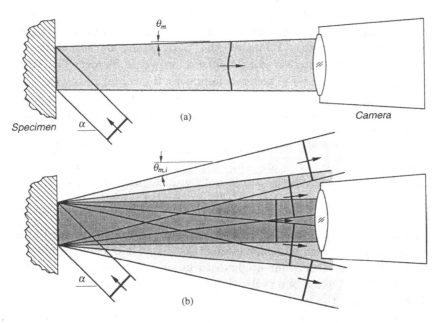

Fig. 6.1   Emergent beams for one incident beam, according to (a) the warped wave front model, for diffraction order $m = -1$, and (b) the Fourier wave front model, for diffraction order $m = -1$.

The Fourier wave front model reduces to the warped wave front model for the special case of an extended field of constant strain. Then, the models predict a single plane wave front for each diffraction order m, except near the specimen boundaries.

6.4 Strain Fields

In an important paper on moiré interferometry, McKelvie pointed out that we must be aware that information describing a strain field can be lost or misrepresented.[2] Obviously, information is lost when, in Fig. 6.1b, angle $\theta_{m,i}$ exceeds the angle of acceptance of the camera lens. Information is misrepresented if mini-orders from the $m = -2$ diffractions enter the camera lens along with those from $m = -1$.

Where the warped wave front model is adequate, angle θ_m in Fig. 6.1a is determined from Eqs. 2.24 and 4.6 as

$$\sin \theta_m = \frac{\lambda f}{2} \varepsilon_x \qquad (6.2)$$

where ε_x is the strain of the deformed specimen. If $\varepsilon_x < 1\%$, which is the usual case for studies by moiré interferometry, the angular deviation is $\theta < |0.44°|$. (If $\varepsilon_x > 1\%$, one would normally subtract off part of the fringe gradient by carrier fringes, whereby θ again becomes very small.) Thus, all the light would be accepted by a practical optical system and no information would be lost. Of course, the argument pertains to observations near the center of the field and to cases where the slope of the specimen surface is zero or small; these conditions could be applied by adjusting the specimen, if necessary.

In the case of the Fourier model, there is potential for $\theta_{m,i}$ to be much larger. McKelvie decomposed the warped wave front for the case of a specimen subjected to a harmonic strain variation

$$\varepsilon_x = \varepsilon_1 \cos 2\pi f_g x \qquad (6.3)$$

where ε_1 is a constant and f_g is the frequency of the strain oscillation. Numerical examples were not undertaken, but it is clear in abstract terms that the mini-order angles can be excessively large when f_g is very large, i.e., when one cycle of the strain oscillation occurs in a tiny distance along the specimen.

Numerical examples were conducted by Hagedorn.[3] For the most severe case considered, a specimen grating of initial frequency f_s was deformed by the strain of Eq. 6.3, where the maximum strain was $\varepsilon_1 = 1\%$ and the frequency of the strain oscillations was $f_g = 0.04 f_s$. Thus one cycle of the strain oscillation occurred within 25 lines of the specimen grating; its period was 20.8 μm/cycle for $f_s = 1200$ lines/mm, or 10.4 μm/cycle for $f_s = 2400$ lines/mm. A full strain excursion, from +1% to −1% strain, would occur in half a cycle, or 5.2 μm in the latter case. The strain gradients were extremely large.

The results are summarized in Table 6.1. The *amplitude ratio* is the amplitude of the ith mini-order beam divided by the amplitude of the $i = 0$ mini-order. The results are tabulated for two frequencies of the virtual reference grating: $f = 2400$ and $f = 4800$ lines/mm. In the

TABLE 6.1 Plane Waves of Fourier Model[†]

Angle of Mini-order, $|\theta_{m,i}|$			Amplitude Ratio
	f, lines/mm		
i	2400	4800	$|a_{m,i}/a_{m,0}|$
0	0	0	1
±1	1.42°	2.83°	0.126
±2	2.83°	5.67°	0.0079
±3	4.25°	8.52°	0.00033
±4	5.67°	11.39°	0.000010

[†] For strain defined by Eq. 6.3, where $\varepsilon_1 = 1\%$, $f_g = 0.04f_s$.
$\lambda = 514.5$ nm, $|m| = 1$, $f = 2f_s$

case of the immersion interferometer, the angles of the mini-orders pertain to the angles in air, as the beams approach the microscope lens.

The amplitude becomes extremely small for the larger angles. Following the reasoning of McKelvie,[2] the mini-orders of extremely small amplitude have negligible importance, i.e., their contributions to the fringe pattern in the film plane are negligible. In the present case, nevertheless, even those mini-order beams with extremely small amplitudes can be accommodated by the camera lens (or microscope lens). For high-resolution moiré interferometry systems, a practical limit for the numerical aperture of the lens is about $NA = 0.3$ (except for Configuration 5, Fig. 5.8, where larger numerical apertures are practical). For $NA = 0.3$, the acceptance angle of the lens is $\theta = \pm17.5°$; for $NA = 0.2$, the acceptance angle is ±11.5°. In both cases, the acceptance angle exceeds the largest angle in Table 6.1. We conclude that no significant information would be lost as a result of the camera lens aperture in the case of $f_g = 0.04f_s$.

The cyclic strain frequencies already considered, and the corresponding strain gradients, are extremely high. Should we consider cases in which f_g is much higher? Probably not. Higher frequencies would mean there are fewer grating lines per cycle. Furthermore, the periods (distance per cycle) would become shorter—shorter than the 20.8 and 10.4 µm periods in the example of Table 6.1. These extremes raise additional issues that influence the accuracy with which detail can be extracted. As the number of specimen grating *lines per strain cycle* is made smaller, the uncertainty inherent in limited sampling becomes larger. In addition, the already small change of fringe order within a period would become smaller, leading to larger errors of intensity measurements and fringe order determinations. If the latter difficulty is circumvented by means of carrier fringes, the problem would change to one of precision, where two nearly equal quantities must be subtracted.

6.5 Boundary Effects, Discontinuities

The Fourier wave front model predicts a sequence of diffracted mini-orders from regions near abrupt changes of grating frequency. Regions near specimen boundaries (where the frequency changes abruptly to zero) and regions near strain discontinuities are in this category. Is information lost or misrepresented? The critical question is whether the moiré fringes represent the specimen grating authentically in such regions.

We investigated boundary effects experimentally in an effort to elucidate the behavior of the moiré fringes. The experiment was designed to increase the likelihood of lost and misrepresented information. Very high sensitivity was employed, corresponding to moiré with 4800 lines/mm. From the results, we conclude there was no measurable fringe distortion attributable to the boundary. The implication is that lost or misrepresented information near boundaries is negligible.

The issue should be revisited when apparatus of higher resolution is developed. Within the current capabilities of high sensitivity moiré interferometry, no departure is detected and no compensation is indicated. We speculate, nevertheless, that if a systematic variation of fringe orders is found in the future, the boundary effect could be compensated by subtraction of no-load fringe orders.

6.5.1 The Experiment

A critical requirement was a specimen grating with known geometrical characteristics of the grating lines. It was satisfied by using an undeformed grating, as free as possible of irregularities of line spacing and straightness.

The specimen was a glass substrate with a thin photoresist coating, and a 1200 lines/mm cross-line grating formed in the coating. The glass was fractured to create an undistorted edge. The process consisted of scoring the glass substrate on its underside and flexing the glass to propagate a crack by brittle fracture. The crack propagated through the photoresist to produce a cleavage that did not involve distortion of the material. An example of the result is shown in Fig. 6.2 for a linear grating, where the undistorted character of the grating is seen to extend to the fracture boundary.

The actual specimen employed a cross-line grating with its lines parallel and perpendicular to the fracture boundary. The apparatus of Fig. 5.1 was used with the immersion interferometer of Figs. 5.4 and 5.9. The virtual reference grating frequency was nominally 4800 lines/mm, which means that diffraction orders $m = \pm 2$ were collected by the microscope lens. The numerical aperture of the lens was a modest $NA = 0.2$. The CCD camera had an image size corresponding to 1.2 μm per pixel.

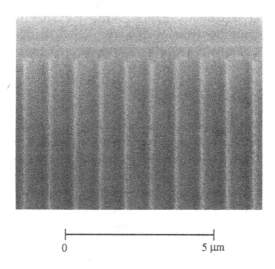

F——————————————————————H
0   5 μm

Fig. 6.2   Scanning electron micrograph of a specimen boundary prepared by brittle fracture. $f_s = 1200$ lines/mm.

Note that the use of a modest numerical aperture increased the likelihood that mini-orders in the second diffraction order would be lost. In addition, use of the second diffraction order increased the likelihood that mini-orders in the first and third diffraction orders would intrude into the microscope. The experiment was designed with these parameters to accentuate the potential fringe distortions.

The results are shown in Fig. 6.3. The fractured boundary of the specimen is at the top of each pattern, parallel to the x axis. Patterns (a) and (b) show the V field, where the grating lines were parallel to the boundary of the specimen. The fringes represent pure rotation in

(a) and rotation plus extension in (b). Pattern (c) shows the U field, with rotation plus extension. These are patterns of two-beam interference, with f = 4800 lines/mm (122,000 lines/in.). No numerical filtering was used. The intensities were binarized near $I = 0$ to depict the paths of the fringes more clearly.

In all cases, the fringes near the specimen boundary have the same characteristics as those far from the boundary. The fringes are essentially straight right up to the boundary, as they should be to represent the specimen grating correctly. There is no systematic disturbance, which means there is no perceptible deviation associated with the Fourier decomposition. Note that such a deviation must be systematic, where *every* fringe oscillates or curves similarly near the boundary.

The fringes represent the specimen grating with high fidelity. For this rather stringent experiment, lost or misrepresented information stemming from a Fourier wave front analysis was negligible.

6.6 Spatial Resolution

The camera lens (or microscope lens) must focus an image of the moiré fringes and the specimen surface on the image plane. The role of the camera lens was discussed in Sec. 2.2.5, where its role was shown to be different for specular and non-specular objects.

Light from a replicated specimen grating, or an etched grating on a polished surface, is specular. Typically the mini-order diffractions of significant amplitude have an angular separation that is small enough to enter the camera lens without intercepting its border or aperture stop. Therefore, the fringes are focused in the image plane (by an aberration-free lens) without spreading or blurring, in correct geometric positions corresponding to the object points.

Light from the specimen boundaries, or from reference marks scratched on the grating, is diffuse. It spreads in all directions, overfills the camera lens, and intercepts the lens boundaries or aperture stop. In the image plane, light from each of these object points is spread in an Airy's disk of diameter s given by

$$s = \frac{1.22\lambda}{NA} \qquad (6.4)$$

where NA is the numerical aperture of the lens and s refers to the size on the scale of the object. Its actual size is sM, where M is the magnification of the image. In the case of microscopic moiré interferometry with a long working distance microscope objective of

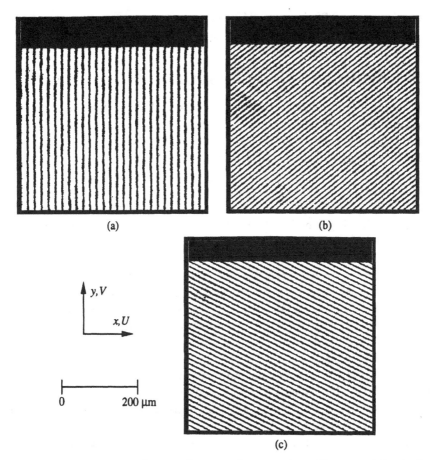

Fig. 6.3   Fringe patterns for a uniform specimen grating. The upper edge is the fracture boundary. $f = 4800$ lines/mm. (a) and (b) V field. (c) U field.

numerical aperture 0.2, Eq. 6.4 yields $s = 6\lambda$. Thus, reference marks and specimen boundaries are spread by about six wavelengths, or about 3 µm. The lens accounts for spreading of light near reference marks, which leads to a small uncertainty of the locations of all points in the field.

6.7 Displacement Resolution

Displacement resolution is the tolerance within which a measurement is assumed to be reliable. By Eq. 4.1, it is proportional to the uncertainty of fringe order N. By visual estimation, fringe orders can be determined to 1/5 or 1/10 of a fringe order at any x,y point. With $f = 2400$ lines/mm, this corresponds to a displacement uncertainty of 0.08 μm or 0.04 μm, respectively. The uncertainty is much smaller when image processing is used.

6.8 Strain and Gage Length

Although displacement fields are often the desired parameter, especially when companion numerical studies are undertaken, strains are frequently extracted from the displacement fields. Then, gage length and accuracy of strain determinations become topics of interest.

Strains are determined from displacement fields by Eqs. 4.2 and 4.3. The equations are applied at any point by extracting the rate of change of fringe order from the moiré pattern. The general procedure was introduced in Sec. 4.11 in connection with a shear specimen and it is demonstrated in Figs. 7.4 and 7.14 in connection with a composite compression specimen. The procedure applies to regions where the strain is known to *vary continuously and smoothly*; and it also applies to analyses of heterogeneous deformations where the investigator wants to average local perturbations and report the global behavior. It is illustrated schematically for ε_x at point A in Fig. 6.4. The procedure consists of these steps: (1) draw a line on the fringe pattern through A parallel to the x axis; (2) plot the curve of N_x versus x for that line; (3) determine the slope $\Delta N_x / \Delta x$ at point A and calculate ε_x by Eq. 4.1. Of course, $\Delta N_x / \Delta x = \partial N_x / \partial x$ at point A.

The term *gage length* means a length on the specimen across which the strains are averaged. It is clear from Fig. 6.4, however, that strains are determined at individual points. No averaging and no gage length is involved when the conditions cited above are in effect.

If it is not known in advance that the strains vary continuously and smoothly, and if fringe-shifting is not employed, only the *average strain* between adjacent fringes can be extracted from the fringe pattern. Then, the gage length is the distance between adjacent fringes. The gage length can be shortened by applying carrier fringes and it can be shortened by employing the centerlines of intensity maxima in addition to intensity minima.

6.8.1 Strain Resolution

Numerical values of strain resolution depends upon the context in which the term is used, upon the nature of the problem, and upon the analysis procedure. It is not uncommon to compare experimental

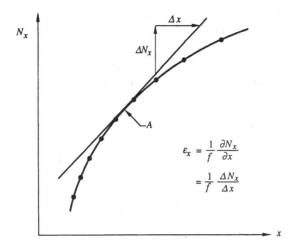

Fig. 6.4 To determine strain, the curve of fringe order versus position is plotted from the moiré data and its slope is measured at any point of interest.

methods by quoting the smallest strain that can be measured reliably. For this purpose, a gage length is usually needed and it is prescribed. For example, an extensometer that measures the strain between two points that are originally 25 mm apart might be compared to a moiré measurement. The average strain in a 25 mm gage length would be measured by both methods.

When moiré fringes are measured to within 1/10 of a fringe order at the two end points of the 25 mm length, the maximum error is 1/5 fringe. The maximum uncertainty of change of length is $1/(5f)$; with $f = 2400$ lines/mm, the uncertainty is 0.08 μm. Then, the uncertainty of the average strain is within $\Delta\varepsilon = 0.08$ μm/25 mm $= 3 \times 10^{-6}$. This is the strain resolution that should be compared to that of the extensometer. In this context, the strain resolution depends upon a prescribed gage length.

Where the strain is known to vary smoothly, strain resolution is determined by the abundance of displacement data. Returning to Fig. 6.4, let the data points be integral fringe orders that are measured and plotted to an accuracy of 1/10 fringe order. If there are many data points in the neighborhood of A, a smoothed curve through A (established, for example, by a least squares fit of the data) would be substantially more accurate than a curve plotted from fewer data points. Of course, the tangent at A would be substantially more accurate, too. In this case, it is the statistical advantage of many data points that improves the displacement resolution, and consequently improves the strain resolution.

The estimate suggested earlier (Sec. 4.1) can be repeated: "regardless of the method used for differentiation --- a loss of an order of magnitude seems to be normal." Thus, it is fortunate that the displacement data usually can be extracted with very high accuracy, so that strains can be determined with reasonable accuracy.

6.9 Dynamic Range

The term *dynamic range* refers to the useful range of response of an instrument or system. For example, the dynamic range of wavelength response of the human eye is typically 400 to 700 nm.

For moiré systems, the dynamic range of relative displacements is very large. Displacement is proportional to fringe order and fringe orders can be measured to within a fraction of a fringe from $N = 0$ to N equals several hundred. An example appears in Fig. 4.7, where 100 fringes appear in (b) and 235 fringes appear in (a); these are enlargements from the same photographic negative which extended over a 76 x 51 mm (3 x 2 in.) field of view, where about 600 well resolved fringes are present.

The dynamic range of strains is large, too. In Fig. 4.7 it extends from $\varepsilon_y = 0$ to $\varepsilon_y = 0.5\%$. In Fig. 7.2 it extends from $\gamma_{xy} = 0$ to $\gamma_{xy} > 1\%$. In addition, the measurement range can be increased substantially with carrier fringes that subtract off uniform strains, while still retaining the original strain resolution capability. As a practical matter, the dynamic range of moiré interferometry greatly exceeds the range needed for measurement of engineering strains.

6.10 Absolute Accuracy

The absolute accuracy of relative displacement measurements depends upon the accuracy of the virtual reference grating frequency, f. It is usually insignificant for deformation analyses, but absolute accuracy can be important for special applications. Examples include moiré interferometry used for strain gage calibration (Chap. 13) and for linear measuring machines. A practical approach is suggested in Chap. 13, whereby the absolute accuracy of the virtual reference grating was adjusted by visual observations to about one part in 10^5.

6.11 Summary

Extremely large strain fluctuations or strain gradients can be accommodated by moiré interferometry; the moiré interference fringes are imaged faithfully by the camera lens. Images of specimen boundaries and reference lines are slightly blurred as a result of the finite aperture of the camera lens.

Displacement resolution is determined by the method employed to extract fringe order data. Where displacements vary smoothly, strains can be determined at individual points; they are not averaged across a gage length. Otherwise, the strains are averaged between adjacent fringes. The dynamic range of moiré interferometry is compatible with large displacements and large strain gradients.

6.12 References

1. J. W. Goodman, *Introduction to Fourier Optics*, McGraw-Hill, New York (1968).

2. J. McKelvie, "On Moiré Interferometry and the Level of Detail That It May Legitimately Reveal," *Optics and Lasers in Engineering*, Vol. 12, pp. 81-99 (1990).

3. G. A. Hagedorn, Mathematics Dept., VPI&SU, personal communications (Feb.-Apr., 1993).

PART II

APPLICATIONS AND ADVANCES

Diverse Applications
Experimental Techniques
Fringe Pattern Analysis

Whereas Part I emphasizes fundamentals and generalizations, Part II emphasizes specific studies and methods of analysis. Applications from diverse fields are illustrated. Specific experimental techniques are introduced in conjunction with the applications. Important details of analysis and interpretation are described.

It is reassuring that the principles of Part I are applied successfully in Part II. An ever-increasing range of applications can be anticipated.

7
Laminated Composites in Compression: Free-edge Effects

7.1 Introduction

This chapter excerpts material from a series of tests of thick laminated composites in compression.[1] As illustrated in Fig. 7.1, the graphite/epoxy specimens were cut from thick-walled cylinders with two different stacking sequences (i.e., the sequence of fiber directions in successive plies of the laminate). They are called quasi-isotropic and cross-ply laminates. In-plane and interlaminar compression tests were conducted, with the compressive loads applied parallel and perpendicular to the plies, respectively. Note that the coordinate system maintains x perpendicular to the plies and y parallel to the plies. The loading fixture of Fig. 4.10 was used.

Moiré interferometry was applied to one face to reveal details of the surface deformations on a ply-by-ply basis. Electrical resistance strain gages were used on the other three faces to measure average strains; they provided a means to monitor the uniformity of compressive loading.

Free-edge effects were investigated. These are strains or stresses on free (unloaded) surfaces of composite bodies that are different from the corresponding strains below the surface in the interior of the body. Fiber composites behave differently near the surface, compared to their behavior in the interior. The reason is that fibers near the surface are less constrained by surrounding material, so they can displace more readily in response to applied loads.

7.2 In-plane Compression, Quasi-isotropic Specimen

The moiré results for the quasi-isotropic specimen are shown in Figs. 7.2 and 7.3. The fringe patterns represent the central region

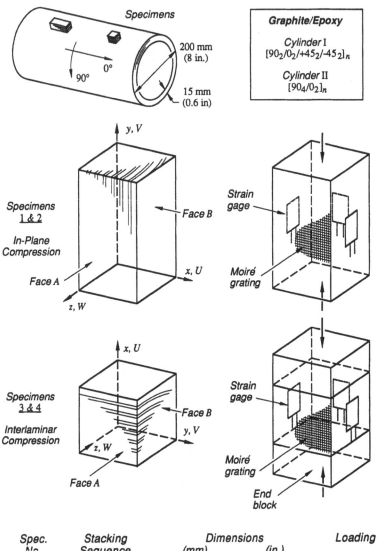

Fig. 7.1   Specimens cut from cylinders I and II.

indicated by the dashed box. They reveal dramatic ply-by-ply variations of displacements and strains. The symbols below the graphs designate the fiber direction in each ply; they represent, from left to right, the 90°, 0°, +45° and –45° fiber directions, respectively.

The compressive strain, ε_y, is obtained from the fringe pattern in Fig. 7.2. It is proportional to the fringe gradient in the y direction (Eq. 4.2), i.e., to $\partial N_y / \partial y$. Here, the vertical distance between adjacent fringes is essentially the same everywhere in the field, which means that ε_y is nearly the same throughout the field. The strain can be determined by measuring the distance between 20 fringes and expressing Eq. 4.2 by finite increments as

$$\varepsilon_y = \frac{1}{f}\left(\frac{\Delta N_y}{\Delta y}\right) = \frac{1}{2400}\left(\frac{-20}{\Delta y}\right) = -0.0027 \qquad (7.1)$$

where Δy is the measured distance (in millimeters) divided by the magnification of the image. The result is a uniform compressive strain of 2700×10^{-6}; it is graphed as a dashed line in Fig. 7.3. It is curious that such a complex pattern can represent a uniform strain.

The gradient of N_y exhibits severe variations in the x direction, and these relate to variations of the shear strain. The shear component $\partial N_y / \partial x$ can be extracted from this pattern. Two techniques are illustrated in Fig. 7.4 for the high shear region encircled in Fig. 7.2. The solid and dashed lines in Fig. 7.4 represent the centers of black and white fringes, respectively. In the graph, the curve of N_y versus x is plotted for line AA. The slope of the curve at any point represents the fringe gradient at that point, as shown for point O. Although finite increments are shown on the tangent line in Fig. 7.4, the slope of the curve at any point is $\partial N_y / \partial x$. In principle, this is an exact method, not an approximate method.

For the second technique, Eq. 4.25a is utilized. It can be rewritten as

$$\frac{\partial N_y}{\partial x} = \frac{1}{\tan \phi}\left(\frac{\partial N_y}{\partial y}\right) \qquad (7.2)$$

where ϕ is the angle of the fringe vector F_y at any point O. The term $\partial N_y / \partial y$ is a constant and its numerical value is already known from Eq. 7.1. Angle ϕ is determined by measuring the angle of the fringe— or more specifically, the angle of the tangent to the fringe at O. This second technique is often more convenient than the first. It too is an exact method, in principle, when $\partial N_y / \partial y$ is constant,

In order to complete the calculation of shear strains, the term $\partial N_x / \partial y$ is needed. It is obtained from the N_x field (with no carrier)

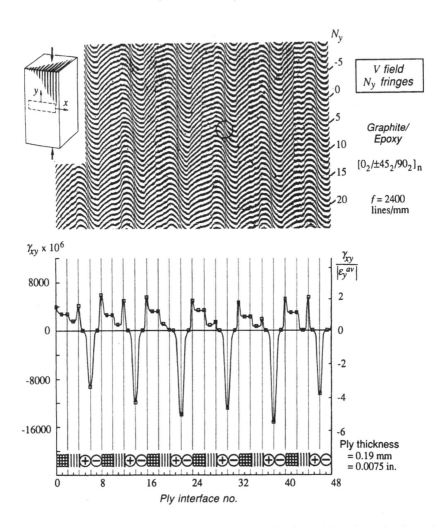

Fig. 7.2   The N_y moiré pattern in the region of the dashed box for a laminated composite in compression. Fringe orders N_y are assigned with an arbitrary zero datum. The graph shows the corresponding interlaminar shear strains acting along a horizontal line. Load = 22.2kN (5000 lb.).

shown in Fig. 7.3. The change of fringe order in the y direction is everywhere zero, or essentially zero, and $\partial N_x / \partial y = 0$.

The shear strain is calculated by Eq. 4.3, with the results given in the graph of Fig. 7.2. Severe shear strains occur near the interface between +45° and −45° plies, with a magnitude about five times greater than the applied compressive strain ε_y. Shears in the 90° plies are essentially constant through the ply thickness, as recognized by the nearly constant slope of N_y fringes across the ply.

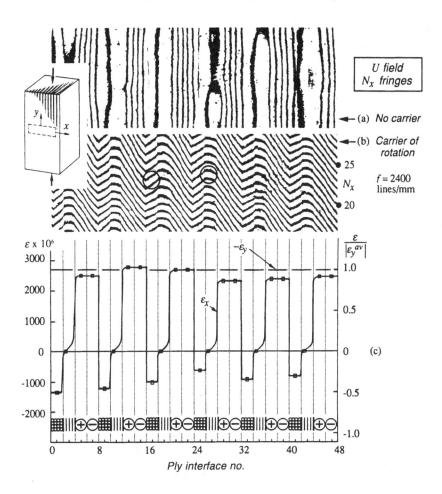

Fig. 7.3 The N_x moiré pattern for the experiment of Fig. 7.2. Carrier fringes were applied to evaluate ε_x. The graph shows the transverse normal strains on a ply-by-ply basis.

Note that nominally equal plies exhibit somewhat different strain levels. These differences reflect small ply-by-ply variations of the material, which is characteristic of composite laminates.

7.2.1 Carrier Fringes

The normal strains ε_x are also determined from the N_x field. These are transverse strains, which are associated with delamination of the composite. It is difficult to interpret the N_x field shown in Fig. 7.3a, particularly to discriminate regions of increasing fringe order from regions of decreasing fringe order.

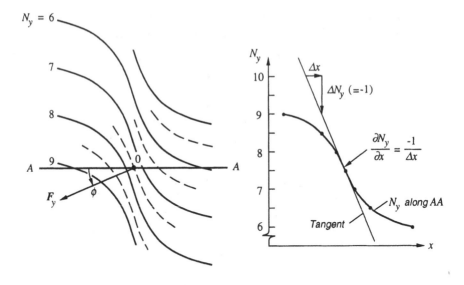

Fig. 7.4   Two procedures for determining $\partial N_y / \partial x$ in the encircled region of Fig. 7.2.

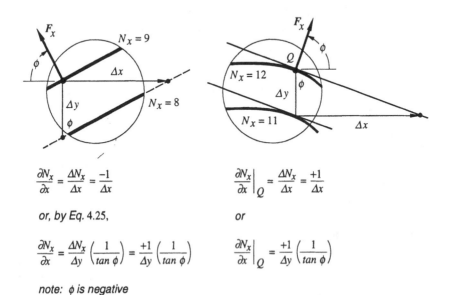

Fig. 7.5   Two procedures for determining $\partial N_x / \partial x$ in the encircled regions of Fig. 7.3.

The issue is clarified by introducing carrier fringes of rotation. The carrier fringes were introduced by applying a tiny rigid-body rotation of the specimen and loading fixture. In the absence of load-induced fringes, the carrier is comprised of uniformly spaced fringes parallel to the x axis (Sec. 4.13). As such, they introduce no fringe gradient in the x direction; $\partial N_x / \partial x = 0$ for the carrier of rotation.

The deformation induced by the compressive load transformed the carrier pattern to that of Fig. 7.3b. Now, $\partial N_x / \partial x$ can be determined unambiguously in each ply. The analysis is illustrated in Fig. 7.5 for the two encircled regions of Fig. 7.3b. The distance Δx can be measured from the fringe pattern in each case. However, it is more convenient to measure Δy and angle ϕ instead; Δy is a constant and it can be determined accurately by measuring the vertical distance between several fringes; ϕ is the angle of the fringe vector, as shown. Angle ϕ and Δy are interrelated in such a way that the results are independent of the magnitude of rotation. Since counterclockwise ϕ is positive, the sign of the derivative and the sign of the strain, ε_x, changes when the slope of the fringe vector changes from positive to negative.

The graph of ε_x versus x is shown in Fig. 7.3. For the 45° plies, ε_x is large and essentially constant across the thickness of the ply, consistent with the nearly constant slope of the corresponding fringes. An extremely abrupt change from tensile to compressive ε_x is seen at interfaces between −45° and 90° plies. The strain is nearly zero in 0° plies.

7.3 In-plane Compression, Cross-ply Specimen

Figure 7.6 depicts the U and V displacement fields for the cross-ply specimen, which has a $[90_4/0_2]_n$ stacking sequence. The U and V patterns extend across the full width of the specimen, while the pattern with carrier fringes shows a smaller central portion. The cross-derivatives of U and V are essentially zero, confirming that shear strain γ_{xy} is essentially zero. The patterns confirm that the cross-ply laminate does not exhibit a γ_{xy} (or τ_{xy}) free-edge effect.

The slowly varying vertical spacing in the V field indicates that the compressive strain ε_y is not constant across the specimen. Instead, ε_y varies smoothly across the width of the specimen, with its average value 9% smaller than its maximum value.

The spacing of fringes in the U field is cyclic, as seen more clearly in the pattern with added carrier fringes of rotation. The carrier fringes are parallel to x and do not influence the derivative with respect to x. The fringe pattern is analyzed by the method of Fig. 7.5.

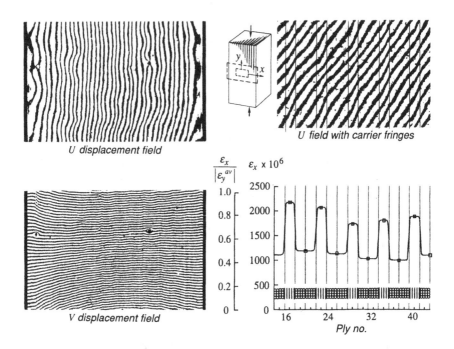

Fig. 7.6   Deformation of face A of the cross-ply specimen.   Load = 22.2 kN (5000 lb.).   The U and V fields indicate nearly zero shear γ_{xy}.   Ply-by-ply variations of ε_x are indicated by fringe slopes in the U field with carrier fringes.

The fringe slope is greater in the 0° plies, leading to the ε_x strain distribution shown by the graph.   The graph pertains to the central part of the specimen, but it is representative inasmuch as substantial point-to-point variations appear within individual plies and in nominally equal plies of the laminate.   Again, such variations are characteristic of composites.   Nevertheless, the transverse strains are consistently larger in the 0° plies, contrary to expectations based upon the Poisson's ratios of individual 0° and 90° plies.   The explanation is that the transverse surface strains measured here represent a *free-edge effect* associated with out-of-plane warpage of the free surface.

7.4   Interlaminar Compression, Quasi-isotropic Specimen

Figures 7.7 and 7.8 show the displacement fields for the quasi-isotropic specimen in interlaminar compression.   The extremely interesting V field (transverse displacement field) shows strong shear strains in repeating cycles.   For analysis, the N_y fringe orders were numbered, starting with an arbitrarily selected zero. The direction of the increasing fringe orders were assigned consistent with the knowledge that transverse strains were tensile, corresponding to a positive Poisson's ratio.

The shear strains along the vertical centerline, CC, were determined by the following technique. First a graph of N_y versus x was constructed along CC; sufficient data were available by using the centers of both the integral and half-order fringes. Note that x should be plotted in specimen dimensions, i.e., the fringe pattern dimensions divided by the magnification. Then, the slope of that curve was measured at various points to determine $\partial N_y / \partial x$. Prior to the introduction of carrier fringes in the U field (Fig 7.8), the load-induced fringes were essentially horizontal, which means $\partial N_x / \partial y$ was negligible. Therefore, $\partial N_y / \partial x$ was multiplied by $1/f$, according to Eq. 4.3, to obtain γ_{xy}.

The interlaminar shear strains at $+45°/-45°$ interfaces are large compared to the applied compressive strains. These occur across the whole width of the face, except near the specimen corners where the peak shears are slightly smaller. These shear strains cannot occur in the interior of the body. Instead, they are free-edge effects, acting in a narrow volume of material near the free surfaces. They can be an important factor leading to material delamination.

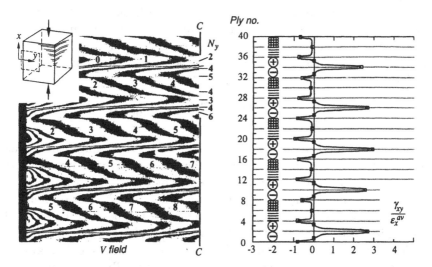

Fig. 7.7 V (or transverse) displacement field for face A of the quasi-isotropic specimen in interlaminar compression. Load = 6.67 kN (1500 lb.). The graph shows the interlaminar distribution of shear strains γ_{xy} along CC, the vertical centerline of face A.

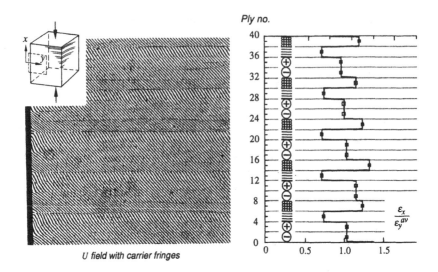

Fig. 7.8 U (or vertical) displacement field with carrier fringes of rotation, corresponding to Fig. 7.7. The graph shows interlaminar compressive strains ε_x on a ply-by-ply basis.

The alert experimentalist will notice another interesting feature in the V field (Fig. 7.7). Fringe orders are consistently lower near the top of the field than near the bottom. This means the top was displaced to the left. Lines that were vertical before loading lean slightly to the left after loading. The loading fixture did not apply a pure compressive displacement to the specimen, but instead, the displacement had a horizontal or shear component, too. The global horizontal displacement is seen as about –5 fringes in 32 plies; globally,

$$\frac{\Delta V}{\Delta x} = \frac{-5(1/2400)\text{mm}}{32(0.19)\text{mm}} = -0.00034$$

Thus, an accidental global shear strain of this magnitude was applied together with the much larger compressive strain. In the absence of this global shear, i.e., under idealized pure compressive loading, the shear strain curve of Fig. 7.7 would be shifted to the right by approximately 0.1. The peak strains would be about 4% higher and shears in the 0° and 90° plies would be essentially zero.

The same fringe pattern, Fig. 7.7, shows that the transverse normal strains, ε_y, are small. They are essentially constant across the entire face of the specimen except near the corners where the fringe gradients, and consequently the ε_y strains, are a little larger.

The compressive strains, ε_x, were determined from Fig. 7.8. The average value is

$$\varepsilon_x^{av} = \frac{1}{f}\frac{\Delta N_x}{\Delta x} = -0.0034$$

where Δx is a multiple of the thickness of the eight-ply sequence and ΔN_x is the change of fringe order in that distance. In this case, the average compressive strain is essentially uniform across the width of the face, i.e., $\Delta N_x / \Delta x$ is constant.

The local strains were determined from measurements of the angular orientation of fringes and their fringe vectors. Equation 4.25a was used, but it was reformulated by interchanging x and y to comply with the notation shown in Fig. 7.8. In derivative form, it becomes

$$\frac{\partial N_x}{\partial x} = \frac{\partial N_x}{\partial y} \tan \phi \qquad (7.3)$$

where ϕ is the acute angle from the horizontal y axis to the fringe vector. The term $\partial N_x / \partial y$ is constant in this case and equal to the carrier fringe frequency, since the load-induced fringes were essentially horizontal before the carrier was introduced. It is determined by measuring a distance Δx corresponding to a ΔN_x of several fringes, e.g., to $\Delta N_x = 20$; again, note that Δx refers to the specimen scale, not the fringe pattern scale. Then, local measurements of ϕ determine the fringe gradient and the compressive strain ε_x. The graph in Fig. 7.8 gives the results, normalized by the average compressive strain. Again the surface strains vary on a ply-by-ply basis, with extremely steep normal strain gradients at interfaces between nonparallel plies.

7.5 Interlaminar Compression, Cross-ply Specimen

Deformation of face A of the cross-ply specimen in interlaminar compression is documented in Fig. 7.9. Two anomalies are evident: the scalloped fringes in the V field which signify γ_{xy} shears near the free edge or corner of the specimen, and the irregular fringe spacings in the U field which signify nonuniform compressive strains ε_x. The latter are investigated in Sec. 7.5.2.

Analysis of the shear strains is a bit more complex than the cases previously encountered. The following paragraphs refer to this edge region—the region of scalloped fringes. Qualitatively, it is seen in Fig. 7.9c that $|\partial N_y/\partial x|$ is largest at the interfaces between crossed plies and diminishes to zero at the midthickness of the ply (or ply group). At the interface, $|\partial N_y / \partial x|$ increases monotonically toward the free edge. The absolute values are indicated because the gradients and the shear strains have opposite signs at the top and bottom of each ply.

In Fig. 7.9b, the gradient $\partial N_x / \partial y$ is zero except very close to the free edge. From there, its magnitude increases monotonically to the free edge, but its sign is opposite to that of $\partial N_y / \partial x$. Their sum reaches its peak value where $\partial N_x / \partial y$ departs from zero. At the free edge, the two gradients are equal in magnitude and opposite in sign and they cancel to yield zero shear strain. The data are consistent with the requirement that the shear stress (and shear strain) must be zero at the free edge.

To evaluate $\partial N_y / \partial x$ from Fig 7.9c, only the fringes in the 90° plies were considered. The fringe pattern was greatly enlarged, the interface lines were marked on the pattern, and vertical lines (parallel to the free edge) were drawn at a series of distances from the edge. A digitizing tablet was used to record the coordinates of intersections of these vertical lines with the integral and half-order fringes in the 90° plies. Then, a curve of fringe order N_y versus position x was plotted for each line and smoothed. The slope of the curve at the interface was measured to provide data points of $\partial N_y / \partial x$. Since the companion derivative $\partial N_x / \partial y$ is zero in the region $y \geq 0.25$ mm, the shear strains were calculated for that region and plotted in (d).

An enlargement of the U field (at the same load level as that for Fig. 7.9c) was used to determine $\partial N_x / \partial y$. Since only one data point was extracted in the range $y < 0.25$ mm, the practical approach was to use Eq. 7.3. Two quantities were required, $\partial N_x / \partial x$ and ϕ, but they could be obtained readily from the fringe pattern. The two cross-derivatives were added and the shear strain was calculated and plotted on the graph. The final step was to draw a smooth curve through the data points.

The results are shown in Fig. 7.9d for four different interfaces. The shear strain (and stress) rises rapidly to its peak value, which occurs very close to the free edge, and then fades to zero at about 1.3 mm from the free edge. The four curves are somewhat different from each other and the differences should be attributed to two factors: experimental error and the inherent variability of the material in nominally equal regions.

Why were data taken only from 90° plies? Equilibrium requires that the shear stresses in the 0° and 90° plies must approach a common value at their interface. Since the shear moduli of 0° and 90° plies are unequal, the shear strains must approach unequal values.

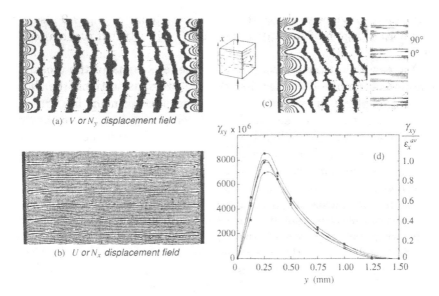

(a) V or N_y displacement field

(b) U or N_x displacement field

(c)

(d)

Fig. 7.9   Deformation of face A of the cross-ply specimen in interlaminar compression. For (a) and (b), the load was 6.67 kN (1500 lb.); for (c) and (d), the load was 13.3 kN (3000 lb.) The graph shows the edge-affected zone near the corner of the specimen. The four curves represent interface shear strains γ_{xy} in four different 90° plies.

The differences are small, but slight discontinuities of fringe paths are visible in Fig. 7.9c near the free edge at the interfaces.

7.5.1 Out-of-plane Displacements

If the plies in the specimen were not bonded, and in the absence of friction, the Poisson effect would cause unequal transverse extensions in the 0° and 90° plies. The 90° plies would bulge, as illustrated in Fig. 7.10a. Continuity is enforced when the plies are bonded. Then the out-of-plane displacements of the specimen surface are continuous and the surface exhibits an array of ridges and valleys. Figure 7.10b represents a deformed cross section; it illustrates the hypothesized shear stresses developed at the interfaces to enforce continuity of the body. The shear strain (and stress) was documented in Fig. 7.9 at the corner of the specimen. The question that arises is whether the same distribution of interface shear persists along the entire width of the specimen.

Figure 7.11 is an interferogram that depicts the surface topology; it documents the array of ridges and valleys. It shows the full width of the specimen. The topography is essentially the same throughout the face, including the corner regions at the ends of the ridges and valleys. It provides evidence that the shears found in Fig. 7.9 also act in a border region of the entire interface, as illustrated schematically in Fig. 7.12.

We conclude that a strong free-edge effect exists in the cross-ply laminates. Unlike the cases discussed in previous sections, the shear strain and stress is zero at the free surface (except near the corners). However, the shears rise rapidly in a border region just inside the free surface and then fade away within a few ply thicknesses. Thus, the shears revealed by the scalloped fringes in Fig. 7.9 also extend behind the free surfaces in a belt region along the perimeter of the interface.

These inferences are qualitative. The magnitude of shears along the perimeter might not be quite the same as those at the corners. Moreover, the ridge-to-valley height (Fig. 7.11) varies randomly over the face and some of the ridges are more pointed than others; these variations are interpreted as anomalies associated with local variations of the material rather than departures from the basic edge-effect phenomenon.

Actually, Fig. 7.11 is taken from a companion study[2] where the stacking sequence was $[90_2/0]_n$ instead of the $[90_4/0_2]_n$ sequence of Fig. 7.9. Nevertheless, it provides the information for the class of problems considered here and it is used to infer the edge effects for both cross-ply specimens.

The out-of-plane contour map was made with a Twyman-Green interferometer (Sec 2.3.7), which was installed in combination with a moiré interferometer as illustrated schematically in Fig. 7.13. Light from the zeroth diffraction order of the specimen grating was used for the out-of-plane observations, and light from the +1 and −1 orders

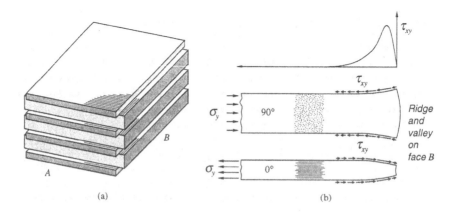

Fig. 7.10   (a) Out-of-plane deformation of an unbonded laminate. (b) Cross-sectional view of deformation and shear stresses in a bonded laminate.

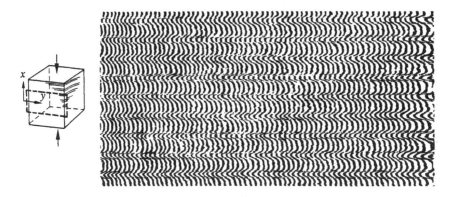

Fig. 7. 11   Surface topography documenting ridges and valleys. The contour interval is 0.257 μm/fringe.

was used for the moiré fields. Incidentally, a pellicle type beam splitter would be a better choice, since the finite thickness of the beam splitter illustrated here can slightly distort the geometry of the specimen.

7.5.2 Normal Strains ε_x, Microscopic Study

Figure 7.9b reveals that the compressive strains ε_x in the 90° plies are greater than those in the 0° plies. Examination with a carrier pattern of rotation revealed fringe curvatures, which indicate nonuniform strains within the plies. Microscopic moiré interferometry was used to quantify the nonuniform strains, i.e., to characterize ε_x in the edge-affected zone.

An immersion interferometer with f = 4800 lines/mm was used and the fringes were processed by the O/DFM method (Chapter 5). Contour maps representing the U displacement field are shown in Fig. 7.14. They were centered on a 0°/90° interface in the central region of face A. The contour intervals are 104 and 26 nm/fringe contour in (a) and (b), respectively, corresponding to 4 and 16 times greater displacement sensitivity than the previous patterns. The technique used for analysis of the patterns will be emphasized in this section.

The load-induced or stress-induced pattern is shown in Fig. 7.14a. The contours are more closely spaced in the 90° ply, which means that the compressive strains are larger than those in the 0° ply. Details of the transition region could be extracted more readily by subtracting off fringes corresponding to the average compressive strain. That was done, approximately, by adding carrier fringes of

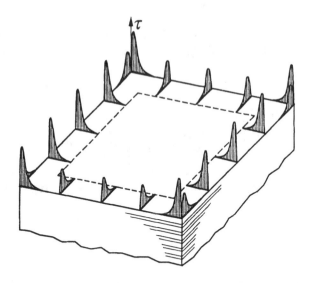

Fig. 7.12   Schematic illustration of shear stress in the border region of the interface.
Shears acting along opposite borders have opposite signs.

extension; in addition, carrier fringes of rotation were introduced to
produce the pattern of Fig. 7.14b. The pattern became more sparse
when the carrier of extension was introduced, and that is why it
became practical to increase the sensitivity by another factor of 4.

The stress-induced contours are given index numbers or contour
orders N_x^σ and the contours representing the stress-induced fringes
plus the carrier fringes are given $N_x^{\sigma,c}$. The carrier fringes of
rotation made it easier to trace the fringe contours across the
interface region. Of course, the fringes of rotation are parallel to line
AA' and therefore, they do not influence the relative fringe orders
along AA'.

The strain distribution along AA' was desired. In the procedure
adopted here, Fig. 7.14b was used to determine the total strain and
then the apparent strain from the carrier fringes was subtracted off.
First, a graph of $N_x^{\sigma,c}$ versus x was plotted, as illustrated in (c); to
facilitate the task, the fringe pattern along AA' was superimposed
upon the x axis as shown and lines were drawn from each fringe
contour to the corresponding $N_x^{\sigma,c}$ position. A smooth curve was
drawn through the plotted points.

Then the gradient $\Delta N_x^{\sigma,c} / \Delta x$ was determined at selected points, as
illustrated for point B. The fictitious strain $\varepsilon_x^{\sigma,c}$ which includes the
apparent strain from the carrier fringes was calculated by

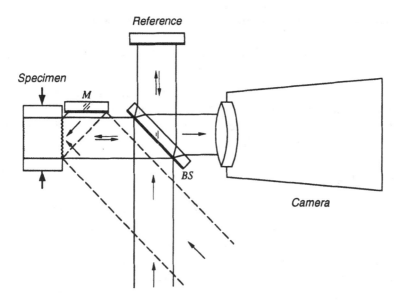

Fig. 7.13   Optical arrangement for in-plane and out-of-plane measurements. The solid lines show the optical paths for Twyman-Green interferometry; the dashed lines show moiré interferometry.

$$\varepsilon_x^{\sigma,c} = \frac{1}{f\beta} M \frac{\Delta N_x^{\sigma,c}}{\Delta x} \qquad (7.4)$$

where M is the magnification of the contour map relative to the specimen, and Δx is measured on the contour map. The result was similar to Fig. 7.14d, but the curve was shifted to the right relative to the coordinate axes; the dots on the curve show the points where data were taken from curve (c).

The stress-induced strain, ε_x^{σ}, is determined by

$$\varepsilon_x^{\sigma} = \varepsilon_x^{\sigma,c} - \varepsilon_x^{c} \qquad (7.5)$$

where ε_x^{c} is the apparent strain from the carrier fringes. Next ε_x^{c} must be evaluated. To do so, the number of contours that cross equal line segments labeled Δx in (a) and (b) were counted. (Any pair of corresponding vertical lines could be used if they both lie in regions where contour orders vary monotonically.) Their difference, adjusted by the sensitivity factor $\beta^{\sigma,c}/\beta^{\sigma} = 4$, was

$$\Delta N_x^{c} = \Delta N_x^{\sigma,c} - 4\left(\Delta N_x^{\sigma}\right) = -11.5 - 4(-22.1) = 76.9$$

The carrier-induced apparent strain in Fig. 7.14b was

$$\varepsilon_x^c = \frac{1}{f\beta} M \frac{\Delta N_x^c}{\Delta x} = \frac{1}{4800(8)} \frac{76.9}{0.2} = 0.0100 \tag{7.6}$$

where $\Delta x / M$ was 0.2 mm. Of course, the carrier pattern and the apparent strain it introduces are both constant over the entire field.

The final step is to subtract ε_x^c according to Eq. 7.5. It was done simply by shifting the strain axis in Fig. 7.14d to obtain the graph of stress-induced strains. The procedure was a highly effective means of extracting the compressive strains in the interface region of the laminate.

In Fig. 7.14d, the experimentally determined strains are drawn as a solid curve and its mirror image is drawn for the next interface as a dashed curve. The symmetry is sketched to convey a qualitative view of the strain distribution. Again, the experimental data show small variations from ply to ply, a condition that is characteristic of the mechanics of composites. The graph reveals a strong strain gradient in the interface region and a modest strain concentration in the 90° ply. The maximum compressive strain is about 17% greater than the strain at the center of the 90° ply, and about 30% greater than the average compressive strain in the multi-ply laminate.

In summary, free-edge effects occur in a belt region near the intersection of the cross-ply interface and the free surface. They are anomalous strains, different from the strains in the interior of the body. Shears are distributed at the ply interface as illustrated schematically in Fig. 7.12. Normal compressive strains occur on the free surface near the ply interface with the distribution of Fig. 7.14d.

7.6 References

1. Y. Guo, D. Post and B. Han, "Thick Composites in Compression: An Experimental Study of Micromechanical Behavior and Smeared Engineering Properties," *J. Composite Materials*, Vol. 26, No. 13, pp. 1930-1944 (1992).

2. D. Post, J. Morton, Y. Wang and F. L. Dai, "Interlaminar Compression of a Thick Composite," *Proc. American Society for Composites*, 4th Annual Meeting (Oct. 1989).

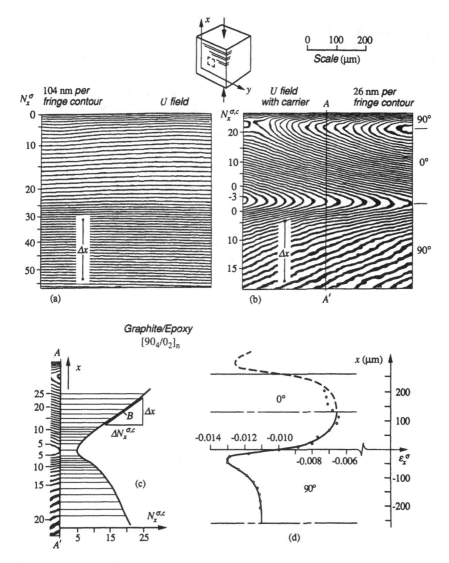

Fig. 7.14   (a) Microscopic moiré interferometry pattern of stress-induced contours, N_x^{σ}. (b) Pattern of load-induced plus carrier-fringe contours, $N_x^{\sigma,c}$. (c) $N_x^{\sigma,c}$ along AA'. (d) Stress-induced compressive strain along AA'.

7.7 Exercises

7.1 Determine the average strain ε_y^{av} from Fig. 7.2. Show or describe the steps in the procedure.

7.2 Determine the shear strain at the center of the circled region in Fig. 7.2 by (a) the graphical procedure of Fig. 7.4 and (b) by the fringe vector procedure. Show or describe all the steps.

7.3 Determine ε_x at the circled region on the left side of Fig. 7.3. (a) and (b) Use both methods outlined in Fig. 7.5. Show or describe all the steps.

7.4 (a) Plot a graph of ε_y versus x across the width of the specimen in Fig. 7.6. (b) State the procedure followed.

7.5 For the graph in Fig. 7.7, it is known from theoretical analyses that $\gamma_{xy} = 0$ at the 0°/90° interfaces. Explain why that result was not obtained in the experiment.

7.6 Describe, in paragraph form, the result depicted in Fig. 7.12.

8

Thermal Stresses Near the Interface of a Bimaterial Joint

8.1 Introduction

The state of elastic strains and stresses in a bimaterial joint subjected to a uniform change of temperature was analyzed by moiré interferometry.[1,2] The specimen configuration is illustrated in Fig. 8.1. The steel and brass plates were joined by a very thin, high-temperature silver-solder film along the mating surfaces. The experimental analysis measured the deformations caused by a change of temperature of $-133°C$. The small gratings at the corners were used to measure the coefficients of thermal expansion

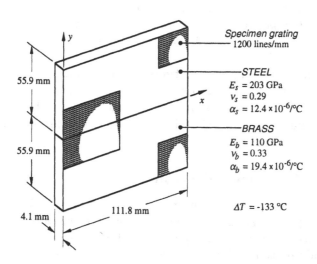

Specimen grating
1200 lines/mm

STEEL
$E_s = 203$ GPa
$v_s = 0.29$
$\alpha_s = 12.4 \times 10^{-6}/°C$

BRASS
$E_b = 110$ GPa
$v_b = 0.33$
$\alpha_b = 19.4 \times 10^{-6}/°C$

$\Delta T = -133 °C$

55.9 mm

55.9 mm

4.1 mm

111.8 mm

Fig. 8.1   Bimetal specimen and specimen gratings.

(contraction) of each material. The large grating that spanned the two materials deformed as a result of two effects: (a) the free thermal contraction of the steel and brass and (b) the state of stress caused by the mutual constraint along the joint interface. The stress-induced strains were determined from the experimental data and the corresponding stresses, σ_x, σ_y and τ_{xy} were evaluated using the known material properties. *Extraordinarily large stress gradients* were detected and these were investigated in detail.

8.2 The Experiments

8.2.1 Experimental Method

A special method was developed for this class of problems.[3] In it, a cross-line grating mold is made on a zero-expansion substrate so that it has the same frequency at room temperature and elevated temperature. The mold is used to replicate the grating on the specimen at elevated temperature. When the specimen cools to room temperature, its thermal deformation is imbedded in the specimen grating.

The mold is also used to tune the moiré interferometer to a null field corresponding to the elevated temperature frequency of the specimen grating. Thus, the moiré pattern observed in the interferometer reveals the absolute change of displacements incurred in cooling from elevated to room temperature.

The method measures the deformation that occurs in the interval between the application of the specimen grating and the subsequent observation of the deformed grating. Thus, the measurements are independent of the state of residual stresses in the specimen. This feature applies to all applications of moiré interferometry. The measurements are independent of any previous state of deformation—deformations prior to the application of the grating—except to the extent that the history of the specimen might alter its mechanical properties.

8.2.2 Relationships and Notation

The following equations of moiré interferometry and elastic thermal stress analysis were used to analyze the experimental results. The results are described in terms of the notation defined here. Additional notation is defined in subsequent sections.

$$U = \frac{1}{f}N_x^t \qquad\qquad V = \frac{1}{f}N_y^t \tag{8.1}$$

$$\varepsilon^t = \varepsilon^\sigma + \varepsilon^\alpha = \varepsilon^\sigma + \alpha\Delta T \tag{8.2}$$

$$\varepsilon_x^\sigma = \frac{\partial U}{\partial x} - \varepsilon^\alpha = \frac{1}{f}\frac{\partial N_x^t}{\partial x} - \varepsilon^\alpha = \frac{1}{f}\frac{\partial N_x^\sigma}{\partial x} \tag{8.3}$$

$$\varepsilon_y^\sigma = \frac{\partial V}{\partial y} - \varepsilon^\alpha = \frac{1}{f}\frac{\partial N_y^t}{\partial y} - \varepsilon^\alpha = \frac{1}{f}\frac{\partial N_y^\sigma}{\partial y} \tag{8.4}$$

$$\gamma_{xy} = \gamma_{xy}^\sigma = \frac{\partial U}{\partial y} + \frac{\partial V}{\partial x} = \frac{1}{f}\left[\frac{\partial N_x^t}{\partial y} + \frac{\partial N_y^t}{\partial x}\right] = \frac{1}{f}\left[\frac{\partial N_x^\sigma}{\partial y} + \frac{\partial N_y^\sigma}{\partial x}\right] \tag{8.5}$$

$$\sigma_x = \frac{E}{1-v^2}\left(\varepsilon_x^\sigma + v\varepsilon_y^\sigma\right) \tag{8.6}$$

$$\sigma_y = \frac{E}{1-v^2}\left(\varepsilon_y^\sigma + v\varepsilon_x^\sigma\right) \tag{8.7}$$

$$\tau_{xy} = G\gamma_{xy} = \frac{E\gamma_{xy}}{2(1+v)} \tag{8.8}$$

where
- N_x^t and N_y^t are fringe orders in the patterns of total displacements
- N_x^σ and N_y^σ are fringe orders in the patterns of stress-induced displacements
- f is 2400 lines/mm (60,960 lines/in.) for the macroscopic analysis
- ε and γ are normal and shear strains
- superscripts t, σ and α represent the total strain, the stress-induced part of the strain, and the free thermal expansion part of the strain, respectively
- α is coefficient of thermal expansion, listed in Fig. 8.1 for the steel and brass
- ΔT is the temperature increment
- σ and τ are normal stress and shear stress
- E is modulus of elasticity, listed in Fig. 8.1
- v is Poisson's ratio, listed in Fig. 8.1, and
- G is shear modulus of elasticity (or modulus of rigidity).

Equations 8.2-8.4 and 8.6-8.8 pertain to the steel and brass portions separately, inasmuch as α, E and ν are different in the two materials. Equations 8.6 and 8.7 are used for the surface of the specimen since there are no normal tractions on the surface. Shear stresses and shear strains are independent of the uniform thermal contraction, so either N^t or N^σ can be used in Eq. 8.5.

8.2.3 Elevated Temperature Replication

The process used to replicate the specimen grating is illustrated in Fig. 8.2. It is different than the standard replication technique of Fig. 4.29 by three minor factors: (a) the mold is a cross-line grating on a zero-expansion substrate; (b) the replication is carried out at elevated temperature in an oven; and (c) the reflective metallic film is applied after replication.

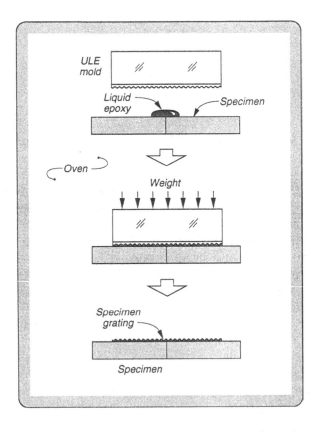

Fig. 8.2   Procedure to produce the specimen grating at elevated temperature. ULE glass is an ultra-low expansion substrate (by Corning Glass Works).

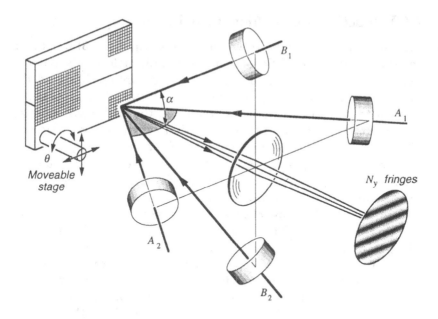

Fig. 8.3 The specimen was moved to view each deformed grating by the moiré interferometer.

The mold was a silicone rubber cross-line grating on a special substrate. It was produced by the procedure described in Sec. 4.9.2, where a photoresist grating was replicated in silicone rubber on a glass substrate. In this case, however, the substrate was a 25 mm (1 in.) thick Ultra-low Expansion (ULE) glass plate (marketed by Corning Glass Works). Its thermal coefficient of expansion is $9 \times 10^{-8}/°C$ in the temperature range of the experiments. The coefficient is two orders of magnitude smaller than that of steel or brass and it can be considered zero for the present purpose.

The steps illustrated in Fig. 8.2 were performed at the oven temperature of 157°C. The specimen and mold were heated to this temperature. Then a small pool of liquid epoxy was poured onto the specimen; the mold was lowered into the pool, but tilted so as to avoid entrapping air bubbles; and a weight was applied to squeeze the epoxy into a thin film. The oven door was opened only briefly to swab the excess epoxy. The epoxy polymerized to a solid in about 1.5 hours and the mold was pried off the specimen after six hours, while both were at the elevated temperature. The separation was clean, because epoxy does not adhere well to silicone rubber. The specimen was allowed to cool to room temperature. Then, an ultra-thin film of aluminum was applied to the grating by evaporation (Fig. 2.10).

8.2.4 Extracting the Coefficients of Expansion

First the interferometer (represented by beams in Fig. 8.3) was tuned to produce null U and V fields, using the ULE mold in place of the specimen. This provided initial conditions corresponding to the specimen grating frequency at the elevated (157°C) temperature. Attention to a practical detail was required; the ULE mold must be rotated 90° with respect to its position when the grating was replicated. The reason is that the lines of the cross-line grating are never precisely perpendicular, and since there is a left-to-right reversal upon replication, compensation for the lack of perpendicularity requires the 90° rotation.

Then the specimen was mounted on a moveable stage so that each of the three gratings could be centered in the moiré interferometer, as indicated in Fig. 8.3. Out-of-plane rotation was monitored by assuring that the zero-order diffractions returned to the source (e.g., to point D in Fig. 4.8). Near the free corners, angular adjustments θ were made to eliminate fringes of rotation, but α was not readjusted. The result is shown in Fig. 8.4a. The fringe gradient $\partial N_y^t / \partial y$ was measured near the free corner and the strain ε^α was determined by Eq. 8.4, where the stress-induced strain ε_y^σ is zero. The thermal coefficient was determined by $\alpha = \varepsilon^\alpha / \Delta T$. This procedure was followed for the free corners in the steel and brass to determine the coefficients listed in Fig. 8.1.

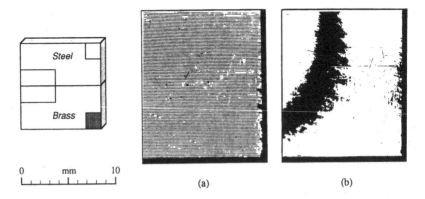

Fig. 8.4   (a) V displacement field near the free corner of the brass. (b) The same V field modified by carrier fringes that nullify the fringe gradient at the free corner.

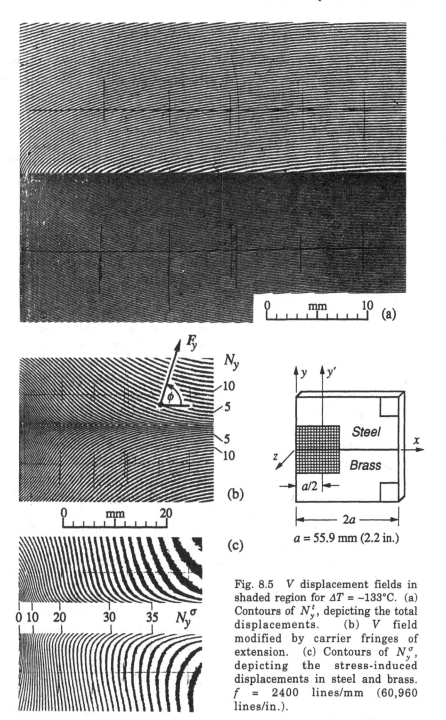

Fig. 8.5 V displacement fields in shaded region for $\Delta T = -133°C$. (a) Contours of N_y^t, depicting the total displacements.   (b) V field modified by carrier fringes of extension. (c) Contours of N_y^σ, depicting the stress-induced displacements in steel and brass. $f = 2400$ lines/mm (60,960 lines/in.).

8.2.5 Fringe Patterns

Next, the large grating that extended across the steel and brass was moved into the field of view of the interferometer. The angular alignment, θ, was adjusted to make $\partial N_y^t / \partial x = 0$ along the interface near the center of the specimen, i.e., the N_y^t fringes were made parallel to the interface. This decision recognizes that displacements must be symmetrical about the vertical centerline. We remember, however, that angular misalignment does not influence the normal or shear strains extracted from the fringe patterns (Sec. 4.12).

The fringe patterns are shown in Figs. 8.5 and 8.6. Figure 8.5a is the total V displacement field, i.e., the N_y^t pattern. The displacements are dominated by the thermal contraction of each part, steel and brass; thus, the gradients $\partial N_y^t / \partial y$ and $\partial N_x^t / \partial x$ are negative everywhere. Of course, the total displacements are continuous across the steel/brass interface. In Fig. 8.5 the stress-induced deformations combine with the free thermal contractions to produce sharp changes of fringe gradients across the interface. The change is emphasized in (b), where carrier fringes of extension were introduced. They have the effect of subtracting off a uniform strain equal to the average (approximately) of the thermal contractions. The change across the interface is rather abrupt. In Fig. 8.6b the carrier fringes are adjusted to prove that the ε_x^t strains in the steel and brass are constant along the interface and identical in the two materials.

In Figs. 8.5c and 8.6c, the carrier fringes introduced a uniform apparent strain equal and opposite to ε^α (or $\alpha\Delta T$), which had the effect of subtracting the free thermal contraction part of the fringe pattern. The result is the patterns of the stress-induced displacements, namely the contour maps of N_y^σ and N_x^σ. The procedure was carried out separately for the upper and lower regions since the magnitudes of α are different for steel and brass. To accomplish the subtraction, the specimen (Fig. 8.3) was translated to view the small grating in the corner of the brass. Then the interferometer was adjusted to produce null U and V fields in the free corner, as shown in Fig. 8.4b. This adjustment canceled the fringes of free thermal contraction. When the specimen was translated back to the main field, the new virtual reference grating frequency produced the stress-induced displacement fields in the brass. The same approach was followed for the steel, to complete the data of Figs. 8.5 and 8.6.

Figure 8.5c shows that the gradients of stress-induced fringes, $\partial N_y^\sigma / \partial y$, have opposite signs in the steel and brass near the interface, and by Eq. 8.4, the strains ε_y^σ have opposite signs. The

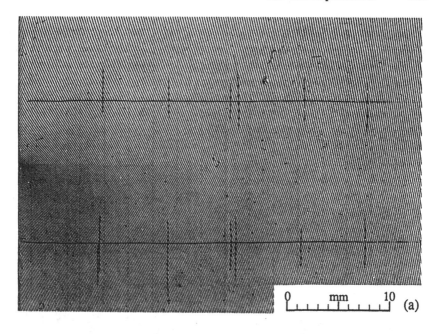

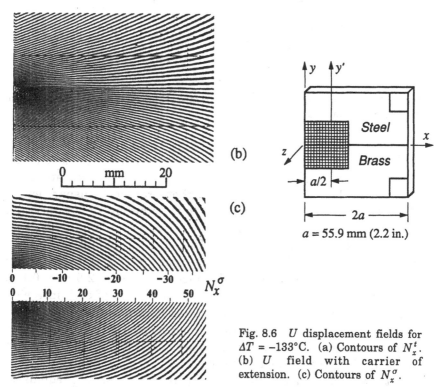

Fig. 8.6 U displacement fields for $\Delta T = -133°C$. (a) Contours of N_x^t. (b) U field with carrier of extension. (c) Contours of N_x^σ.

fringe orders marked in Fig. 8.6c indicate that the stress-induced strain, ε_x^σ, along the interface is compressive in the steel and tensile in the brass; this is consistent with the greater thermal contraction of the brass and the opposing restraint by steel.

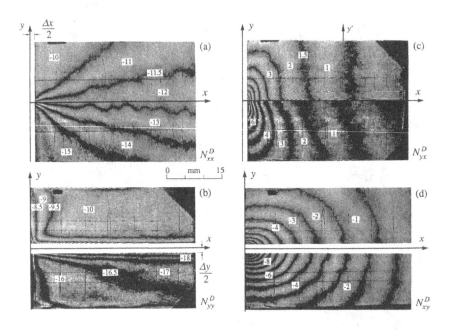

Fig. 8.7   Contours of displacement derivatives by mechanical differentiation. $\Delta x = \Delta y = 2.5$ mm (0.1 in.).

Figure 8.7 shows the corresponding fringes of mechanical differentiation (Sec. 4.22). Using (a) as an example, the pattern was constructed by superimposing two photographic transparencies of Fig. 8.6a, with a shift of one relative to the other by a finite increment Δx. When shifted, the dark lines of the two patterns interweave and create a geometric moiré effect. The same procedure was used to portray all four derivative fields, i.e., shifting each pattern N_x^t and N_y^t by increments Δx and Δy. The increments were 2.5 mm (0.1 in.) in all cases, and the data pertains to the midpoint of each increment. Therefore, zones of width $\Delta x / 2$ and $\pm \Delta y / 2$ are blank in the figures because the parent patterns did not overlap in these zones and no data exist there. These blank zones are only 1.25 mm (0.05 in.) wide, or about 1% of the specimen length.

At any point in the pattern, the fringe order N_{xx}^D equals the difference, ΔN_x^t, of fringe orders N_x^t in the two superimposed transparencies (Sec. 4.22). Thus,

$$N_{xx}^D = \Delta N_x^t = \Delta x \frac{\Delta N_x^t}{\Delta x} \approx \Delta x \frac{\partial N_x^t}{\partial x} \qquad (8.9)$$

The double subscript refers, first, to the displacement component of the fringe pattern and, second, to the direction of the shift. Accordingly, N_{xx}^D is proportional to the finite-increment approximation of the differential. By Eq. 8.3,

$$\varepsilon_x^\sigma \approx \frac{1}{f \Delta x} N_{xx}^D - \varepsilon^\alpha \qquad (8.10)$$

Thus, the total strains were determined from fringe orders by $\varepsilon_x^t = N_{xx}^D / f \Delta x = 164 \times 10^{-6} N_{xx}^D$, and the stress-induced strains by $\varepsilon_x^\sigma = 164 \times 10^{-6} N_{xx}^D - \varepsilon^\alpha$ Note that ε^α is negative for both materials. Similar relationships apply for all four derivatives. The numbers shown on the fringe patterns are N^D values for each derivative field. The finite increment approximation must exactly equal the derivative at one point in the interval and it is very effective at all points in regions where the strain gradient in the interval Δx is either linear or small. In this case, it is a useful means of extracting the derivatives on a whole field basis everywhere except in the immediate vicinity of the interface.

An obvious feature of the strain distribution is the irregular nature of fringes in Fig. 8.7a. A theoretical elasticity solution would prescribe smooth contours, so the irregularities must be associated with some facet of the experiment. They are attributed to small variations of the physical properties of the materials, namely thermal coefficient of expansion, modulus of elasticity and Poisson's ratio. These variations might be heightened in a heat-affected zone near the silver-soldered interface, where the fringe irregularities are most severe. The fringes were smoothed for data interpretation, but such anomalies illustrate the vagaries of real materials.

The largest departure from a smooth N_{xx}^D fringe occurs near the interface in the brass, where the departure is 1/4 fringe order. If a variation of thermal coefficient of expansion in the brass was fully responsible for this irregularity, the local value of α would have changed by $3 \times 10^{-7}/°C$, or by 1.5% of its global value.

8.3 Stress Analysis

The displacement patterns of Fig. 8.5 and 8.6 were analyzed in accordance with Eqs. 8.3-8.5. A detailed analysis was performed along the y' axis at the quarter-width of the specimen. The results for stress-induced strains are plotted in Fig. 8.8. Corresponding strains are substantially different in sign and magnitude for the steel and brass. Near the interface ($y' \approx 0$), the difference of strains ε_x^σ in steel and brass is almost equal to the difference of free thermal contractions; since ε_x^t at the interface is the same in both materials (Fig. 8.6), the difference should be exactly equal. The ε_x^σ and γ_{xy} curves are nearly linear and data extraction was relatively easy. Special attention was given to the determination of the ε_y^σ curves and their values near the interface. These data were extracted from Fig. 8.5a by (a) measuring fringe positions on an enlarged pattern using a digitizing table, (b) calculating the incremental derivative $\Delta N_y^t / \Delta y$ between fringes, and (c) calculating the best fit of a second-order polynomial curve to the high gradient portion of the data. The ε_y strains were not extrapolated to the interface, but remain undetermined, for now, in the narrow zone bounded by $y \approx \pm 100$ μm (0.004 in.), which will be called the *interface zone*.

The corresponding stresses were calculated from the strains by Eqs. 8.6-8.8, which are valid over the free surface of the specimen, where $\sigma_z = 0$. These stresses are plotted in Fig. 8.9. The stress distributions are remarkably similar in the steel and brass. The

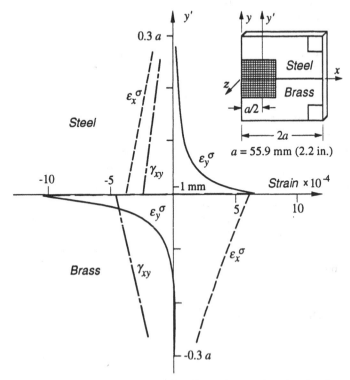

Fig. 8.8   The distribution of stress-induced strains at the y' axis along the free surface.

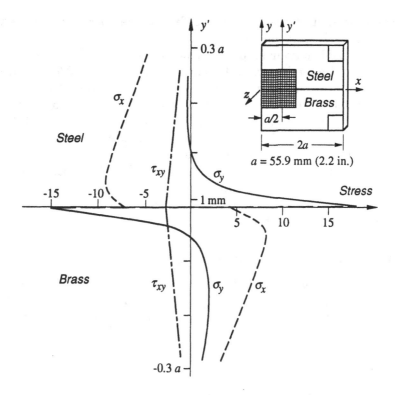

Fig. 8.9   The distribution of stresses along the y' axis. The units on the stress scale are psi when the numbers are multiplied by 1000 and MPa when multiplied by 6.9.

shear stresses are nearly symmetrical and exhibit equal magnitudes at the interface. Normal stresses σ_x have similar distributions, but opposite signs.   The difference between interface values does not violate equilibrium. Normal stresses σ_y exhibit similar distributions, but with opposite signs.   These stresses were not extrapolated through the interface zone; if they were, a violation of σ_y equilibrium would appear. The σ_y stresses in the interface zone will be discussed in subsequent sections.

A detailed analysis of the fringe patterns, Fig. 8.5 and 8.6, was also performed along the length of the interface zone. The stresses plotted in Fig. 8.10 represent σ_x and τ_{xy} *at the interface* and the peak values of σ_y, which occur *near the interface*. They illustrate that the stress disturbance at the specimen corner is highly localized—within about 1% of the specimen length for the normal stresses and 3% for the shear stress. Elsewhere, the normal stresses are constant and the shear stresses vary linearly along the interface length. Of course, these represent stresses at the surface of the specimen, where the measurements were made.

Whole-field contour maps of the surface stress fields are shown in Fig. 8.11. The mechanical differentiation data of Fig. 8.7 were used, except in the high gradient region.   For the shear stress determination, the graphical process of Fig. 3.10 was used to add the cross-derivatives (Fig. 8.7c and d) and construct the map of shear strains.   Equations 8.6–8.8 were used to calculate the stresses. In the high gradient region, stresses were calculated using data from the moiré patterns of Figs. 8.5 and 8.6. Figure 8.11b illustrates that σ_y is very small except near the interface and near the corner. Near the interface and away from the free corner, the σ_y distribution is essentially independent of x.

8.3.1 σ_y within the Interface Zone

In order to assess the distribution of σ_y in the immediate vicinity of the interface, the sharpness of the chevron-like fringes of Fig. 8.5b must be evaluated. The fringes are clearly delineated outside the

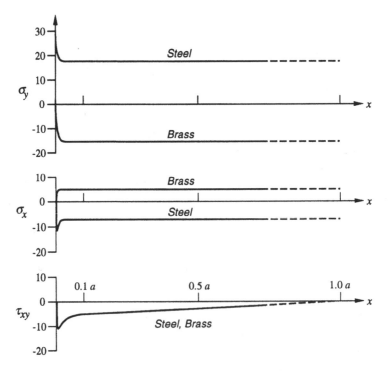

Fig. 8.10   The distribution of peak stresses near the interface along the free surface. Stress units are specified in Fig. 8.9.

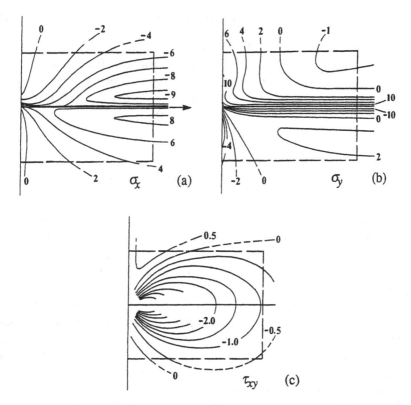

Fig. 8.11   Contour maps of surface stress distributions in 30 x 40 mm portion of the specimen. The contour numbers denote stresses in psi when multiplied by 1000 and MPa when multiplied by 6.9.

interface zone and the corresponding stresses, σ_y, are established in Fig. 8.9. Within the tiny interface zone, however, the paths of the fringes are inconclusive. If the clearly delineated fringes in the steel and brass are extrapolated to the interface with a monotonically increasing inclination, (i.e., increasing ϕ), or else with a constant inclination, then the peak values of σ_y in both materials would occur at the interface, $y = 0$. A physically inadmissible condition would result, where σ_y at each interface point would have two distinct values and where vertical equilibrium would be violated.

On the other hand, if the fringes change their course and *bend toward the interface*, the curve of σ_y versus y' would reach peak positive and negative values at some small distance from the interface. The curve would then turn sharply away from the peaks and progress in the interface zone with finite slopes between the peak values. The relationship between the angle ϕ of the fringe vector (Fig. 8.5b) and σ_y is derived in Ref. 4, with the result

$$\sigma_y = k \tan \phi - k_1 \qquad (8.11)$$

where k and k_1 are constants. Equation 8.11 applies in the interface zone as well as outside the zone. Thus, the *continuation* of the σ_y curve of Fig. 8.9 into the interface zone depends upon the course of the chevron-like fringes in that zone. Additional experimental evidence was sought by microscopic moiré interferometry to ascertain the course of the fringes.

8.4 Experiments by Microscopic Moiré Interferometry

The same specimen as that of Fig. 8.1 was used, except its width and height were cut to half the previous dimensions. The specimen grating was applied at elevated temperature, as before, and observed at room temperature to reveal the thermal deformation. The specimen grating was applied at the center as depicted in Fig. 8.12. This change from the quarter-width position of y' is permissible since σ_y is independent of x in these regions. Replication of the specimen grating in the small region of interest was achieved with a much smaller grating thickness, approximately 2 μm thickness in this case. Accordingly, the shear lag that occurs through the grating thickness in the vicinity of high strain gradients is now expected to affect a very narrow zone, where the effect can be significant within ±2 μm and vanishes by about ±5 μm from the interface.

The resulting fringe contours are shown in Fig. 8.12, which represent the thermal deformations plus carrier fringes. The carrier fringes were applied again to emphasize the changes near the interface. Carrier fringes of extension were applied to subtract the average (approximately) of the fringe gradients in the steel and brass, as before. Carrier fringes of rotation were applied to create a vertical bias in the pattern and thereby delineate the fringe contours across the interface. The interface is indicated by the broken line in Fig. 8.12. Its location relative to the contour map was determined by microscopic inspection.

The patterns show a distinct decrease of $|\phi|$ in a 50 μm (0.002 in.) zone, which proves that the curves of σ_y for the steel and brass turn around from their peak values and move toward each other. The distribution of σ_y is plotted in Fig. 8.13b, where the vertical scale is greatly magnified. The axis is marked y', but the results apply for most of the interface zone as indicated in Figs. 8.10 and 8.11b. The dashed line recognizes the remaining uncertainty in the high gradient region. The region of uncertainty is now diminished by an order of magnitude, and a monotonic variation of σ_y within the region

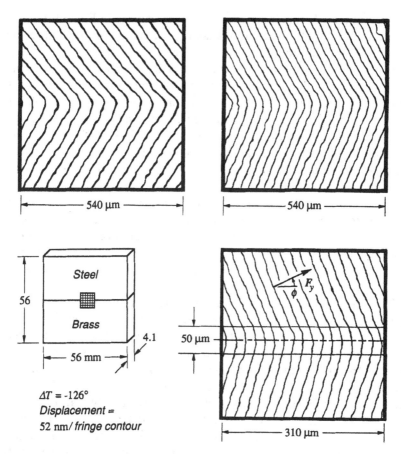

Fig. 8.12   Contour maps by microscopic moiré interferometry depicting N_y^* plus carrier fringes.

can be rationalized with substantial confidence. The microscopic analysis indicates a σ_y curve near the interface that is always single-valued, i.e., the curve has finite slope, and the potential equilibrium imbalance is eliminated. The stress gradients are very high, but finite. The microscopic analysis proves the continuity of the σ_y stresses in the interface region. It documents that the peak stresses occur near the interface (but not on it), and it documents a severe stress gradient across the interface.

8.5 Comments

Note that $\partial N_y^* / \partial x$ is not affected by the fringe curvature in the interface zone. Note that there is no corresponding uncertainty for the N_x fringes. Accordingly, the σ_x and τ_{xy} curves in Fig. 8.9 extend to the interface, $y' = 0$. Note, too, that the stress magnitudes all remain within the elastic range for these materials.

It is interesting to observe the similarity of stress distributions in Fig. 8.9, 8.10, and 8.11 for the steel and brass elements. When the bimaterial body experiences a temperature change, one element tends to expand relative to the other. The dominant constraint that restricts the relative expansion is a system of shear forces. At each point on the interface the shear forces acting on the steel and brass are equal in magnitude and opposite in direction. If we consider the steel and brass parts separately, and if we consider *only* these shear forces, we see that the parts have identical geometry and exactly opposite force systems. Since the stresses in an elastic body depend only upon the forces and body geometry, these forces generate equal stress distributions in the steel and brass, but of opposite signs. Indeed, the results in Figs. 8.9–8.11 show approximately equal and opposite stress distributions in the steel and brass.

The theoretical singularities predicted in the literature cannot occur in real joints that maintain their structural integrity. Nevertheless, it is seen that the experimentally determined stresses feature rapidly rising tensile and compressive stresses on opposite sides of the interface. Then they reach finite peaks, followed by an extremely rapid decrease of the normal stress $|\sigma_y|$. Their magnitudes decrease precipitously in an extremely narrow zone, a zone of less than 50 μm (0.002 in.) width. This very strong tension/compression stress system occurs along the entire length of the interface and it can be assumed to prevail along the entire perimeter of the joint. It is a free-edge effect, acting in a narrow, shallow belt of material near the junction of the interface and free surface of the body.[2] It is like the free-edge effect found in Fig. 7.14, but since the transition of material properties is much more abrupt in the steel/brass specimen, the edge-effect is more severe.

A localized but strong disturbance was observed in the corner region where two free edges intersected. The experiments represent an analysis of a fully three-dimensional, elastic, bimaterial thermal stress problem. Both macroscopic and microscopic moiré interferometry were required to extract the stress distribution.

The presence of solder as a third material does not seem to be significant. We accept that the silver solder behaves elastically at these stress levels. The turn-around points of the σ_y curve (Fig. 8.13b) are at least three times more distant from the centerline of the solder material than the boundary of the solder. Therefore, based on St. Venant's principle, essentially the same results would be expected in the turn-around region for a theoretical bonding agent of zero thickness.

Although the tests addressed thermal stresses, the results have much broader implications. In many cases, the stresses in bimaterial bodies subjected to mechanical loading differ from those in corresponding thermal stress problems merely by a constant. The mechanism responsible for theoretical singularities in both cases is the differential expansion at the bimaterial junction, caused by unequal thermal expansions in one case and unequal Poisson's ratios in the other.

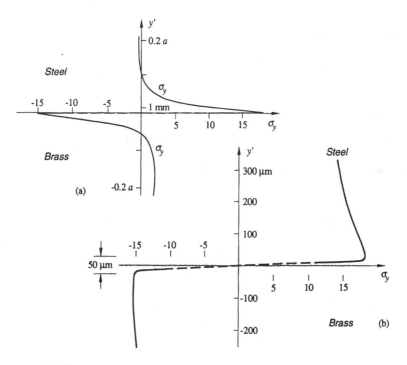

Fig. 8.13   (a) The σ_y stresses along y'. (b) Expanded vertical scale showing the σ_y peaks near the interface and a severe stress gradient between the peaks. The numbers denote stresses in psi when multiplied by 1000 and MPa when multiplied by 6.9.

8.6 References

1. J. D. Wood, M.-Y. Tsai, D. Post, J. Morton, V. J. Parks and F. P. Gerstle, Jr., "Thermal Strains in a Bimaterial Joint: An Experimental and Numerical Analysis," *Proc. 1989 SEM Spring Conference on Experimental Mechanics*, pp. 543-551, Society for Experimental Mechanics, Bethel, Connecticut (1989).

2. D. Post, J. D. Wood, B. Han, V. J. Parks and F. P. Gerstle, Jr., "Thermal Stresses in a Bimaterial Joint: An Experimental Analysis," ASME *J. Applied Mechanics*, Vol. 61, No. 1, pp. 192-198 (1994).

3. D. Post and J. D. Wood, "Determination of Thermal Strains by Moiré Interferometry," *Experimental Mechanics*, Vol. 29, No. 3, pp. 318-322 (1989).

4. J. D. Wood, "Determination of Thermal Strains in a Bimaterial Joint," Ph.D. Dissertation, Virginia Polytechnic Institute and State University, Blacksburg, Virginia (1992).

8.7 Exercises

8.1. Section 8.2.4 prescribes a 90° rotation of the ULE mold to adjust the interferometer for null fields. Explain with words and sketches why the rotation is required.

8.2 Plot ε_y^σ versus y'' from Fig. 8.5a, where y'' lies at $x = 15$ mm.

8.3 Plot σ_y versus y'' using data from Figs. 8.1, 8.5 and 8.6; y'' lies at $x = 15$ mm.

8.4 Plot ε_x^σ versus y'' from Fig. 8.6; y'' lies at $x = 15$ mm.

8.5 Plot σ_x versus y'' using data from Figs. 8.1, 8.5, and 8.6; y'' lies at $x = 15$ mm.

8.6 Plot ε_y^t and ε_y^σ versus y'' from Fig. 8.7; y'' lies at $x = 15$ mm.

8.7 Determine the frequency of carrier fringes in Fig. 8.5b.

8.8 Plot γ_{xy} versus x'' from Fig. 8.7; x'' lies at $y = 5$ mm.

8.9 Plot τ_{xy} versus x'', from Fig. 8.7; x'' lies at $y = 5$ mm.

9
Textile Composites

9.1 Introduction

Although it was projected more than a decade ago that laminated composite materials would become the material of choice, aluminum is still the primary structural material for aircraft. High manufacturing costs, poor out-of-plane properties, and low damage tolerance[1] have been limiting factors for laminated composites. In response to these shortcomings, a new generation of composite materials is being evaluated. This class of materials is called advanced textile composites. They have the potential for superior through-the-thickness properties, damage tolerance,[2] and cost effectiveness.

Advanced textile composites are differentiated from other textile composites by their high fiber volume fraction and by the use of advanced fibers such as graphite and aramid. Textile composites have various types of fiber architecture, including braids, weaves, knits, and stitched fabrics. They can be made so that the fiber reinforcement forms a three-dimensional network giving desired properties in specific directions, much like laminates but additionally providing superior through-the-thickness properties. Since textile composites are geometrically complex, mathematical modeling has proved to be difficult. With the currently small data base and inherent modeling difficulties, the potential use of advanced textile composites hinges on a focused, comprehensive experimental effort. Moiré interferometry provides a valuable tool to investigate how textile composites deform and how failures initiate and propagate. The work presented here is part of an ongoing comprehensive moiré study of structural textile composites conducted at the NASA Langley Research Center.[3-5]

9.2 Woven Textile Composites

A 2-D plain weave textile composite in tension was investigated to determine the degree of inhomogeneity of deformation. The material system consisted of AS4 graphite fibers in a 3501-6 epoxy resin matrix. The plain weave architecture is illustrated in Fig. 9.1, where the smallest repeating unit of one layer or ply is shown. The plain weave architecture consists of yarns in two orthogonal directions that alternately pass over and under each other. The yarns are large bundles of fibers, i.e., they contain a great number of individual fibers. The yarns that emerge perpendicular to the weaving loom are called warp yarns while the parallel yarns are called weft or fill. Composite panels were produced by resin transfer molding, whereby fiber preforms consisting of ten layers of plain weave fabric were impregnated with resin and cured with heat and pressure. The panels were consolidated textile composites with a fiber volume of about 60%. The designation 2-D, or two dimensional, means that no fibers are present to connect successive layers of fabric. For the material tested, the smallest repeating unit, known as the unit cell, is about 2.5 mm (0.1 in.) square. The unit cells do not stack in an orderly fashion but rather they nest randomly. By examining cross-sections of the consolidated composite it was determined that both the warp

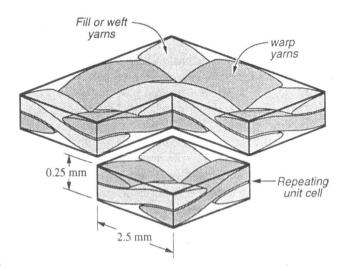

Fig. 9.1   A plain weave textile composite specimen was evaluated using moiré interferometry. The fabric architecture and smallest unit cell are shown.

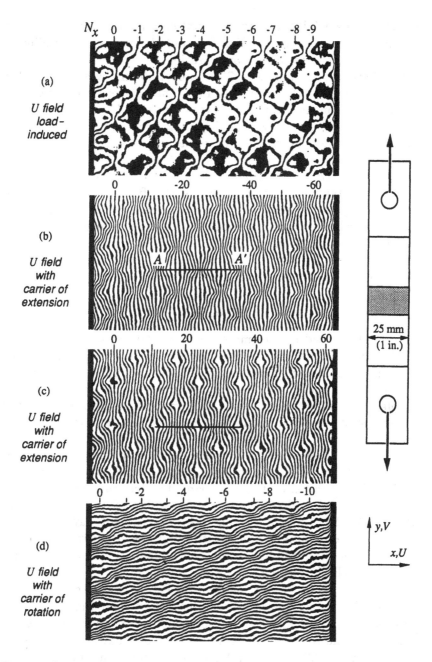

Fig. 9.2   Patterns depicting the transverse displacement field for the plain weave tensile specimen.   Various carrier patterns have been added to aid in data reduction. f = 2400 lines/mm.

and weft yarns exhibit the same amount of crimp (or curvature). The thickness is exaggerated in Fig. 9.1 and the yarns (or bundles of fibers) actually have a flatter configuration.

The tensile specimen was 25.4 mm (1.00 in.) wide and had a nominal thickness of 2.5 mm (0.10 in.). The heterogeneous nature of the deformation is best illustrated by the U, or transverse displacement field shown in Fig. 9.2, revealing the Poisson contraction. The load-induced fringe pattern is shown in Fig. 9.2 a. More detail is evident and potential ambiguities are eliminated by adding carrier fringes. The fringe pattern in (b) is the U field with a carrier of extension[6] added, whereby the fringe orders decrease monotonically in the positive x direction. In (c), the carrier causes the fringe orders to increase with x. The same deformation with a carrier of rotation is shown in (d). The location of the zero-order fringe was chosen arbitrarily in each case. Useful concepts of data reduction can be expressed with this example.

9.2.1 Data Reduction and Results

The procedure used to determine the transverse strain ε_x along line AA' (Fig. 9.2b) is described. The fringes are sufficiently close together in (b) and (c) that their gradient $\partial N_x / \partial x$ can be approximated by the finite increment $\Delta N_x / \Delta x$. An enlarged image of the fringe pattern was placed on a digitizing tablet. The center of each black fringe was located visually using a magnifying eyepiece with a cross-hair. The digitizer recorded the x coordinate of each fringe along AA' and stored the coordinates and corresponding fringe orders in a personal computer. A simple algorithm calculated Δx and the fringe gradient $\Delta N_x / \Delta x$ for each pair of neighboring fringes and its value was assigned to the midpoint $\Delta x / 2$. The algorithm proceeded by subtracting the uniform carrier-induced gradient $\Delta N_x^c / \Delta x$ from each measured gradient and multiplying the difference by $1/f$ to obtain the load-induced strain at each midpoint. These values are plotted as circles in Fig. 9.3.

The same procedure was followed for Fig. 9.2c and the individual values are plotted as squares. It is interesting to note that the fringes are closest together in (c) at the locations where they are furthest apart in (b), and thus provide data points where they are most useful. The curve in Fig.9.3 was drawn through both sets of data points.

The scatter shows strain deviations of about 100 µm/m. The strain distribution from one unit cell to the next is much less consistent. The lack of repeatability is characteristic of composites, wherein nominally equal structures exhibit somewhat different deformations.

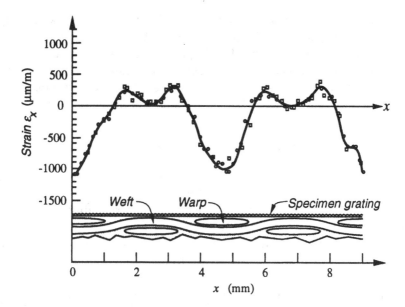

Fig. 9.3   Strain distribution along line AA', which lies at the center of one transverse yarn.

In the present case, the strain deviations at corresponding points of different unit cells are much larger than the scatter from experimental measurements, and efforts to refine the procedure would not be justified.

The curve in Fig. 9.3 reveals that the strain changes from positive to negative in a cyclic manner which corresponds to the fiber architecture.

The result can be seen in Fig. 9.2d, too, which is the U field pattern with a carrier of rotation. Where the fringe slope is positive the strain is compressive, where the slope is zero (i.e., where the fringe is horizontal) the strain is zero, and where the slope is negative the strain is tensile. The average slope is positive and likewise the average strain is compressive. The quantitative data could be extracted from (d), using the angles of fringe vectors and Eq. 4.25. However, the approach described above is easier for these patterns.

Figure 9.3 includes a cross-sectional representation of the portion of the specimen that corresponds to the strain plot. Compressive strains occur in regions where the fibers aligned in the y (or loading) direction are closest to the surface. The tensile strains are located above the regions where the fibers aligned in the x (or transverse) direction are near the surface. The Poisson contraction causes the transverse yarns to bend. As the yarns (fiber bundles) snake through the composite the outside radii undergo tension, while the inside radii undergo compression. The highest tensile strains appear to be located where transverse yarns exhibit the most curvature. This result was not anticipated before the test, but the behavior of the material was rationalized after analysis of the fringe patterns.

9.3 Braided Textile Composites

A study of 2-D triaxial braided textile composite materials was performed to determine how architecture parameters influence their mechanical properties. Braided composites are being assessed as materials for primary aircraft structures, including stiffeners and stringers. Moiré interferometry is being used in this program for quantitative evaluation of the strain inhomogeneities that occur with tensile loading.

The composite consisted of AS4 graphite fibers and 1895 epoxy resin matrix. The fiber preform was produced by triaxial braiding to achieve the architecture illustrated in Fig. 9.4. In the fabric, each braider yarn passes over two opposing braiders and then under two. The axial yarns are covered above and below with braider yarns. Successive layers of fabric nest randomly. The result is a layered preform with no through-the-thickness reinforcement. The preforms were flattened and border stitched to avoid yarn shifting. Then the

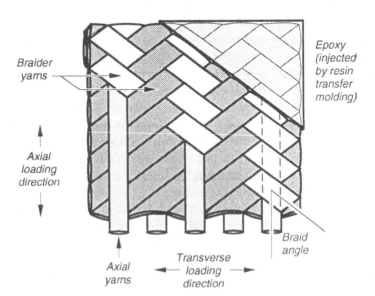

Fig. 9.4   The 2-D triaxial braid fiber architecture.

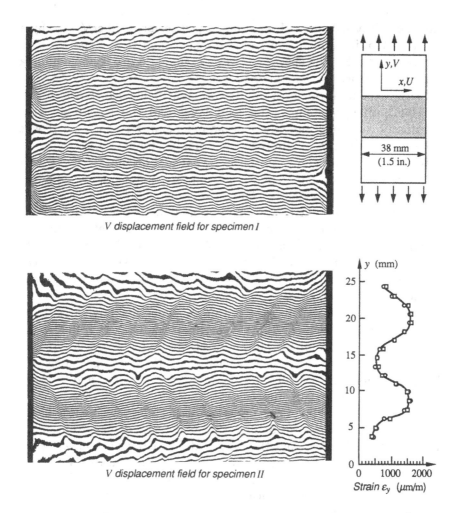

Fig. 9.5 Displacement fields for two 2-D triaxial braids under transverse tensile
loading. The strains along the vertical direction exhibit cyclic variations.
$f = 2400$ lines/mm.

resin transfer molding process was used to impregnate the yarns
with the epoxy matrix material and cure the composite. The braider
yarns snake through the composite in a nominally sinusoidal
manner while the axial yarns exhibit only minor curvatures.

Two transverse tensile specimens were tested; transverse means
that the tensile loads were applied perpendicular to the axial yarns.
Specimen I had 6k braider yarns (6,000 fibers/yarn) with a braid
angle of 70°, and 30k axial yarns. Specimen II had 15k braiders at 70°
and 75k axial yarns. The textile architecture of the specimens was

the same, but specimen II was coarser, in terms of fibers/yarn, by a factor of 2.5; in terms of unit cell size, specimen II was coarser by a factor of about 2. The specimens were 38 mm (1.5 in.) wide and had a nominal thickness of 3.2 mm (0.13 in.).

The moiré patterns of the V displacement fields are shown in Fig. 9.5 for both specimens. The load level and the thickness were nominally the same, but the average extension for specimen II was nearly 10% larger. Whereas the displacement field would be represented by straight, uniformly spaced horizontal fringes for uniform extension, the fringe spacings are not constant and the fringes exhibit waviness. The scale and location of the fiber architecture is evident in the displacement fields. The alternating regions of high and low fringe densities indicates that the normal strain ε_y varies in a cyclic manner. The graph in Fig. 9.5 shows the variation along a vertical line near the center of specimen II. The cyclic strain distribution is a global response that extends over most of the specimen, but changes near the free edges.

Both of the specimens exhibit this response although it is more pronounced for specimen II, the specimen with large architecture. The ratio of maximum to minimum strain ε_y for specimen II was about twice as great as that for specimen I. For both specimens, the high strains occur over the axial yarns (which are aligned transverse to the loading direction).

Since the unit cells of the top layer do not align with those of the underlying layers, we rationalize that the cyclic deformation of the top layer is different from the deformation in the next lower layer. An implication is that large shear strains must act between the layers to accommodate their relative movements.

In addition to the high normal strains ε_y in regions over the axial yarns, high shear strains occurred between the braider yarns in the same regions. Figure 9.6 shows an enlarged portion of the V and U fields for specimen II, with point A located in a high-strain region. Notice the anomaly along a 70° angle through point A, where the fringes are perturbed in both fields. This is the region directly above the adjoining edges of two neighboring braider yarns. Although high shears are evident along the x,y coordinates, one can anticipate even higher shears along the 70° direction, i.e., along the edges of the braider yarns. These off-axis strains can be determined once the strains in the x,y coordinate system are evaluated.

The methods illustrated in Fig. 7.5 were used to determine the x,y strains. Tangent lines were drawn (Fig. 9.6) and the increments Δx and Δy were measured for each field. Then, the four derivatives ($\partial N_y / \partial y$, $\partial N_y / \partial x$, etc.) were determined and the strains were calculated

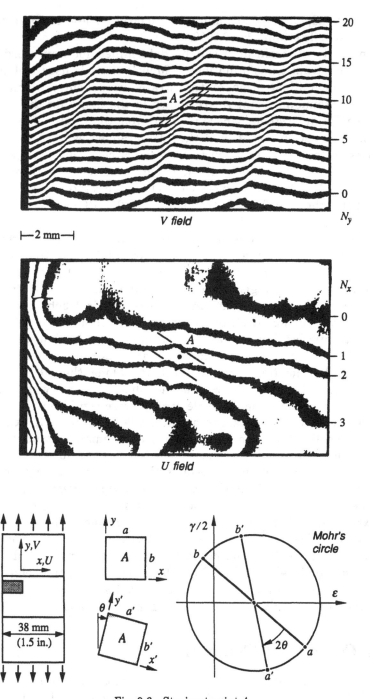

Fig. 9.6 Strain at point A.

as $\varepsilon_y = 0.0023$, $\varepsilon_x = -0.0003$, and $\gamma_{xy} = -0.0022$. These compare to the average tensile strain $\varepsilon_y^{av} = 0.0010$.

The state of strain at A can be determined for any coordinate system by strain transformations. Mohr's circle is used in Fig. 9.6 for graphical illustration of the strain components. Elementary blocks surrounding point A are shown in the figure for the x,y and x',y' coordinate systems, where y' is 70° from the x axis. The strains ε_y and γ_{xy} exist on face a, and they are graphed as a on the circle. Strains on face b are graphed as b on the circle. The corresponding strains for faces a' and b' are found by a 2θ rotation where θ is depicted by the elementary blocks as a clockwise rotation of 90° − 70° = 20°. Thus, the transformed strains are $\varepsilon_{y'} = 0.0013$, $\varepsilon_{x'} = -0.0007$, and $\gamma_{x'y'} = -0.0034$.

The shear strain along the 70° braid angle is nearly the maximum shear strain at A, and it is about 50% larger than ε_y at A. The maximum shear strain is also influenced by the registration (nesting) or lack of registration of underlying layers, so it does not necessarily occur at the 70° angle. The physical behavior of the material was rationalized based on evidence from the moiré tests. Under tensile loading the braider yarns attempt to slide past each other, but shear stresses in the matrix constrain the movement. Upon further loading, cracks developed along the 70° direction (represented by face b') where the shear strains were nearly maximum and the normal tensile strains were significant.

The braid with the larger architecture, specimen II, exhibited damage earlier than the specimen with the smaller structure. In fact, damage was not seen in specimen I until the load was twice the load of damage initiation in specimen II. The initial damage for both materials was in the form of cracks between the braider yarns.

The unusual strain distributions found in this series of tests was unexpected, and at the time of the experiments they were not predicted. The results served as a starting point for computer modeling to predict the mechanical behavior of such materials. The desire was to develop models that adequately predict the behavior in the interior of the textile composite as well as at the surface, and account for the sensitivity to the scale of the unit cell of the composite.

9.4 Open Hole Tension

A series of tests were performed to determine the strain distributions in 2-D braided open hole tensile specimens. The main objectives were to determine the distribution and severity of the strain concentration, and to document damage initiation and progression. Two architectures were tested, both 2-D triaxial braids. Specimen I had 15k braider yarns, 36k axial yarns and a 45° braid angle; specimen II had 15k braider yarns, 75k axial yarns, and a braid angle of 70°. The tensile loads were applied in the direction of the axial yarns. Both specimens were 50.8 mm (2.00 in.) wide and had a nominal thickness of 3.2 mm (0.13 in.), with a hole diameter of 9.53 mm (0.375 in.).

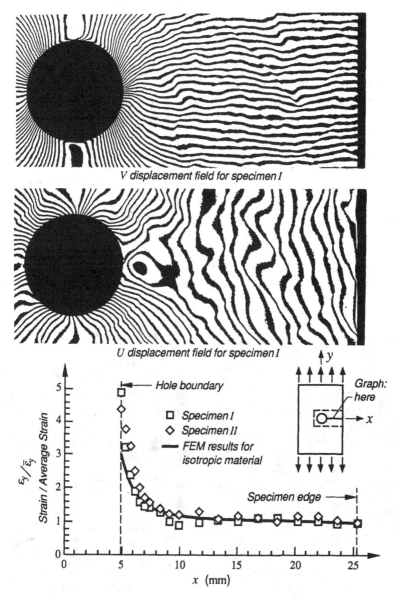

Fig. 9.7   The displacement fields around a hole in a triaxial braid composite. f = 2400 lines/mm. Hole diameter = 9.53 mm (0.375 in.).

Figure 9.7 shows the displacement fields around the hole for specimen I. The strain concentration is evidenced by high fringe gradients near the hole. The graph shows the ε_y strain distribution along the horizontal centerline of the specimen. Both of the specimens are represented. The two specimens, and also a Finite Element Method (FEM) solution for an isotropic material, exhibit a similar global response. The braided materials, however, show a higher strain concentration. In a series of tests with different specimens, the strain concentration factors varied between 3 and 6, depending upon the location of the hole with respect to the unit cell.

In axial tension experiments on braided materials without holes, the maximum ε_y strain was about twice that of the average ε_y strain. These strain variations occur within each repeating unit cell. The strain concentration factors found in the moiré tests are sensitive to the position of the hole relative to the unit cell. When the hole boundary coincides with a compliant part of the unit cell, the strain concentration rises. Of course, the maximum strain on the outer surface might not be as high as the strain at the hole boundary for an interior layer. The problem is very complex. For moiré measurements on composite laminates (not textiles), Boeman[7] found that the free-edge effects along the cylindrical surface of a hole are very different from the free-edge effects at a straight boundary.

As the load level increased, cracks propagated from the hole along the braid angles. The onset of cracking did not lead to catastrophic

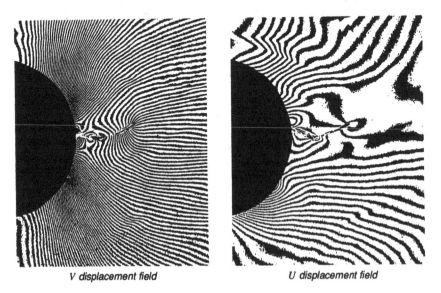

V displacement field   U displacement field

Fig. 9.8   Crack formation at the hole (2-D triaxial braid, specimen II).

failure of the specimen; instead, cracking occurred over an extended load range. The first cracks, which occurred in the matrix of the braider yarns, were detected well before half of the ultimate load. The axial yarns continued to carry a major part of the tensile load. Figure 9.8 shows the fringe patterns of an enlarged region of specimen II where a crack had formed. Surface cracks occurred between braider yarns and followed paths along the junctions between two braider yarns. In each case, the surface crack would terminate at the intersection of the next axial yarn, which acted as a temporary crack stopper. Subsequent loading would produce crack extension until global failure occurred. Surface cracks are illustrated in the figure, but in some cases the first crack occurred in an interior layer of fabric.

9.5 Compact Moiré Interferometer

From an experimentalist's point of view, these experiments are unique because of the instrumentation. The compact moiré interferometer described below was used, and it proved to be very effective. Two such interferometers were put into serious practice at federal research laboratories.[8]

These experiments on textile composite materials were conducted in a materials testing laboratory rather than an optics laboratory. A large capacity screw driven testing machine was employed to load the specimens. The moiré interferometer rested on a platform attached to the testing machine.

The interferometer is illustrated in Fig. 9.9. Its design is based on the optical arrangement of Fig. 4.20, which is shown again in Fig. 9.10 with an identification of elements corresponding to those in Fig. 9.9. The instrument has a circular field of view of 45 mm (1.8 in.) diameter. The overall dimensions are 255 mm (10 in.) width, 280 mm (11 in.) height, and 180 mm (7 in.) depth. It creates a virtual reference grating of 2400 lines/mm (60,960 lines/in.).

Light from a helium-neon laser (632.8 nm) is brought to the interferometer by an optical fiber (1), which terminates at the fiber chuck (2). The diverging light passes through a concave lens (3) which increases the divergence. The cone of light is directed by mirror (4) through a hole in the support structure and redirected by mirror (5) to the central axis of the system. There it is directed by a 45° double-sided mirror (6), it is rendered parallel by the collimating lens (7), and it reaches the 1200 lines/mm crossed line diffraction grating (8). The collimated beam strikes the diffraction grating normal to its surface. It is divided into four first-order diffracted

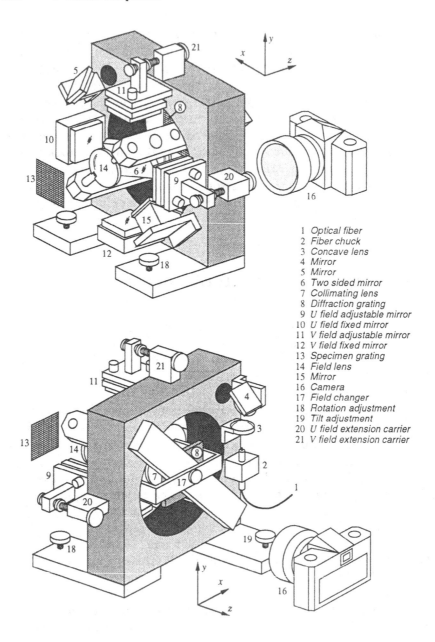

1 Optical fiber
2 Fiber chuck
3 Concave lens
4 Mirror
5 Mirror
6 Two sided mirror
7 Collimating lens
8 Diffraction grating
9 U field adjustable mirror
10 U field fixed mirror
11 V field adjustable mirror
12 V field fixed mirror
13 Specimen grating
14 Field lens
15 Mirror
16 Camera
17 Field changer
18 Rotation adjustment
19 Tilt adjustment
20 U field extension carrier
21 V field extension carrier

Fig. 9.9   The compact moiré interferometer used in these experiments.

beams, plus the zero-order (which is not used); the diffraction angles for the second-order and higher-order beams would exceed 90° (Eq. 2.54) and these diffractions do not exist. The four beams are redirected by four mirrors (9, 10, 11, 12) toward the specimen grating

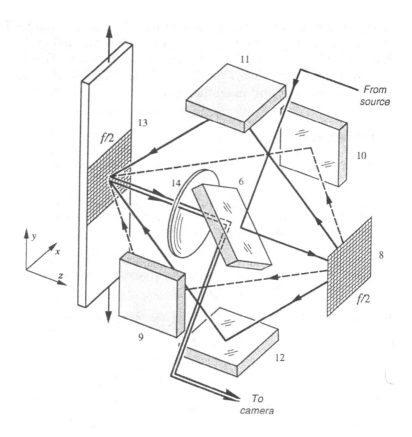

Fig. 9.10   Optical arrangement.

(13). Mirrors (9) and (10) form the U field virtual reference grating
and (11) and (12) form the V field reference grating. A baffle (17), can
be swiveled to block the beams of either the U field or the V field. The
incident beams are diffracted by the specimen grating to emerge
along the z axis. The emergent light is collected by a field lens (14), it
is reflected at the mirrorized back side of the 45° two-sided mirror (6),
and it propagates toward the camera after reflection by mirror (15).
The camera, which is focused on the specimen surface, captures the
fringe pattern. The use of a field lens in the imaging system is
illustrated in Fig. 2.25. The interferometer includes all the usual
features for moiré interferometry, including provisions to control
carrier fringes of extension and rotation.   Carrier fringes of
extension are introduced by (20) and (21), which also permit precise
recovery of the initial field when the carrier is removed.

A 10 mW laser was used with a 2 m length of optical fiber. The typical photographic exposure time was 1/60 sec using Kodak Technical Pan 35 mm film with 1:1 image magnification. The 35 mm camera was equipped with a long focal-length lens and extension bellows to achieve the unit magnification.

9.6 References

1. R. M. Jones, *Mechanics of Composite Materials*, McGraw Hill, New York (1975).

2. F. Scardino, "An Introduction to Textile Composites and Their Structures," *Textile Structural Composites*, T. W. Chou and F. K. Ko, editors, Elsevier Science Publishing Co., New York, pp. 1-26 (1989).

3. P. G. Ifju, J. E. Masters and W. C. Jackson, "Using Moiré Interferometry to Aid in Standard Test Method Development for Textile Composite Materials," *Composites Science and Technology*, Vol. 58, pp. 155-163 (1995).

4. R. A. Naik, P. G. Ifju and J. E. Masters, "Effect of Fiber Architecture Parameters on Deformation Fields and Elastic Moduli of 2-D Braided Composites," *Journal of Composite Materials*, Vol. 28, No. 7, pp. 656-681 (1994).

5. P. G. Ifju, "Shear Testing of Textile Composite Materials," *ASTM Journal of Composites Technology and Research*, Vol. 17, No. 3, pp. 199-204 (1995).

6. Y. Guo, D. Post and R. Czarnek, "The Magic of Carrier Patterns in Moiré Interferometry," *Experimental Mechanics*, Vol. 29, No. 2, pp. 169-173 (1989).

7. R. G. Boeman, *Interlaminar Deformations on the Cylindrical Surface of a Hole in Laminated Composites: An Experimental Study*, Center for Composite Materials and Structures, Report 91-07, Virginia Polytechnic Institute and State University, Blacksburg, Virginia (1991).

8. D. Mollenhauer, P. G. Ifju and B. Han, "A Compact Versatile Moiré Interferometer," *Journal of Optics and Lasers in Engineering*, Vol. 23, pp. 29-40 (1995).

9.7 Exercises

9.1 Determine the frequency of the carrier pattern (a) in Fig. 9.2b and (b) in Fig. 9.2c.

9.2 Referring to Fig. 9.2, determine ε_x at the center of AA'. Describe the measurements and procedure.

9.3 Referring to Fig. 9.5, explain why ε_y is considered essentially independent of x.

9.4 Determine the magnitude and direction of the maximum tensile strain at point A in Fig. 9.6.

9.5 (a) Determine the peak strains, ε_y, at both the left and right sides of the hole in Fig. 9.7. (b) Why are they unequal?

9.6 From Fig. 9.8, determine the crack-opening displacement (V displacement) at the point located $a/3$ from the crack tip, where a is the crack length.

9.7 List advantages of the system of Fig. 9.9 relative to the system of Fig. 4.8.

10
Thermal Deformations in Electronic Packaging

10.1 Introduction

An understanding and documentation of the micromechanics of electronic devices is critical to the industry, and moiré interferometry is taking a leadership role for experimental analyses. In the industry, the technology dealing with the interconnections of electronic components is called *electronic packaging*. Thermal strains are the major cause of fatigue failures of interconnections. The strains result from the mismatch of the thermal expansions of the various elements comprising the package. As the components and structures are made smaller, the thermal gradients increase and the strain concentrations become more serious. Hence, there is a continuously increasing activity in experimental analysis, both for specific studies and for guidance of numerical programs.

The problems are very complex since the structures have complicated geometry and the assemblies are composed of various materials. Nevertheless, the thermal strain techniques introduced in Chapter 8 have proved versatile and effective. The equations of elastic thermal strains are given in Chapter 8. In the present chapter, the patterns without carrier fringes represent total strains, which are the stress-induced strains plus $\alpha\Delta T$. This chapter illustrates several applications and describes specific techniques developed for the studies. In many cases, macroscopic observations by moiré interferometry are followed by microscopic studies of critical details by the methods of Chapter 5.

The examples will deal with *first level* and *second level* packaging.[1] At the first level, active components in the form of large scale integrated circuits on silicon *chips* are interconnected with *chip carriers*. The chip carriers provide electrical pathways to complete electrical circuits within the chip, and to channel input and output signals between larger circuit boards and the active chip. Second level packaging involves the electrical interconnections between the large circuit boards and any number of chip carriers.

10.2 Grating Replication

In the usual procedure for thermal loading (Sec. 8.2), a cross-line grating is replicated on the specimen at an elevated temperature, the grating deforms as the specimen cools to room temperature, and the deformation is recorded at room temperature.

Electronic components require a special technique to replicate the specimen grating because the cross-section usually has such a complex geometry that the procedure for removing excess epoxy cannot be used inside an oven while maintaining a constant

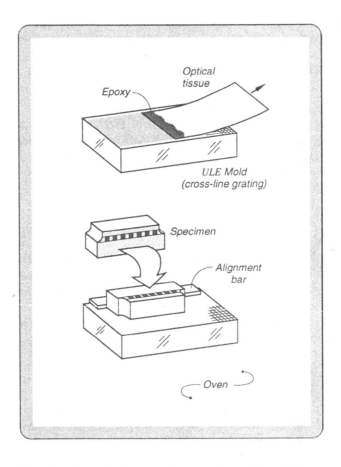

Fig. 10.1   Procedure to replicate a specimen grating at elevated temperature. A thin layer of epoxy is spread on the heated mold by dragging an optical tissue.

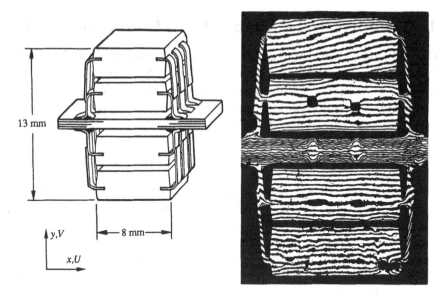

Fig. 10.2   V displacement field of a stacked memory module. The specimen grating was replicated at 82°C and the replication procedure produced very clean edges. Courtesy of Y. Guo (IBM Corp.).

temperature.[2]   The excess epoxy is critical inasmuch as it could reinforce the specimen and change the local strain distribution.

Figure 10.1 illustrates an effective scheme. First, an epoxy grating mold is produced on a ULE glass substrate by replication from a silicon rubber cross-line grating. The mold is double coated by ultra thin layers of aluminum which later act as a parting agent. A tiny amount of liquid high-temperature-curing epoxy is poured onto the heated mold; the viscosity of the epoxy is extremely low at elevated temperature. Then, a lintless optical tissue (a lens tissue) is dragged over the surface of the mold, as illustrated in Fig. 10.1. The tissue spreads the epoxy and absorbs the excess, to produce a very thin layer of epoxy on the mold. The specimen is pressed gently into the epoxy, and it is plied off after the epoxy has polymerized (in 1 to 2 hours). Before polymerization at elevated temperature, the surface tension of the epoxy pulls the excess epoxy away from the edges of the specimen and the normal cleaning procedure is not needed. The result is a specimen grating with very clean edges.

An example is illustrated in Fig. 10.2. The specimen is a stacked memory module. Its cross-section was ground and polished to produce a flat, smooth surface. The specimen grating was replicated at 82°C and the fringes were recorded at room temperature. Very clean edges of the specimen grating are evident.

10.3 Controlled-Collapse Chip Connection (C4)

Controlled-collapse chip connection technology, called C4 technology, has evolved through the years as a reliable and high volume assembly technique for increasingly complex chips. In the technique, solder bumps are deposited on wettable metal terminals on the active chip. A matching footprint of solder-wettable terminals is on the chip carrier. The chip and chip carrier are stacked in registration and all the solder connections are made simultaneously by reflowing the solder.[3] The assembly is subjected to repeated thermal cycles in its normal service. The thermal expansion mismatch between the chip and chip carrier is a critical parameter because it creates deformations and stresses at each solder joint, which can cause electrical and mechanical failures.

The dominant deformation of C4 joints is shear strains produced by the thermal expansion mismatch. Normally, the magnitude of the shear strain on each solder bump is proportional to its distance from the *neutral point*, which is at the geometric center in the case of a symmetrical device. Consequently, the size of the device and the coefficients of thermal expansion of the materials must be restricted to prevent premature failure. However, a redistribution of strains and a dramatic improvement in the fatigue life of C4 assemblies were achieved by filling the gaps between the solder bumps with an epoxy encapsulant (which has a coefficient of thermal expansion nearly equal to that of solder). Thermal strains were measured and analyzed for packages of this type.

10.3.1 Ceramic Chip Carrier

In this first level package, an active silicon chip is attached to a ceramic chip carrier by C4 interconnections. A specimen grating was replicated on the cross section of an assembly at 122°C and the deformation incurred during cooling was observed at room temperature ($\Delta T = -100$°C). Figure 10.3 depicts the U and V displacement fields obtained by macroscopic moiré interferometry. The relative horizontal displacements between the chip and ceramic substrate can be determined from the U field pattern. The largest relative displacement occurred at the end of the assembly and it was about 0.4 μm. The corresponding average shear strain is 0.4%, which was determined by dividing the relative displacement by the height of the interconnection layer (100 μm). The term $\partial N_y / \partial x$ was neglected since it is obviously very small. The average shear strain is zero at the center and it increases gradually until it rises rapidly near the

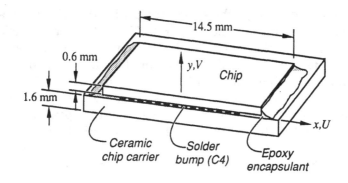

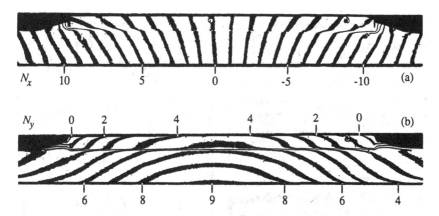

Fig. 10.3   *U* and *V* displacement fields for a ceramic chip carrier assembly, induced by thermal loading of $\Delta T = -100°C$. $f = 2400$ lines/mm.

end of the assembly.   The nonlinear distribution confirmed the expected behavior, wherein shears in an adhesive joint are concentrated at the ends.

Whereas shear strains dominate in the case of no encapsulant,[4] the patterns indicate relatively low shears, but they show that substantial bending occurred.   The *U* field shows greater compressive strains at the bottom of the assembly than at the top. From the *V* field, the relative vertical displacement between the center and the end of the chip was about 1.8 μm.   This bending was ascribed to the presence of the encapsulant, which increased the stiffness of the interconnection between the chip and ceramic, and thus increased the coupling between them.   As the coupling increases, the flexural deformation increases and the shear deformation diminishes. The shear strains are attenuated.

Microscopic moiré interferometry was applied to the same specimen to record localized deformations.   The *U* and *V* displacement fields around the solder bump with the largest distance from the neutral point are shown in Fig. 10.4, where carrier fringes of extension and rotation were added. The contour interval used here was 17 nm for the U field and 35 nm for the V field, which corresponds to fringe multiplication factors, β, of 12 and 6, respectively. In the V field, the closely spaced contours in the encapsulant represent the thermal expansion of the epoxy and solder bump normal to the chip surface. Although the location of the solder bump is evident in the U field, the presence of the epoxy significantly reduced the shear strain concentration that would otherwise occur at this location.

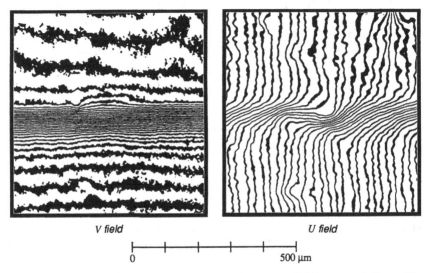

V field   U field

├───┼───┼───┼───┤
0   500 μm

Fig. 10.4   Microscopic (a) U and (b) V displacement fields around the solder bump that was most distant from the center, for the specimen of Fig. 10.3. Fringes of extension and rotation were added. The contour interval is 17 nm and 35 nm for the U and V fields, respectively.

10.3.2 Organic Chip Carrier

With the improvement afforded by the epoxy encapsulant, the C4 technique was extended for an organic chip carrier package. The chip carrier was a glass/epoxy laminate, which has a much larger coefficient of thermal expansion than the ceramic chip carrier considered above.

In Fig. 10.5, (a) and (b) depict the U and V displacement fields of the assembly, where the deformations were induced by thermal loading of $\Delta T = -60°C$. The smaller ΔT was used because the thermal expansion of the laminate was so much larger.

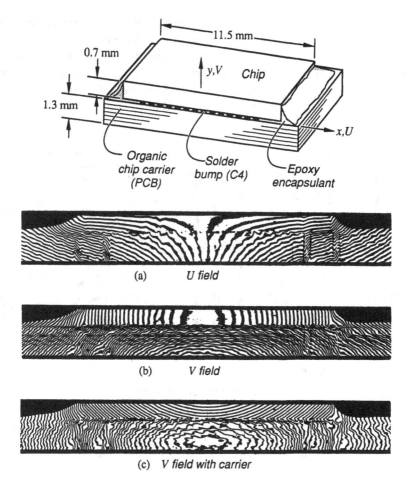

Fig. 10.5 Moiré fringes for an organic chip carrier assembly, induced by thermal loading of $\Delta T = -60°C$; (a) U field. (b) V field. (c) V field with carrier fringes of extension.

The zigzag nature of fringes in the organic chip carrier is not caused by optical noise. Instead, it represents the heterogeneous deformation of the multi-ply glass/epoxy composite, which also has holes for electrical connections.

The patterns report very significant bending of the assembly. The U field shows compressive strains at the bottom of the assembly and essentially zero strain at the top. The V field shows curvature of the active chip, with 11 µm greater upwards displacement at the center relative to the ends of the chip. The curvature greatly exceeds that of the ceramic package. The increased bending can be attributed to the larger coefficient of thermal expansion of the PCB material and its smaller bending stiffness.

Figure 10.5c shows the V displacement field with carrier fringes of extension. The carrier makes it more obvious that the fringe gradients are nearly the same at the top and bottom of the assembly, which means that the curvatures of the chip and chip carrier are nearly the same. Thus, there was a high degree of coupling, which moved the highest shear strains to the end of the joint beyond the last solder bump.

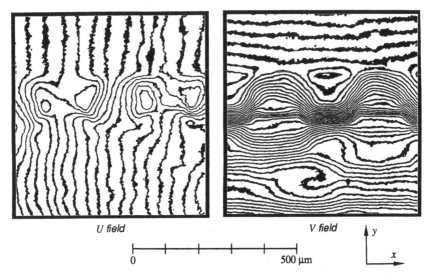

U field   V field

0   500 μm

Fig. 10.6   Microscopic U and V displacement fields around two solder bumps that were most distant from the neutral point, for the specimen of Fig. 10.5. Carrier fringes of extension and rotation were added. The contour interval is 52 nm.

Figure 10.6 shows the microscopic moiré fringe contours around the two solder bumps with the largest distance from the neutral point. Carrier fringes of extension and rotation were added to minimize the complexity of the patterns. The contour interval was 52 nm. The patterns show the details of deformation of the solder bumps. The largest shear strains were approximately 0.7% in the solder-epoxy layer. Unlike Fig. 10.4, the normal strains (ε_y) vary substantially in the solder-epoxy layer, such that the stress-induced ε_y is also of the order of 0.6%.

10.4 Solder Ball Connection

The solder ball connection (SBC) is a second level interconnection technology in which multilayer ceramic modules are connected to a

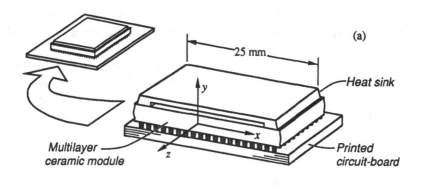

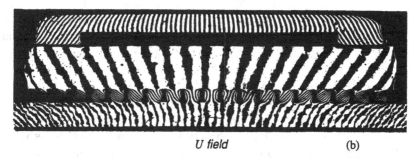

U field   (b)

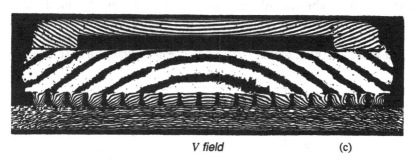

V field   (c)

Fig. 10.7   (a) Schematic diagram of the solder ball connection assembly. (b) U
and (c) V displacement fields, induced by thermal loading of $\Delta T = -60°C$.
Courtesy of Y. Guo (IBM Corp.).

printed circuit board (PCB). The ceramic module is a subassembly of
one or more active chips on a ceramic chip carrier. The solder ball
connection is an array of nonhomogeneous solder columns. They are
nonhomogeneous inasmuch as they consist of two materials, a (90
tin/10 lead) high temperature solder ball and eutectic solder fillets
which have a lower melting point. The eutectic solder connects the
solder balls to the module and to the printed circuit board during the
reflowing process. Low cycle thermal fatigue tests have revealed

failures of the solder fillets. Thermal deformations were measured by moiré to acquire a better understanding of the thermomechanical behavior.

The specimen is a strip cut from a 25 mm SBC assembly, as illustrated in Fig. 10.7a. The ceramic module is 2.8 mm thick and the printed circuit board is 1.6 mm thick. An aluminum heat sink is attached to the module by an epoxy adhesive. The solder balls have a diameter of 0.89 mm and a pitch of 1.27 mm. The actual stand-off between the module and the board is 0.94 mm after the soldering process. Thus, the columns are an order of magnitude longer than those of first level packaging.

A specimen grating was replicated at 82°C and the deformation was measured at room temperature ($\Delta T = -60°C$). Before replicating the specimen grating, the specimen was kept at the elevated temperature for about 3 hours to let the solder joints creep toward a stress-free state.

In Fig. 10.7, (b) and (c) depict the U and V displacement fields of the assembly. The relative horizontal displacements between the top and bottom of the solder columns were determined from the U field pattern and they are plotted in Fig. 10.8. If the solder columns were totally relaxed, there would be no mechanical force applied by the solder. Then, the module and printed circuit board would deform freely and their relative displacements would be the distance from the neutral point multiplied by $\Delta\alpha\Delta T$, where $\Delta\alpha$ is the mismatch of the coefficients of thermal expansion. These relative displacements

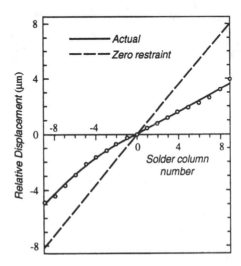

Fig. 10.8   Relative horizontal displacements across the height of the solder columns. The dashed line shows the relative displacements for unrestrained thermal expansion.

are shown by the dashed line in Fig. 10.8. The difference in the curves indicates that the assembly is mechanically constrained by stresses remaining in the solder columns.

Figure 10.9 illustrates another example of a SBC package. In this case, smaller solder balls (250 µm) were used for higher density interconnections. Microscopic moiré interferometry was utilized to determine the thermal deformation of a single solder column (most distant from the neutral point) subjected to $\Delta T = -60°C$. The microscopic U and V fields are shown in Fig. 10.9, where the contour interval is 52 nm ($\beta = 4$). The shear-dominant deformations are basically the same as those found in Fig. 10.7 for the larger columns. The nature of the deformation is governed largely by the aspect ratio of the column, rather than its size.

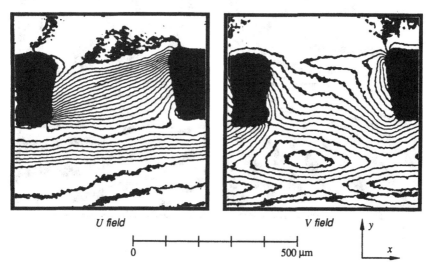

U field   V field y

0   500 µm x

Fig. 10.9   Microscopic moiré fringe contours for a smaller solder ball interconnection, induced by a thermal loading of $\Delta T = -60°C$. 52 nm per fringe contour.

10.5 Low Temperature Application

Very few electronic packages are designed to *operate* at low temperatures, but all must survive the cold that might be encountered in storage, shipping, or during power outages. Another factor becomes important at low temperatures, namely the nonlinear mechanical properties of some of the materials used in electronic packaging, especially solders and adhesives. The interconnection

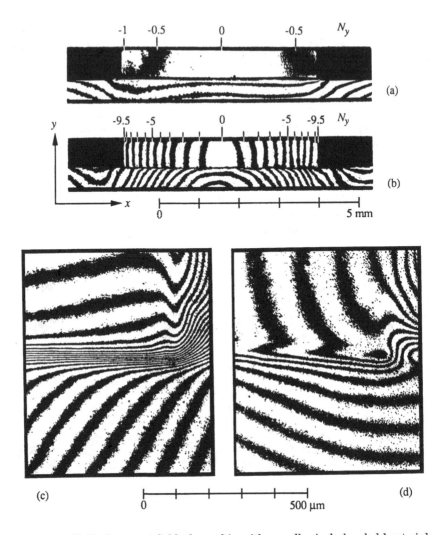

Fig. 10.10 V displacement fields for a chip with an adhesively bonded heat sink, induced by cooling the assembly from (a) 80°C to 20°C and (b) 80°C to –40°C; $f = 2400$ lines/mm. (c) and (d) U and V fields, respectively, at –40°C, near the corner of the chip; $f = 4800$ lines/mm, $\beta = 1$.

materials restrain the free thermal expansion (or contraction) of the major components, and their nonlinear strength, stiffness, and expansivity must be taken into account to ensure circuit dependability.

A compelling example of nonlinearity is illustrated in Fig. 10.10. The specimen is an active silicon chip with an aluminum heat sink adhesively bonded to its lower surface. Macroscopic patterns for the V field are shown in (a) and (b). In (a) the temperature excursion is

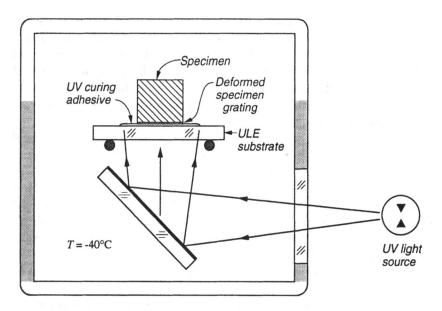

Fig. 10.11 Arrangement to replicate the deformed grating on a ULE substrate at –40°C. The adhesive is polymerized by ultraviolet light.

$\Delta T = -60°C$, from 80°C to 20°C. In (b), $\Delta T = -120°C$, from 80°C to –40°C. If all the materials had constant properties, independent of temperature, the deformation would be twice that in (a). However, the bending deformation of the chip is an order of magnitude larger in (b). The strong nonlinearity of the structural behavior is ascribed to the temperature dependent property of the adhesive layer. Below 0°C, Young's modulus of the adhesive is an order of magnitude greater than at room temperature. Consequently, the coupling and bending were greatly increased at the low temperature.

The deformation at –40°C was investigated by microscopic moiré interferometry. In Fig. 10.10, (c) and (d) depict the U and V fields, respectively, near the corner of the bonded joint. The U field shows a large relative displacement of the chip and heat sink.   The accumulated shear strain in the adhesive reached 3.9%, which was close to its ductility limit.

10.5.1 Experimental Technique: Replication of Deformed Grating

Normally the specimen grating is replicated with epoxies that cure at room temperature or elevated temperature. Such materials do not polymerize at low temperatures.   Instead, adhesives that are polymerized by ultraviolet (UV) light have been effective for replication at low temperatures. The UV irradiation takes place in a cold chamber as illustrated in Fig. 10.11. It is obvious that the normal procedure—whereby the mold is coated with an opaque film of aluminum to act as a parting agent—cannot be used since the coating would block the UV light. Consequently, another procedure was developed for low temperature tests.

First, a specimen grating is applied to the specimen in the normal way (Fig. 10.1) at either room temperature or elevated temperature. Then, the specimen is cooled to the desired low temperature and the deformed specimen grating is copied, or replicated, at that temperature by the method of Fig. 10.11. The replica is on a (nearly) zero-expansion substrate, so it does not change when it is brought to room temperature. Thus, the stable replica is interrogated at room temperature in a moiré interferometer.

The technique follows the more general concept of Sec. 4.21.1. The silicone rubber specimen gratings suggested there are not as practical for electronic components since it is more difficult to produce very thin gratings with very clean edges. Instead, the following procedure was used for the experiment of Fig. 10.10(b-d):

- the specimen grating was replicated in epoxy at 80°C (Fig. 10.1) and the specimen was cooled to room temperature
- another layer of evaporated aluminum was applied
- a thin layer of UV curing adhesive was spread with an optical tissue (Fig. 10.1) on a *smooth* ULE glass plate
- the specimen grating was pressed lightly into the liquid adhesive and the mold and specimen were transferred into a cold chamber
- when thermal equilibrium at the desired low temperature was reached, the adhesive was polymerized by UV light (Fig. 10.11)
- the specimen was gently pried off, the ULE substrate was heated to room temperature, and the deformed grating was viewed in the moiré interferometer. The interferometer was tuned with the original ULE grating mold, which represented the undeformed specimen grating at 80°C.

The high quality of the patterns in Fig. 10.10(b-d) corroborates our expectation that replication of deformed gratings is a viable and attractive technique. It is reassuring to see that this applies to both macroscopic and microscopic moiré interferometry.

10.6 Coefficient of Thermal Expansion

Knowledge of the coefficients of thermal expansion (CTE) is extremely important for accurate predictions of thermal strains in

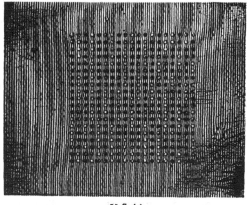

U field

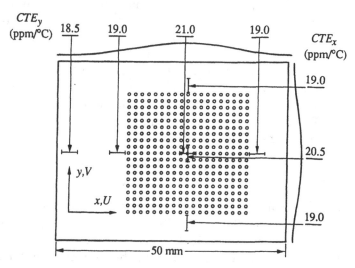

Fig. 10.12   Coefficients of thermal expansion for a multilayer printed circuit board. (a) U field, f = 2400 lines/mm. (b) Distribution of CTE. Courtesy of Y. Guo (IBM Corp.).

interconnections between active components, chip carriers, and printed circuit boards. Moiré interferometry is an exceptionally attractive technique for the measurements since the overall CTE and the local CTE can be determined across the entire body. Figure 10.12 illustrates its application to a multilayer printed circuit board.

A multilayer circuit board is a very complex laminate containing layers of metal conductors separated by sheets of insulating material. In the case of Fig. 10.12, the insulators are a woven-glass/epoxy laminate and the conductors are copper. Each layer of copper is patterned into numerous ribbons such that each ribbon provides an electrical pathway between specific points in the circuit board. After the layers are laminated together, an array of holes is drilled and their cylindrical surface is made conductive by copper plating. They are called *plated through holes*, which interconnect various interior conductors and also provide sockets or pads for surface-mounted components. In this case, the holes are filled with solder, which serves as pads for solder ball connections.

As a result of the complex internal geometry of copper and the presence of plated through holes, the CTE varies across the circuit board. For the 50 mm board shown in the figure, a cross-line specimen grating was applied at elevated temperature across the entire surface and the deformed grating was viewed at room temperature. The change of temperature was $\Delta T = -50°C$. The fringe pattern is the U field, but both the U and V fields were recorded to survey the thermal expansion. The closely spaced fringes map the thermal expansion. A pattern of broad low-contrast fringes can be discerned, too, and these are supermoiré fringes that occur from the interaction of the moiré fringes and the texture of the woven fabric. They should be disregarded. The effect of the texture could have been eliminated, if desired, by the double replication procedure (Sec. 4.9.1).

The coefficient of thermal expansion, in a temperature range $T_2 - T_1 = \Delta T$, is determined from the moiré fringes by

$$CTE(\text{ppm}/°C) = \frac{1}{f} \frac{\Delta N_x}{\Delta x \Delta T} \times 10^6 \qquad (10.1)$$

where Δx is any gage length across which ΔN_x is determined, and ΔT is measured on the Celsius scale. Of course, an equivalent expression applies for the V field. The accuracy is governed by the accuracy of the individual parameters, but for this material, an accuracy of ±0.3 ppm/°C is readily achieved when the gage length is $\Delta x = 5$mm and ΔT is 50°C. The accuracy increases proportionally with increased gage length, i.e., the potential error is inversely proportional to the gage length.

Measurements were made at several locations in the U and V fields and the corresponding CTE values are shown in the figure. They varied from 19.0 to 21.0 ppm/°C in the printed circuit region. Since surface mount components are assembled over different locations of the printed circuit board, this variation must be included in numerical models for stress predictions.

10.7 References

1. J. W. Dally, *Packaging of Electronic Systems: A Mechanical Engineering Approach*, McGraw Hill Publishing Co., New York (1990).

2. Y. Guo, W. T. Chen and C. K. Lim, "Experimental Determination of Thermal Strains in Semiconductor Packaging Using Moiré Interferometry," *Proc. 1992 Joint ASME/JSME Conference on Electronic Packaging*, American Society of Mechanical Engineers, New York, pp. 779-784 (1992).

3. N. Koopman, T. Reiley and P. Totta, "Chip-To-Package Interconnections," Chap 6, *Microelectronics Packaging Handbook*, R. Tummala and E. Rymaszewski, Editors, Van Nostrand Reinhold, New York, pp. 366-391 (1989).

4. J. Clementi, J. McCreary, T. M. Niu, J. Palomaki, J. Varcoe and G. Hill, "Flip-chip Encapsulation on Ceramic Substrates, "*Proceedings of 43rd Electronic Components & Technology Conference*, IEEE, New York, pp. 175-181, (1993).

10.8 Exercises

10.1 Referring to Fig. 10.7, sketch the deformed shape of the aluminum heat sink and the ceramic module.   Exaggerate the deformation.   (b) On separate diagrams, show the major tractions acting on each part. (c) What assumptions were made?

10.2 Figure 10.9 is for a solder ball interconnection at the right side of an assembly of a ceramic module (at top) and a PCB (at bottom). (a) Plot a graph of the shear strain γ_{xy} versus y along the vertical centerline.   (b) Describe the distribution of shear strains in the ceramic module and the PCB.   Explain why the shear strains in the PCB are high.

10.3 In Fig. 10.9, the condition $\gamma_{xy} = 0$ appears to be violated at the free boundary near the upper-right corner of the solder interconnection. What is the probable explanation?

Illustrative Solution

At the free boundary parallel to the y axis, there are no external tractions and therefore the shear stress τ_{xy} and the corresponding shear strain γ_{xy} must be zero. In this case $\partial U/\partial y$ is large at the boundary, while $\partial V/\partial x$ appears to be zero. The probable reason is that a tiny piece of the specimen grating has been chipped off in that region and the pattern does not extend to the actual boundary of the specimen. If the fringes of the V field extended to the boundary, they would hook downwards, such that $\partial V/\partial x = -\partial U/\partial y$ at the boundary. Then, γ_{xy} would be zero.

10.4 Determine the location and magnitude of the maximum shear strain in Fig. 10.10(c-d).

11

Advanced Composites Studies

11.1 Introduction

Brief descriptions of several studies are presented in this chapter. Diverse experiments are described where the common thread is composite materials. They address the mechanics of highly complex bodies. These examples were chosen to demonstrate the broad scope of moiré interferometry and to express the virtue of whole field observations. These are not in-depth reports. Instead, numerous fringe patterns are shown and discussed to familiarize the reader with real-world analysis.

11.2 Ceramic Composites

Figure 11.1 documents a special phenomenon. Although the tensile specimen has cracked—with cracks so distinct that light can be seen through the otherwise opaque material—it continues to resist the tensile load. The matrix has failed in this case, but vertical fibers bridge the gap and carry the load.

The material is a ceramic laminate with silicon carbide fibers in a silicon nitride matrix. It is a cross-ply laminate with 11 plies of alternating 0° and 90° fiber directions, designated [(0/90)$_5$/0]. The 0° plies are vertical, in the direction of the tensile load. The specimen width is 19 mm (0.75 in.) and it has three open holes of 6.3 mm (0.25 in.) diameter. Steel extensions were cemented to the specimen with an epoxy adhesive and the extensions were fitted with pin connections for loading in a universal testing machine. A moiré interferometer and the related optical system was mounted on the testing machine, too, as described in Ref. 1.

In Fig. 11.1a, b, the average tensile strain was about 240 x 10^{-6}. No cracks were present and the strain distribution was orderly. The U field shows the transverse deformation associated with Poisson contraction. The longitudinal deformation is shown in (c) for a higher load level. Matrix cracks occurred at the upper and lower holes, with consequent major redistributions of displacement fields near the cracks. At this stage, the average strain near the central hole was about 300×10^{-6}. At a higher load, a matrix crack occurred at the central hole, as seen in (d). The local strain near the central crack actually reduced to about 200×10^{-6}.

After cracks are formed, the $0°$ fibers carry the tensile load across the gap. Slippage occurs at the fiber-matrix interface; the matrix stresses relax in part, and the matrix recedes back from the crack plane. The matrix stresses do not fully relax, since friction at the interface transfers some of the fiber strain to the matrix until, at some distance from the original crack plane, strain equalization in the fibers and matrix is complete.

The fringe orders are assigned in Fig. 11.1 starting from an arbitrary datum in each case. Black contours are given integral fringe orders in the V fields and white contours are integral orders in the U field. The crack opening cannot be determined from these patterns since there is no path for a continuous fringe count from one side of the crack to the other. It would be possible to measure the crack opening width by attaching an artificial bridge according to the method of Sec. 4.16. The bridge can take the shape illustrated in Fig. 11.1d and it can be made from a low stiffness plastic.

11.2.1 Delamination of Specimen Gratings

Slippage *must occur* between the fibers and matrix in order for the fibers to span the gap. Without slippage, the fibers in the gap would be required to stretch from an initial length of zero, which is obviously impossible.

The same consideration applies to a specimen grating when a grating of finite thickness is used for fracture studies. If the grating adheres to the specimen sufficiently well, and if the grating material is sufficiently brittle, then the grating will crack together with the specimen and convey the specimen deformation. Otherwise, the grating will delaminate along the crack border and span the gap, at least momentarily prior to complete separation of the specimen. Such action would lead to relaxation and curling of the grating in the delamination zone and cause an absence of data near the crack border. Tiny zones where this occurred are visible in Fig. 11.1c near the upper crack. For fracture studies, replicated gratings should be made to adhere as strongly as possible and to be as thin as possible. Zero-thickness gratings (Sec. 4.9.5) would be most suitable.

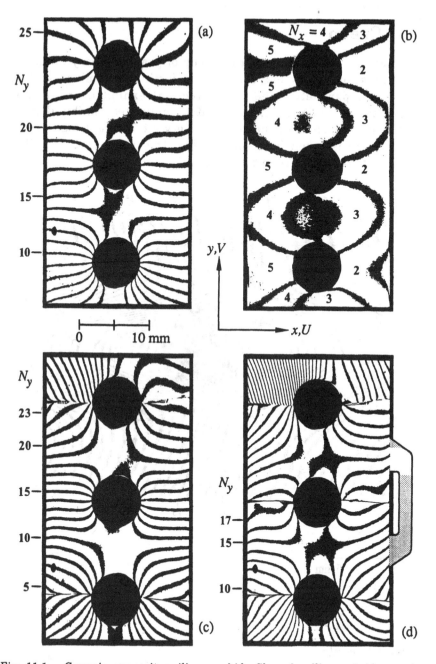

Fig. 11.1   Ceramic composite: silicon carbide fibers in silicon nitride matrix, [(0/90)$_5$/0]. (a), (c) and (d) V field for increasing tensile loads. (b) U field for same load as (a). f = 2400 lines/mm.

11.3 Creep

Tuttle and Graesser[2] investigated creep of graphite/epoxy laminates in compression. They found that fiber-dominated laminates (those with fibers parallel to the load direction) exhibited small creep deformations, while matrix-dominated laminates exhibited much greater time-dependent behavior. The very interesting fringe pattern in Fig. 11.2a is taken from that study. The fringe pattern will be discussed here, rather than the creep behavior.

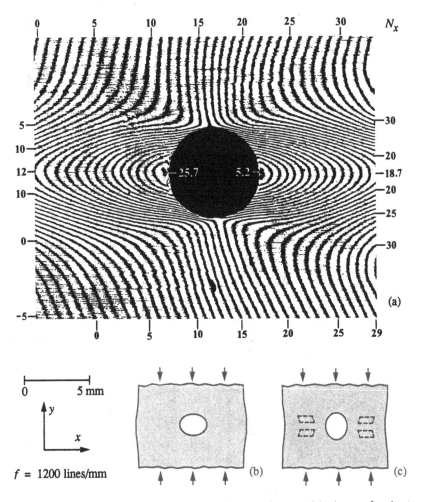

Fig. 11.2   (a) Transverse displacement field for $[90]_{48}$ graphite/epoxy laminate. Courtesy of M. E. Tuttle (Univ. of Washington). (b) Deformation for isotropic material. (c) Actual deformation, exaggerated.

The specimen was a 25.4 mm (1 in.) wide coupon with a 6.3 mm (0.25 in.) hole loaded in compression. It was machined from a 48 ply unidirectional laminate of IM7-8551 graphite/epoxy; the compressive load was applied perpendicular to the plies, defining the laminate as $[90]_{48}$. It was highly matrix dominated since the fibers resisted the stresses in the Poisson direction rather than the load direction. A cross-line specimen grating of 600 lines/mm was applied by replication and an optical system similar to that of Fig. 4.8 was used to form virtual reference gratings of $f = 1200$ lines/mm.

Figure 11.2 maps the transverse displacement field, N_x. It is a remarkable pattern—it is counterintuitive! With a focus on isotropic materials, one would expect a compression specimen to assume the shape sketched in Fig. 11.2b, where the hole elongates horizontally and the sides bulge outwards. The actual deformation was the opposite. The hole contracted horizontally and the sides curved inwards, as sketched in (c).

Remember that positive displacements are displacements in the +x direction. Accordingly, the boundary point on the *right* side of the hole moved to the right by $U_R = 5.2/f$ (relative to an arbitrary datum point of zero fringe order), while the point on the *left* side of the hole moved to the right by $U_L = 25.7f$. The left boundary point moved closer to the right boundary point. The relative displacement was $-20.5/f$, or -17 μm, and the horizontal diameter actually decreased.

Similarly, the width of the specimen increased by $34/f$ at locations above and below the hole. Along the centerline, the width increased by only $(18.7 - 12.0)/f$, or $6.7/f$. The width at the centerline is actually smaller than that above and below by $27.3/f$, i.e. by 23 μm, as depicted in Fig. 11.1c.

High shear strains occur in regions above and below the horizontal centerline, where the cross-derivatives $|\partial N_x / \partial y|$ are large. These shears cause local deformations as indicated schematically in Fig. 11.2c. Of course, the corresponding deformation is consistent with the global shape sketched in (c).

11.4 Metal-Matrix Composite

The specimen material in this study[3] was a boron/aluminum composite—boron fibers in an aluminum matrix. Boron fibers have a relatively large diameter, 140 μm in this case, or about 20 times the diameter of graphite fibers. The consequence for our work is substantial. Fiber-matrix interactions can be observed by the macromechanics methods of Chapter 4. Material behavior on the micromechanics level can be viewed across the large field afforded by macromechanics techniques.

The specimen was a tensile coupon with a central slot. It was 19 mm (0.75 in.) wide, 1.5 mm (0.06 in.) thick, with a slot length of 5.5 mm (0.22 in.). The laminate had 8 plies with a $[\pm45/0_2]_s$ stacking sequence.

A variation of the $\pm45°$ method of Sec. 4.7.10 was used. Moiré fringe patterns were recorded for the x, y, x' and y' fields. Figure 11.3 displays the fringe pattern for the x' field; it is a contour map of the displacements in the x' direction, which is the direction of the fibers in the outer ply. The contour interval is $1/(\sqrt{2}f_s)$, where $f_s = 1200$ lines/mm, or 0.59 μm (23 μin.) per fringe order. Referring to Fig. 4.24a, the virtual reference grating was produced by beams AO and CO.

The zigzag nature of the fringes is not optical noise, but real information. It depicts the larger shear displacements in the matrix and smaller shear displacements in the fibers. An interesting micromechanical feature is highlighted by cross-hatching between two neighboring fringes, each of constant fringe order. It shows that large displacements of fibers occur in blocks of several fibers.

The graph shows the variation of shear strains along the y' axis. The shear strains were concentrated in the aluminum matrix material. The spikes indicate large plastic shear strains between randomly spaced blocks of fibers. These zones of anomalous plastic slip did not extend along the entire length of the interface between neighboring fibers; instead, the extreme strains gradually diminished, and sometimes reappeared at interfaces between nearby fibers.

The global deformation of the outer 45° ply was accomplished largely by local shear displacements in the fiber direction. We can surmise that deformation of the next inner ply was accomplished largely by local shears in *its* fiber direction. However, the directions were 90° apart. In order to accommodate different local strains in adjacent plies, the aluminum matrix between plies must undergo severe shear strains, too. The micromechanics of multidirectional laminates is very complicated, with strains varying through the thickness as well as in the planes of the plies.[3]

11.5 Residual Stresses

Gascoigne[4] investigated residual stresses in graphite/epoxy cross-ply laminates. For each ply, the thermal coefficient of expansion is small in the fiber direction and much larger in the perpendicular direction.

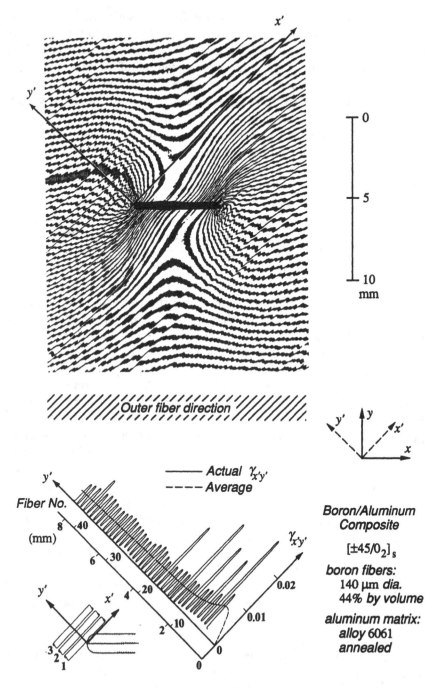

Fig. 11.3 N_x' displacement field, mapping the displacements parallel to the fibers of the outer ply. The shear strains are concentrated in the matrix between fibers.

A cross-ply laminate is relatively stress-free at its elevated fabrication temperature. Upon cooling, the free thermal contraction of each ply is constrained by its neighboring ply(s) and a system of residual stresses is developed. An analogous system of residual stresses results from the thermal expansion mismatch of the fiber and matrix materials; these fiber-level stresses were not addressed, but the ply-level residuals were evaluated.

Gascoigne sliced thick three-ply laminates into their individual plies to eliminate the mutual constraint of neighboring cross-plies. He cut through the 0°/90° interfaces with a thin diamond-impregnated circular saw. He also sliced the thick ply into a thin sub-ply and found no additional relief of stresses, demonstrating that residual stresses are fully released by cutting along the interface. He used moiré interferometry by applying the specimen grating to the laminate and then slicing it to relieve the constraints. The deformation measured in the slice was taken to be equal and opposite to that previously caused by locked-in stresses. Strains were determined from the moiré patterns and the corresponding residual stresses were calculated by the elastic orthotropic stress-strain relations. In principle, the residual stresses can be determined for any laminate by cutting through the interfaces to release the inter-ply constraints.

The technique is illustrated by Fig. 11.4. The specimen was a thick laminate of graphite/epoxy IM7/8551/7A with stacking sequence $[90_{20}/0_{20}/90_{20}]$. It was machined to the dimensions given in the figure and a 1200 lines/mm cross-line grating was replicated on the surface. The moiré interferometer was adjusted for a null field before slicing the specimen. It was determined that the grating mold produced the same null field and the mold was used for subsequent adjustments. The V field obtained from a 0.5 mm slice is shown in the figure.

The fringe pattern is dramatic evidence of high residual stresses in the outer ply. The maximum normal strain occurred at the hole boundary along the horizontal diameter. It was 0.45% compression, indicating a residual *tensile* strain of that magnitude in the original specimen. The corresponding stress was calculated as 35 MPa (5100 psi) using the elastic orthotropic properties of the material and data from the U and V fields.

This residual tensile stress is severe, accounting for about 60% of the transverse strength of the unidirectional material. By evaluating the results of alternative fabrication techniques, moiré interferometry can make valuable contributions to achieve superior composite structures.

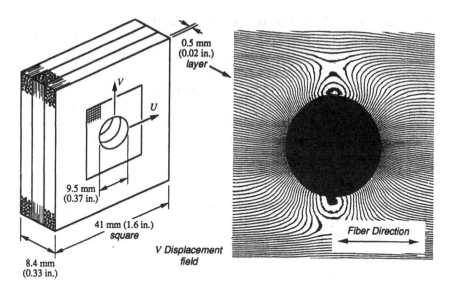

Fig. 11.4   Deformation of the outer ply after a slice was cut from the specimen to remove the constraint of its neighboring ply. f = 2400 lines/mm.

11.6 Thermal Strains, Residual Strains

The method of Chapter 8 was used to determine thermal strains in the graphite/epoxy specimen illustrated in Fig. 11.5. The specimen grating was applied at an elevated temperature of 121°C (250°F). The grating deformed together with the specimen as the specimen cooled, and the deformation was observed at room temperature (24°C). The thermal strains introduced by the −97°C increment were determined from the in-plane displacements.

If we assume, as we did in the previous section, that the laminate solidified and hardened at its curing temperature, that it was essentially stress-free at its 121°C curing temperature, and that there was no creep or strain relaxation as it cooled to room temperature, then the measured strains represent the residual strains produced in the fabrication process. These assumptions are reasonable for some materials and fabrication processes. Thus, the method can sometimes be used to determine residual strains of laminates and it can always be used to determine steady-state thermal strains.

The specimen was a block cut from a thick-walled cylinder with a $[0/90_2]_n$ ply sequence. The room temperature displacement fields are shown in Fig. 11.5 for the region of the dashed box. The optical system illustrated in Fig. 7.13 was used, where moiré interferometry with f = 2400 lines/mm provided the U and V fields and Twyman-Green interferometry provided the W field.

The fringe patterns suggest *déjà vu*—that we have seen them before. We have not, but the patterns are remarkably similar to those of the $[0_2/90_4]_n$ laminate in interlaminar compression, shown in Figs. 7.9 and 7.11. The techniques discussed in Sec. 7.5 for data reduction are the same.

Some of the signs are different. Here the fringes of the U field show thermal contraction or compressive strain, whereas Fig. 7.9a results from Poisson expansion or tensile strain. The strains are compressive in the V field for both cases. The ridges and valleys of the W field are similar in shape but they have opposite signs; the valleys occur at 0° plies in the thermal contraction case. (The carrier fringes had opposite signs in the two experiments.) Another difference is the general waviness of fringes in some regions of the V field. We conjecture that the waves are associated with nonuniformities of the laminate, especially variations of the thickness of resin-rich zones between plies and the greater thermal contraction in the thicker parts.

Nevertheless, the similarities are abundant. The scalloped fringes of the U field signify a strong free-edge effect, such as that graphed in Fig. 7.9d. The periodic variations of $\partial N_y / \partial y$ in the V field are associated with the edge effect, as in the case of interlaminar compression. Judging from the results of the microscopic study of Sec. 7.5.2 (and also the bimaterial study of Chapter 8), an ε_y strain concentration would be expected near the cross-ply interfaces. The common feature is the protrusion of plies relative to their neighbors. The strain distributions that result are similar, regardless of whether the protrusions are caused by Poisson expansion or thermal expansion.

Global values of the thermal coefficients of expansion can be determined easily from these patterns. The average coefficients are α_x = 1.7 x 10^{-6}/°C and α_y = 21.9 x 10^{-6}/°C.

11.6.1 Micromechanics: Thermal Deformation of Boron/Aluminum

Moiré interferometry can elucidate the micromechanics of composites at the fiber level if the fibers are large. This section exhibits experimental results for a *unidirectional* boron/aluminum composite subjected to thermal loading.

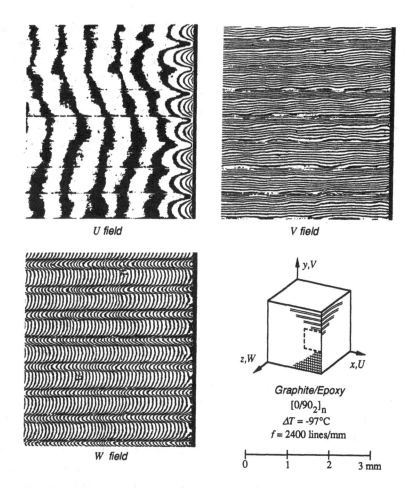

U field

V field

W field

Graphite/Epoxy
$[0/90_2]_n$
$\Delta T = -97°C$
$f = 2400$ lines/mm

0   1   2   3 mm

Fig. 11.5   Temperature-induced deformation.   The contour intervals are 0.417 μm and 0.257 μm per fringe order for the in-plane and out-of-plane displacements, respectively.

The methods of Chapters 5 and 8 were employed.   A surface perpendicular to the fibers was ground and polished, and a cross-line grating was applied by replication at 132°C, using a mold on a ULE glass substrate.   The microscopic moiré apparatus of Fig. 5.1 was tuned to $f = 4800$ lines/mm, using the same mold, to establish a null field.   Then the deformed specimen was installed and observed at room temperature, whereby $\Delta T = -110°C$.

The results are shown in Fig. 11.6 for a hexagonal array of fibers in (a) and (b), and a square array in (c) and (d).   The fringe multiplication factor is $\beta = 6$, providing a contour interval of 35 nm/contour.   The dominant feature is the more closely spaced fringes in the aluminum, consistent with its higher coefficient of thermal expansion. The patterns show compressive strains everywhere for this case of negative ΔT. However, there are areas of aluminum where lower fringe densities persist. Such areas are most noticeable in the ligaments between fibers along vertical rows in (a) and (b), and along horizontal rows in (d). The free thermal contraction ($\alpha \Delta T$) of aluminum in these areas was constrained by the fibers, with the result that the stress-induced part of the deformation was tensile. Although the patterns of total deformation are orderly, with monotonic variations of fringe contour numbers, the *stress-induced* deformations show the heterogeneous nature of the composite. When $\alpha \Delta T$ is subtracted for each material, strain discontinuities at the fiber/matrix boundaries become evident.

11.7 Simulation

Sometimes it is instructive to simulate a body by an enlarged or simplified model. Fig. 11.7 is an example. It models a unidirectional composite in compression. It enforces a square array of fibers in which the central section represents a unit cell that is reasonably free of specimen boundary effects. It is a great enlargement of normal composites, which means that a given strain level is represented by many more displacement fringes.

The model was made by aligning glass rods in a fixture and pouring liquid epoxy into a box that surrounded the fixture. After the epoxy cured, the ends with glass rods were cut to plane, smooth faces by a diamond impregnated circular saw. A cross-line grating of $f_s =$ 1200 lines/mm was replicated on the face and the model was loaded in compression by means of the fixture of Fig. 4.10. Note that residual stresses were generated by the polymerization shrinkage of the epoxy around the glass rods, but no cracks or delaminations occurred. The deformation measured by the grating is that which occurred after the grating was applied, so the patterns are independent of the initial condition of residual stresses.

Detailed measurements can be made from the patterns, but some generalizations are evident. From the U field we notice that the change of fringe order across the specimen width is greater at the top and bottom than it is in the region of the rods. Thus, the glass rods reduce the transverse (Poisson) extension of the specimen. Curiously, the transverse strain in the horizontal ligament region directly above and below the rods is much larger than the average transverse strain, and much larger than the transverse strain in the rods. As a consequence, strong shear strains must develop near the

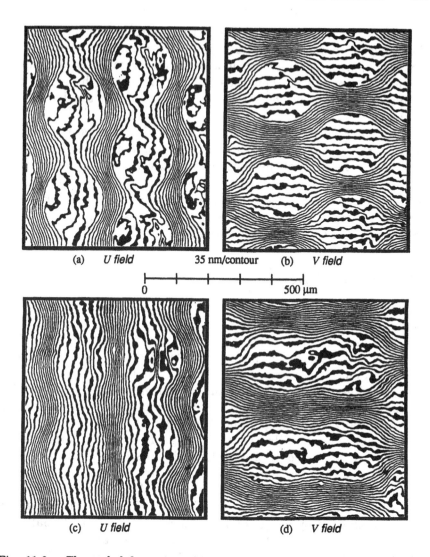

Fig. 11.6   Thermal deformation of boron/aluminum composite.   (a) and (b) hexagonal array of fibers.   (c) and (d) square array.

interface to maintain continuity of the material. Here, these shears are combined with strong compression across the interface, but in the case of a tension test, the tensile stresses would combine with the shear stresses to hasten interface failures. Curiously, too, the transverse strain changes from tension in the ligament to a weak compression in the low strain region centered between four rods.

From the V field, it is clear that the largest compressive strain occurs in the matrix in the horizontal ligament region directly above and below the rods. It is about four times larger than the strain in the adjacent lowest strain regions. The maximum compressive strain is approximately twice the average compressive strain. In the vertical ligament (between rods on its left and right sides) the compressive strain is about half of that present in the horizontal ligament.

Tiny *black holes* and *white holes* (Sec. 4.19) can be seen in the V field near the boundary above and below some of the rods. These result from the *slope* of the specimen surface where the out-of-plane displacement (the Poisson extension) of the matrix is constrained by the stiff glass rods. Factors that would eliminate the blank zones include use of a camera lens with a larger solid angle of light acceptance, and observations at lower load levels.

11.8 Wavy Plies

Among the problems encountered in the fabrication of thick composite structures, wavy plies is recognized as an important issue. Wavy plies occur inadvertently in the fabrication process. For example, when successive layers are wound around a mandrel, the tension in outer layers can cause inner layers to warp into an array of wavy plies. Their influence on the strain distribution and the strength of thick composite rings has been investigated experimentally by Gascoigne and Abdallah.[5]

The specimen is illustrated in Fig. 11.8. The laminate is comprised of repeated groups of four plies each, with the fibers of three plies in the circumferential or hoop direction and the fibers of the fourth ply in the axial direction, perpendicular to the diagram. The stacking sequence is designated as $[(90_3/0)_{20}/90_3]$. Two regions were investigated by moiré interferometry: (A) where essentially no ply waviness was present and (B) where severe waviness occurred. The ring was loaded by external hydrostatic pressure of 34.5 MPa (5000 psi) by a unique test fixture comprised of a robust frame and a rubber bladder capable of being pressurized to 140 MPa (20,000 psi).

Fringe patterns for region (A) are shown in Fig. 11.9. Data were extracted for determination of strains along a radial line near the center of the field. The fringes of the V field have an essentially uniform spacing in the hoop direction, yielding a uniform compressive strain of -0.25%. The hoop strain is shown in the graph as ε_y. The small anomalies in the V field are characteristic of composite materials and they were disregarded in the analysis. Note, too, that the graph does not extend beyond ply group 17; the reason is

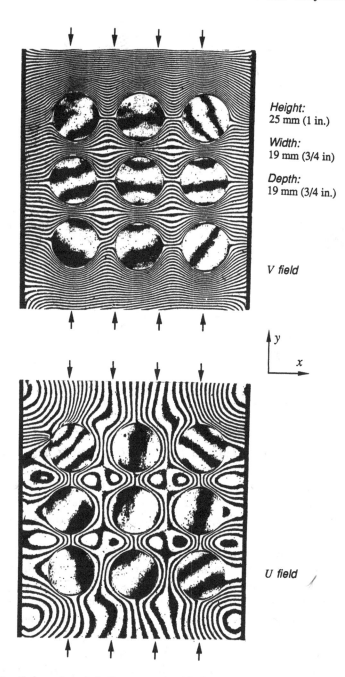

Height:
25 mm (1 in.)

Width:
19 mm (3/4 in)

Depth:
19 mm (3/4 in.)

V field

U field

Fig. 11.7   Enlarged model of a composite block in compression. Material: glass rods in an epoxy matrix. f = 2400 lines/mm.

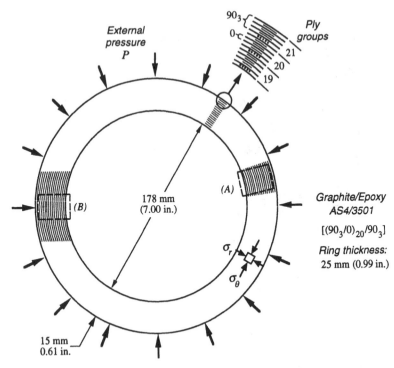

Fig. 11.8   Composite ring subjected to external pressure, P. Regular plies at (A) and wavy plies at (B) were investigated.

that the loading fixture blocked part of the incoming light near the outer perimeter of the ring.

The U field is shown with carrier fringes of rotation. The carrier was used to transform the pattern to one that is easily interpreted by the method of Fig. 7.5. The corresponding graph of radial strains is labeled ε_x.

The strong undulations of ε_x mimic those of Fig. 7.6, where the compressive stresses in the coupon correspond to the compressive hoop stresses, σ_θ, in the ring. The 0° plies of Fig. 7.6 correspond to the hoop plies here, and it is seen that they both exhibit maximum strains (on an absolute scale). The axial plies here correspond to the 90° plies of Fig. 7.6, and they both exhibit the minimum strains in each ply group. In the ring, the radial compressive strains force the entire curve downwards. As in the case of Fig. 7.6, we should remember that these transverse strains, ε_x, represent free-edge effects, and they should not be taken as the strain distribution inside the ring.

The influence of wavy plies in region (B) is revealed in Fig. 11.10. Now the spacing of fringes in the V field is highly irregular. The most closely spaced fringes occur at the arrows, near inflection points in the waves. The strain is $\varepsilon_y = -0.50\%$ at those locations, which is twice as large as the hoop strain found in Fig. 11.9. The waves caused a strain concentration by a factor of 2.

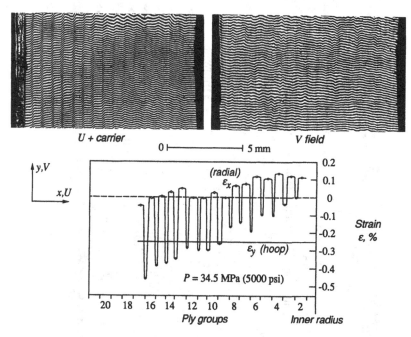

Fig. 11.9   Fringe patterns in region (A) and graph of surface strains along a radial line. $f = 2400$ lines/mm.   Courtesy of H. E. Gascoigne (California Polytechnic Univ.) and M. G. Abdallah (Hercules Aerospace Co.).

Whereas shear strain γ_{xy} was very low in region (A), strong shears are present in the wavy plies. The closely spaced fringes in the U field (Fig. 11.10) signify large values of $\partial U / \partial y$. The largest value occurs at the arrow, again near an inflection point. The gradient $\partial V / \partial x$ is nearly zero in the same part of the V field. By Eq. 4.3, the shear strain is nearly 1%. Notice that the large shears near the inflection points occur across the full width of the ply group. The waves are responsible for substantially elevated compressive strains and large shears in the same location. Pressure tests to failure confirmed a severe degradation of the ultimate strength of rings with wavy plies.

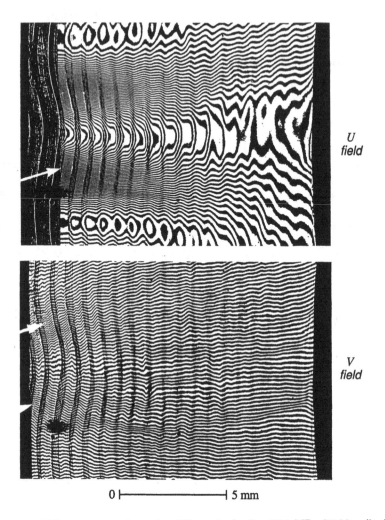

U field

V field

0 ⊢————————⊣ 5 mm

Fig. 11.10   Fringe patterns in region (*B*), again for *P* = 34.5 MPa (5000 psi). The wavy plies caused large increases in compressive and shear strains. Courtesy of H. E. Gascoigne and M. G. Abdallah.

11.9 References

1. Y. Guo and P. Ifju, "A Practical Moiré Interferometry System for Testing Machine Applications," *Experimental Techniques*, Vol. 15, No. 1, pp. 29-31 (1991).

2. M. E. Tuttle and D. L. Graesser, "Compression Creep of Graphite/Epoxy Laminates Monitored Using Moiré Interferometry," *Optics and Lasers in Engineering*, Vol. 12, No. 2, pp. 151-172 (1990).

3. D. Post, Y. Guo and R. Czarnek, "Deformation Analysis of Boron/Aluminum Specimens by Moiré Interferometry," *Metal Matrix Composites: Testing Analysis, and Failure Modes*, ASTM STP 1032, W. S. Johnson, editor, American Society for Testing and Materials, Philadelphia, pp. 161-170 (1989).

4. H. E. Gascoigne, "Residual Surface Stresses in Laminated Cross-ply Fiber-epoxy Composite Materials," *Proc. VII International Congress on Experimental Mechanics*, pp. 1077-1084, Society for Experimental Mechanics, Bethel, Connecticut (1992).

5. H. E. Gascoigne and M. G. Abdallah, "Displacements and Strains in Thick-walled Composite Rings Subjected to External Pressure Using Moiré Interferometry," *Proc. 2nd International Conference on Photomechanics and Speckle Metrology*, Vol. 1554B, pp. 315-322, SPIE, Bellingham, Washington (1991).

11.10   Exercises

11.1   In Fig. 11.2, let $x = y = 0$ be located at the center of the hole. Determine (a) $\partial N_x / \partial x$ and (b) $\partial N_x / \partial y$ at the point $x = 5$ mm, $y = 2.5$ mm.

11.2   Reproduce the pattern of Fig. 11.3 (using an office-type copying machine). Assign fringe orders $N_{x'}$ by marking numbers on the copy.

11.3   In the fringe pattern of Fig. 11.3, the shaded zone lies between two neighboring fringes of fringe orders Q and $Q + 1$. The paths of these fringes seem ambiguous. How can we prove that each of these irregular fringes is a contour of constant fringe order?

11.4   Using Fig. 11.4 as an example, explain (in paragraph form) the method of residual strain measurements.

11.5   Using Fig. 11.6 (a and b) as an example, explain (in paragraph form) the method of steady-state thermal strain measurements.

11.6   Make an enlarged image of the V field of Fig. 11.7. (a) Mark fringe orders on the pattern. (b) Plot graphs of displacements V and strains ε_y along the vertical centerline. (c) List any assumptions you made.

11.7   The arrow in the U field of Fig. 11.10 points to a region of high strain. Determine (a) ε_x, (b) ε_y and (c) γ_{xy} in that region.

12
Metallurgy, Fracture, Dynamic Loading

12.1 Introduction

A broad range of studies is introduced to emphasize the diverse capabilities of moiré interferometry. This work was performed during the last several years at six different photomechanics laboratories. They are at the Idaho National Engineering Laboratory (INEL), University of Strathclyde, University of Washington, Rockwell Science Center, IBM Corporation and Virginia Polytechnic Institute and State University (VPI&SU).

12.2 Metallurgy

12.2.1 Elastic-plastic Joint

The specimen was a tensile coupon of two materials: commercially pure titanium and a titanium alloy containing 6% aluminum and 4% vanadium.[1] The materials were diffusion bonded, which is a joining process that requires clean surfaces, very high contact pressures and modest temperatures to create a metallurgical bond. Unlike welding, no filler material is used.

The patterns of Fig. 12.1 show residual deformations after the tensile load was removed. The titanium alloy has a yield strength about 2.4 times higher than that of the pure titanium and yielding occurred in the pure material while the alloy remained elastic. Shear bands appear in the yielded material. Substantial point-to-point variations of strain occur, resulting from a random distribution of plastic shear strains. We can notice bending of the tensile specimen, which is evidenced by a closer spacing of fringes on the right side of the V field.

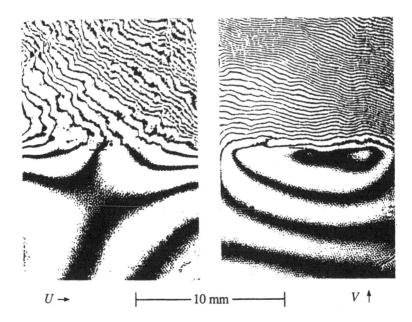

$U \rightarrow$ |————— 10 mm —————| $V \uparrow$

Fig. 12.1   Residual deformation in a tensile member. Plastic strains are developed on one side of a welded joint between pure titanium (top) and a 6Al-4V titanium alloy. $f = 2400$ lines/mm. Courtesy of J. S. Epstein (INEL).

12.2.2 Weld Defects

Figure 12.2 is a contour map of vertical displacements of a 304 stainless steel tensile specimen.[2] The specimen was made from two parts that were welded together using a nickel filler material. The pattern exhibits regions where the weld was successful and other regions where the two sides are not fused together. A lack of fusion is established by a discontinuity of the fringe count across the weld. For example, in the weld region located slightly to the left of center, $N_y = 14$ above the junction and $N_y = 7.5$ below it. The opening or gap at that point is $6.5f$, or 2.7 μm (107 μin.).

12.2.3 Micromechanics: Grain Deformations

A preliminary test of large grain titanium in *elastic* tension is reported in Figs. 12.3 and 12.4. The loading fixture applied direct plus flexural forces, resulting in tension on the outside face of the specimen. The specimen cross-section was 6 mm by 5 mm, its yield point stress was 1450 MPa (210,000 psi), and its nominal elastic modulus was 97 GPa (14×10^6 psi). Numerous grain sizes were present, up to a maximum of 400 μm.

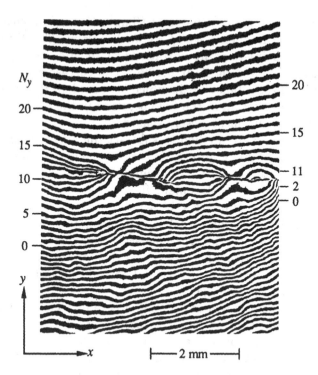

Fig. 12.2 Weld defects in a stainless steel tension specimen. f = 2400 lines/mm. Courtesy of S. A. Chavez (INEL).

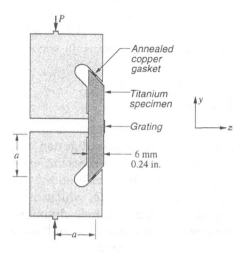

Fig. 12.3 Specimen and loading fixture.

The microscopic moiré scheme of Chapter 5 was employed. The reference grating frequency was $f = 4800$ lines/mm. The V field (Fig. 12.4a) reveals point-to-point variations of tensile strain, with an average of $\varepsilon_y = 0.0076$ m/m, or 0.76%. The corresponding average stress was 730 MPa (106,000 psi), which is about half the yield stress. The largest tensile strain in the field occurred in the circled region A, where the fringes are most closely spaced.

Carrier fringes were applied to subtract the average strain, with the result shown in (b). Figure 12.4c is the same pattern with a fringe multiplication factor of $\beta = 8$. The contour interval is 26 nm/contour. The corresponding U displacement field is shown in (d), where the average transverse strain was subtracted off.

The multiplied patterns reveal strong anomalies. Whereas elastic strains are uniform on a macroscopic scale, they are seen to vary on a microscopic scale. The anomalous, or extra, strain in region A is $\varepsilon_y = 0.39\%$. The total tensile strain is $\varepsilon_y = 1.15\%$ at that location, or about 50% greater than the average.

The largest shear strain in the field occurs in region B (where the anomalous tensile strain is zero). Inspection of the U and V patterns prior to the application of carrier fringes reveals that the cross-derivatives are both negative at B, which means that their contributions to the shear strain are additive. The shear strain at B is $\gamma_{xy} = -0.59\%$.

The strains are elastic. The load was removed after the patterns of Fig. 12.4 were recorded and the displacement fields for zero load were displayed with $\beta = 8$. They showed no permanent deformation, thus verifying reversibility and elastic deformation.

The anomalous elastic strains were large, approximately 50% (normal) and 75% (shear) of the average tensile strain. The micromechanics of polygranular materials warrants more detailed investigation.

12.2.4 Tube-plate Joints

This section is based upon a study performed at the University of Strathclyde, Scotland.[3] It was an investigation of the elastic-plastic deformation that occurs during the process of expanding tubes into mating holes in thick plates. An interference fit is required to seal the joint between the tube and plate, which means that the tube diameter would be larger than the hole diameter if their mutual constraint was removed. Tube-plate assemblies are used in heat exchangers, including those in nuclear power plants, and effective sealing is essential.

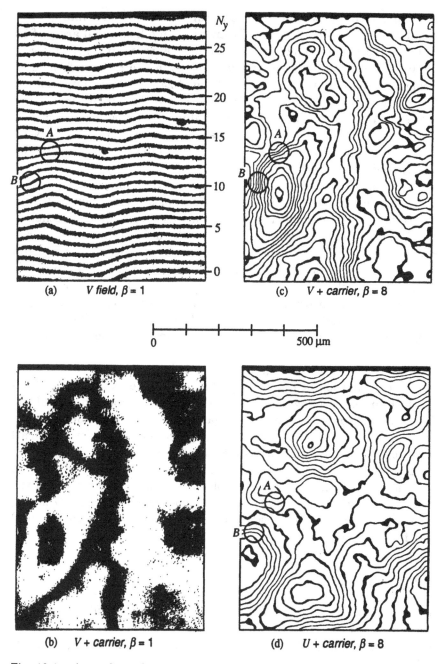

Fig. 12.4   Anomalous elastic deformation of large grain titanium in a tensile field. In (b, c, d) the average strain was subtracted by carrier fringes. $f = 4800$ lines/mm. Contour interval = 26 nm.

The specimen was a titanium tube plate of 127 mm (5 in.) diameter and 70 mm (2.75 in.) thickness, with a hexagonal array of 19 holes of 15.9 mm (0.627 in.) diameter. A cross-line grating was replicated on the surface and viewed by a moiré interferometer designed according to the scheme of Fig. 4.24b.[3,4] Deformation of the tube plate was monitored during a program in which steel tubes were inserted and expanded sequentially. The tube expansion tool was electrically driven. It consisted of three rollers that pressed against the interior surface of the tube to enlarge its diameter by plastic deformation.

A remarkable aspect of this work is that the moiré patterns were recorded *during* the mechanical deformation process. Figure 12.5 is an example. The vibration environment was severe, but nevertheless the fringe pattern is distinct. Although the deformation process was quasi-static (i.e., wave propagation was not investigated), the environment was akin to that encountered in dynamic analyses.

A 5 mW helium-neon laser was used. The fringe photographs were recorded at 5 frames per second by a 35 mm camera equipped with a power winder and a large capacity film cassette The exposure time was 1/2000 second on Kodak Tri-X film, processed for an ASA rating of 1500.

The upper hole in Fig. 12.5 was being expanded during the exposure. The blank regions are shadows of the expansion tool. A slight blurring of fringes is noticeable below the upper hole and the blur was attributed by the investigators to fringe movement during the exposure time. The pattern represents the V (or vertical) displacement field at an early stage of tube expansion.

Figure 12.6 is the U displacement field after the tubes were expanded in all the neighboring holes. Strains on the surface of the tube plate reached $\varepsilon_x \approx 0.009$ below the upper hole, or nearly 1% strain. The contour interval in these displacement fields is 1.05 μm (41 μin.) per fringe order.

12.3 Fracture

Advances in the field of *fracture mechanics* are critically dependent upon detailed knowledge of the behavior of real materials. Examples of the role of moiré interferometry are given in the following sections.

12.3.1 Plastic Wake, Damage Wake

Inelastic deformation appears to accompany the fracture process, to some degree, even for very brittle materials. It is recognized as

Fig. 12.5 *V* displacement field in a titanium tube plate. The pattern was recorded while a steel tube was being expanded for a tight fit in the plate. Mechanical vibrations were severe. Diameter of holes is 15.9 mm (0.627 in.). Courtesy of C. Walker and J. McKelvie (Univ. of Strathclyde).

Fig. 12.6 *U* displacement field after tubes in the surrounding holes were fully expanded. Courtesy of C. Walker and J. McKelvie.

permanent deformation in the neighborhood of the crack path, and for metals it is known as the *plastic wake*.

Figure 12.7 is an example. The specimen was an edge-notched coupon of high strength aluminum 7075-T6, of 1.52 mm (0.060 in.) thickness. It was subjected to tensile fatigue loading until a crack initiated and extended to the length shown in the figure.[5] The patterns represent the U and V displacement fields that remained after the specimen was fully unloaded. This deformation stems from two causes. One is the plastic strains near the crack path and the other is elastic strains that are present because the crack cannot close when the loads are removed. The plastic deformation along the crack path prevents a return to the original specimen geometry and prevents the complete closure of the crack. Thus, the fringes at some

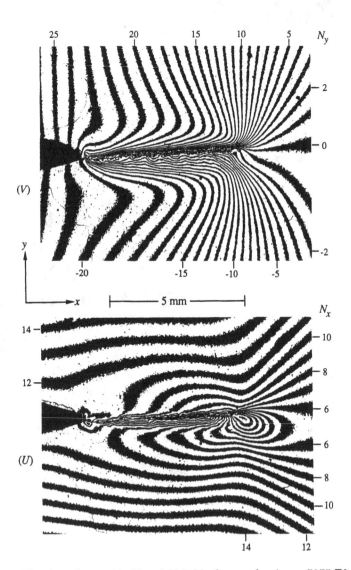

Fig. 12.7   Plastic wake seen in V and U fields for an aluminum 7075-T6 fatigue specimen. $f = 2400$ lines/mm.

distance from the crack result from rigid-body rotation of the upper portion of the specimen relative to the lower portion, and also from elastic strains required to accommodate the new geometry.

The V pattern documents positive $\partial N_y / \partial y$ in the wake region, signifying a tensile ε_y and a permanent stretching of material. Perturbations of fringes along the crack border occur at local contact points, where the ε_y strains are slightly reduced. We can rationalize that the tensile ε_y strains must be permanent (i.e., plastic strains) because otherwise compressive strains would appear along the crack border as a result of the contact stresses.

The U pattern reveals positive $\partial N_x / \partial y$ on both sides of the crack. Shear strains are present in the wake region; below the crack they are caused by movement of material points toward the right (relative to the adjacent elastic material) and above the crack, to the left.

Fracture specialists recognize the asymmetry seen here as a condition called *shear lips*. Nicoletto[6] interpreted these patterns in terms of the ratio of plastic zone size to specimen thickness, and concluded that the asymmetry of the plastic wake is caused by shear lips near the specimen surface. The crack faces are not perpendicular to the specimen surface. Instead, they are warped such that they intersect the specimen surface at an oblique angle. The complex three-dimensional state of stress that causes the shear lips is responsible for the asymmetrical deformation found at the outside surface of the specimen.

Similar wakes are seen in Fig. 12.8 for nonmetals, and they are called *damage wakes*. The materials are (a) silicon carbide (SiC), (b) alumina (Al_2O_3) and (c) concrete. These ceramics and concrete are markedly different from each other, but they all exhibit a damage wake of permanent ε_y strains. The patterns in (a) and (b) are taken from Ref. 7; for these, $f = 2400$ lines/mm. The moiré pattern in (c), for concrete, is from Ref. 8; $f = 1200$ lines/mm.

A curious feature appears in Fig. 12.8a, where the wake extends along only one portion of the crack length. In this case the crack was present in the specimen prior to application of the specimen grating. Then the crack length was increased to study the damage wake. Of course, the deformation that occurred before the grating was applied is not revealed by the fringe pattern.

12.3.2 The J Integral

The J integral has evolved as an important concept to characterize crack extension under elastic-plastic conditions. From the viewpoint of physical measurements, it circumvents the difficult task of extracting data from the region of plastic deformations. Instead, it utilizes data taken along a path in the elastic material that surrounds the plastic region.

Figure 12.9 illustrates the analysis.[9] The patterns map the U and V displacement fields in an edge-notched specimen, loaded statically in tension. The material was aluminum 2024-T3, 2mm (0.80 in.) thick. The J integral requires knowledge of the four derivatives of displacement $(\partial V/\partial y, \partial V/\partial x, \partial U/\partial x, \partial U/\partial y)$ along a closed path surrounding the crack tip, as drawn in the figure. In this case, two data paths were chosen, the small rectangular path near the crack

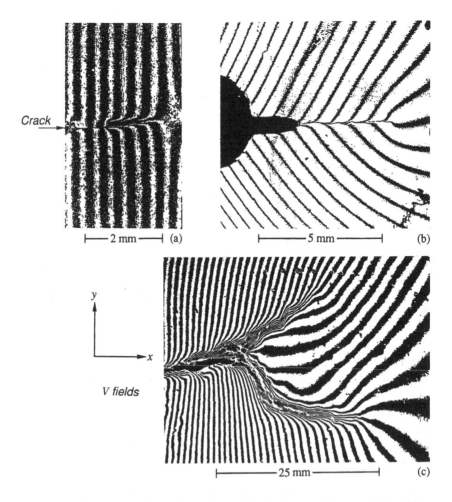

Fig. 12.8   Damage wakes of permanent strain in (a) silicon carbide, (b) pure alumina and (c) concrete. (a) and (b) courtesy of J. S. Epstein (INEL). (c) courtesy of A. S. Kobayashi and Z. K. Guo (Univ. of Washington).

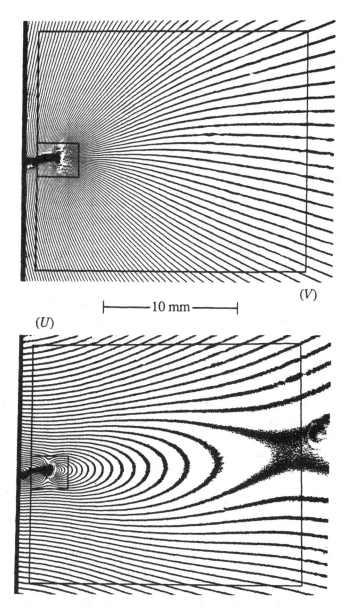

├─────10 mm─────┤

(V)

(U)

Fig. 12.9   Alternative integration paths for calculation of the J integral. Courtesy of M. S. Dadkhah (Rockwell Science Center).

tip and the large rectangular path relatively far from the tip; the J integrals calculated from these data were nearly the same.

The region near a crack tip is usually masked by a black hole (Sec. 4.19.1), where relatively large slopes of the specimen surface direct light away from the camera lens. In many cases, plastic deformation near the crack tip, and in the plastic wake behind it, causes discrete movements of each metal grain relative to its neighbors (e.g., grain rotations), which produce highly irregular, closely spaced moiré fringes. These effects are seen in Fig. 12.7, where they are not masked by a black hole. A beauty of the elastic-plastic fracture parameters lies in their avoidance of the moiré data in these intractable zones.

12.3.3 Micromechanics: Crack Arrest

Microscopic moiré interferometry was applied to study the fracture mechanics of multilayer ceramic/metal composites. The specimen is illustrated in Fig. 12.10. Alternating layers of alumina (a high modulus brittle ceramic) and nearly pure aluminum (a ductile metal) were laminated by diffusion bonding to fabricate a beam. A cross-line specimen grating was applied prior to the application of loads by a four-point pure-bending fixture. The arrangement was deflection controlled rather that load controlled, whereby the displacement at the loading points was increased in small increments.

The patterns in Fig. 12.10 correspond to an intermediate stage in the fracture process. As the flexural deflection increased, a first crack occurred at A in the ceramic (alumina). The crack was arrested by the aluminum layer, where extensive plastic deformation developed to accommodate the crack tip opening displacement. Subsequently, a second crack occurred at B and a third at C, each for successively larger beam deflections. Referring to the drawing of the specimen, the first crack occurred in layer (1) and the second and third cracks in layer (2). Additional beam deflection caused cracks in layers (3) and (4). The patterns show layers (1) and (2) of ceramic and the adjacent metal layers.

The patterns offer a wealth of information. From the U field, the crack opening displacements can be determined by $COD = \Delta N_x^* / f\beta$, where ΔN_x^* is the change of fringe contour numbers across the crack. At the tip of crack A, the COD is 4.1 μm; near the center of cracks B and C, it is 1.04 μm and 0.62 μm, respectively. The cracks relieved the strain ε_x in the alumina almost completely, as noted from the nearly horizontal fringes ($\partial U / \partial x \approx 0$). The stiff ceramic constrains the ε_x strain in the metal near the interfaces (where the fringes are nearly horizontal), except near the crack tip where there is a very large ε_x strain concentration. In addition, strong shear strains are seen to occur near the ceramic/metal interfaces, since $\partial U / \partial y$ is large while $\partial V / \partial x$ is small.

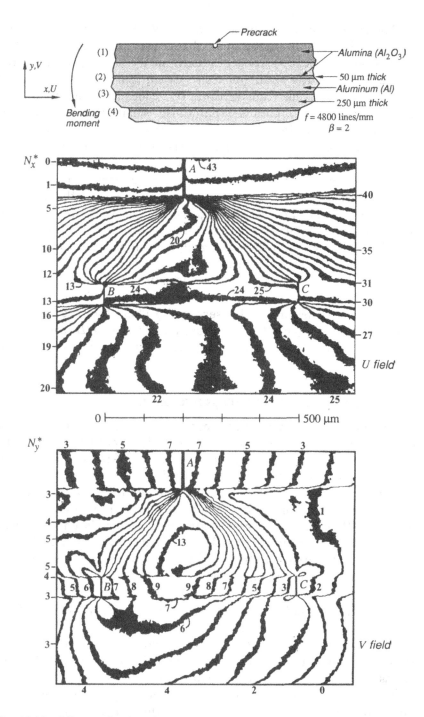

Fig. 12.10   Micromechanics of fracture in a multilayer ceramic/metal composite subjected to pure bending.

The fringe contour numbers in the V field suggest material discontinuities, but except near the crack tips, the material is continuous. Instead, strong gradients of $\partial V/\partial y$ appear near the interfaces, where the gradients are so strong that the fringe contours are not fully resolved. Tensile strain (ε_y) concentrations occur near the interfaces. It is possible that these tensile strains represent effects that occur in the interior of the specimen, in combination with free surface effects (Chap. 7); analysis is needed to discriminate between them.

Intense concentrations of normal and shear strains occur in the aluminum near the crack tips. In some local zones, the displacement data is masked by the convergence of fringe contours. However, the data needed for fracture parameter studies remains clear, including assessment of the J integral, determination of plastic zone size, and data for model verification.

These contour maps cannot discriminate interface cracks perpendicular to the cracks in the ceramic. Some interface separation must occur, or else severe slip along the interface must occur, since the adjacent metal must absorb the *COD* over a *finite* length; otherwise the strain concentration in the metal would be infinite.

12.3.4 Interior Strains, Stress Freezing

Throughout the book we have been concerned with deformation measurements on external surfaces of a body. Very few methods are available for measurements inside the body. Of these, the methods that are used most are imbedded strain gages for point-by-point strain measurements and three-dimensional photoelasticity for extended-field measurements. Models are used in both cases, which precludes the analysis of real engineering materials and generally restricts the applications to the linear elastic realm. Poisson's ratio is usually not matched to that of the prototype material, a condition that must be considered when interpreting the experimental results; for stress-freezing photoelasticity, $v \approx 0.5$.

With three-dimensional photoelasticity we measure elastic stress or strain parameters that are averaged across the finite thickness of a *slice* isolated from the model.[10,11] Averaging is implicit for the two techniques that are available: scattered light photoelasticity and stress-freezing photoelasticity. Sometimes the average value is not desired, but instead, the deformation at a specific plane inside the body is needed. In that case, a hybrid technique of moiré interferometry and stress-freezing photoelasticity can be used. The hybrid technique[12] was developed for measurements in fracture mechanics, but it can be utilized, too, for experimental determination of interior deformations in various three-dimensional elasticity problems.

A diphase property of certain model materials—notably epoxies—is used for stress freezing. The diphase characteristic can be explained by an analogy to an elastic sponge whose pores are filled with a resin. At room temperature, both the sponge and resin act elastically and contribute to a relatively high modulus of elasticity. At an elevated *critical temperature*, the resin acts essentially as a liquid; the sponge acts as an elastic body, but with substantially reduced stiffness. If external loads are applied at the critical temperature, the body deforms elastically. Then, as the body is cooled to room temperature, the resin hardens in the deformed configuration. The external loads can be removed, but the deformation of the body remains enforced by the resin. Furthermore, the body can be cut into numerous pieces without relieving the original condition of elastic deformation that is locked into the sponge. All the internal forces are self-equilibrated in each structural element of the body, where local forces of the sponge are resisted by opposite local forces exerted by the hardened resin. The internal forces, or stresses, are locked-in or frozen, and the process is called *stress freezing*. The process is reversible, too. If the deformed body is heated to critical temperature again, this time without the external loads, the resin liquefies and allows the sponge to relax to its stress-free condition. When cooled to room temperature, the body is in its original undeformed configuration. The procedure is called *annealing*.

When the diphase material is viewed as self-equilibrating on a molecular scale, we can visualize that cutting the body does not relieve its deformation. In practice, no change of the deformation is found, even after several years of storage at room temperature.

Moiré interferometry was coupled with stress-freezing by C. W. Smith and his students for studies of linear elastic fracture mechanics. They investigated the variation of stress-intensity factors through the thickness of cracked specimens, where the near-tip behavior conformed to nearly plane strain conditions in the interior and approximately plane stress near the free surfaces.[13,14] Epoxy specimens with *artificial cracks* or *natural cracks* were heated to the critical temperature, the external loads were applied, and the specimens were cooled slowly to room temperature. Each specimen was cut into several thin slices that represented different depths through the thickness of the specimen. A moiré grating was applied to the surface of each slice and then the slices were annealed to remove the deformation. At room temperature, the moiré interferometry pattern revealed the relaxation, which was equal and opposite to the deformation field previously locked into the specimen at the location of the slice. Figure 12.11 is an example of the V displacement field after annealing, for an artificial crack. In a series of experiments, the data analyzed from each set of slices revealed the variation of stress-intensity factor through the thickness of the specimen.

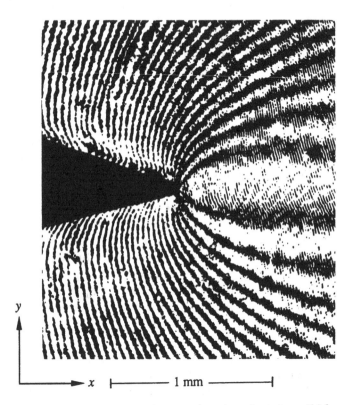

Fig. 12.11 V displacement field along an interior plane in a thick specimen with an artificial crack. $f = 1200$ lines/mm.

12.4 Dynamic Moiré Interferometry

12.4.1 Equipment

Dynamic loading, where the effects of wave propagation in the body are investigated, requires very short exposure times to prevent blurring of the fringe patterns. *Pulsed lasers* that produce intense bursts of light of 10 to 100 nanoseconds (10^{-9} s) duration are fully adequate with respect to coherence, intensity, and exposure time.

A problem arises, however, because such lasers cannot provide the continuous illumination needed to align and adjust the moiré interferometer. In practice, continuous wave helium-neon lasers have been used for the initial setup and pulsed ruby lasers for subsequent dynamic photography; their wavelengths are 632.8 nm and 694.3 nm, respectively. Use of two light sources with different frequencies introduces complications, but a practical approach is to use a double-frequency grating.

The narrow beams from the ruby laser and the helium-neon laser are made coincident by means of a beam splitter (or a removable mirror). The dual-frequency specimen grating, or grating mold, is made in the normal way with an apparatus such as that sketched in Fig. 4.32. Each laser is used, in turn, to expose the photosensitive plate. If the system is arranged for f_s = 1200 lines/mm with the helium-neon laser, the coincident ruby laser beams will produce f_s = 1094 lines/mm. The superimposed gratings will have reduced diffraction efficiency, but otherwise they will perform effectively. A normal moiré interferometer can be used (e.g., Fig. 4.4a); after adjusting it with the helium-neon source, the ruby laser is used to record the dynamic events. Since the angle of incidence α is fixed, the ruby laser produces a virtual reference grating frequency of 2188 lines/mm, given by $f = (2 \sin \alpha)/\lambda$ (Eq. 4.4).

A complete system of this sort is reported by Deason and Ward of INEL.[15] It utilizes a Q-switched ruby laser as a multiple-pulse light source, a high-speed rotating-mirror camera to capture successive moiré patterns, and a triggering and timing system to synchronize the laser pulses with the dynamic event.

Looking ahead, it is likely that developments in laser and laser diode technology will circumvent the need for two separate lasers. Coherent light sources that provide the persistent viewing of the fringes needed for setup and also the pulsed output needed for dynamic recording could then be used with many different types of moiré interferometers.

12.4.2 Wave Propagation in a Composite Laminate

A unique dynamic experiment was undertaken at the Idaho National Engineering Laboratory, and the results are fascinating.[16] The specimen was a 48 ply graphite-epoxy laminate with alternating 0° and 90° plies. The ply thickness was approximately 0.13 mm (0.005 in.). The specimen was impacted at its end with an air-driven ram to induce a compressive wave that propagated along its length. The effect of the dynamic compressive loading was recorded by moiré interferometry.

A dual grating was applied along the edge of the specimen, where the 0° and 90° plies were exposed. Instead of being superimposed, the gratings made with the helium-neon and ruby lasers were located side-by-side. The initial field was adjusted to give an array of carrier fringes, which were subsequently subtracted off by superimposing the no-load fringe pattern (prior to the arrival of the compressive wave) upon the load-induced pattern, and optically filtering the pair as described in Sec. 4.17-4.18.

The result is shown in Fig. 12.12 for a portion of the specimen edge. The specimen was impacted from the left side and the compressive wave was traveling to the right. This frame was exposed 108 µs after the impact. The pattern exhibits noise as a result of all the processing, but the basic shape of the fringes is clear. They exhibit a strong zigzag shape, in step with the alternating 0/90 plies.

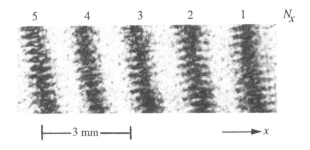

Fig. 12.12   Dynamic wave propagation in a $[0/90]_n$ laminate. The compressive wave is traveling from left to right. The zigzag fringes indicate interlaminar shear strains, γ_{xy}. $f = 2188$ lines/mm. Courtesy of J. S. Epstein (INEL).

However $\partial N_x / \partial x$ is the same in 0° and 90° plies, so ε_x does not change at the ply boundaries. The compressive strain is independent of the fiber directions. In this respect, the dynamic result is the same as the corresponding static strain shown in Fig. 7.6, where the V field indicates continuity of the compressive strain across the 0/90 plies.

On the other hand, the V field of Fig. 7.6 does not exhibit zigzag fringes. For static loading, the 0/90 laminate does not exhibit γ_{xy} shear strains. The dynamic case is different. The severe zigzag shape of fringes documents a strong $\partial N_x / \partial y$, which is indicative of large shear strains.

The explanation appears to lie with the velocity of sound. Normally, the compressive wave travels with the velocity of sound in the material, but in this case the velocity in the 0° plies is nearly three times that in the 90° plies. The compressive wave in the 0° plies tries to advance relative to the 90° plies. To succeed and advance independently, the interface bond must be broken. Instead, shear stresses and strains were developed at the interfaces, which retarded the wave velocity in the 0° plies and increased the wave velocity in the 90° plies, such that the compressive wave propagated at an intermediate velocity. It is important to realize that interlaminar shear is a consequence of dynamic loading, and to document it experimentally. Unlike free-edge effects, these interlaminar shears can be expected to exist throughout the specimen thickness. They can be expected, too, to play a role in structural delamination under impact loading.

12.5 References

1. J. S. Epstein, S. M. Graham, K. E. Perry and W. G. Reuter, "Displacement and Strain Fields for a Bimetallic Strip Under Remote Tension," *J. Applied Mechanics*, ASME (submitted for publication).

2. S. A. Chavez, V. A. Deason and J. S. Epstein, "Use of Moiré Interferometry in Weldments," *Proc. Int. Conf. on Trends in Welding Research*, S. A. David, editor, American Society for Metals, Metals Park, Ohio, pp. 533-537 (1986).

3. C. A. Walker, A. McDonach, P. MacKenzie and J. McKelvie, "Dynamic Moiré Measurement of Strains Induced in a Titanium Tube Plate During Rolling of a Series of Tubes," *Experimental Mechanics*, Vol. 25, No. 1, pp. 1-5 (1985).

4. A. McDonach, J. McKelvie and C. A. Walker, "Stress Analysis of Fibrous Composites using Moiré Interferometry," *Optics and Lasers in Engineering*, Vol. 1, No. 2, pp. 185-205 (1980).

5. D. Post, R. Czarnek and C. W. Smith, "Patterns of U and V Displacement Fields Around Cracks by Moiré Interferometry," *Application of Fracture Mechanics to Materials and Structures*, G. C. Sih, E. Sommer and W. Dehl, editors, Martinus Nijhoff Publishers, Boston, pp. 699-708 (1984).

6. G. Nicoletto, "Moiré Interferometric Fringe Patterns about Crack Tips: Experimental Observations and Numerical Simulations," *Optics and Lasers in Engineering*, Vol. 12, No. 2, pp. 135-150 (1990).

7. J. S. Epstein and M. S. Dadkhah, "Moiré Interferometry in Fracture Mechanics," *Experimental Techniques in Fracture*, Vol. III, J. S. Epstein, editor, VHC Publishers, Inc., New York, pp. 427-508 (1993).

8. Z. K. Guo, "Experimental and Numerical Characterization of the Fracture Behavior of Quasi-brittle Materials," Ph.D. Thesis, University of Washington, Seattle, Washington (1993).

9. M. S. Dadkhah and A. S. Kobayashi, "HRR Field of a Moving Crack, An Experimental Analysis," *Engineering Fracture Mechanics*, Vol. 34, pp. 253-262 (1989).

10. C. A. Burger, "Photoelasticity," Chap. 5, *Handbook on Experimental Mechanics*, 2nd edition, A. S. Kobayashi, editor, VCH Publishers, New York (1993).

11. D. Post, "Photoelasticity," Chap. 6, *Manual on Experimental Stress Analysis*," 5th edition, J. F. Doyle and J. W. Philips, editors, Society for Experimental Mechanics, Bethel, Connecticut (1989).

12. C. W. Smith, "Use of Optical Methods in Stress Analysis of Three-Dimensional Cracked Body Problems," *Optical Engineering*, Vol. 21, No. 4, pp. 696-703 (1982).

13. C. W. Smith, "Measurements of Three-dimensional Effects in Fracture Mechanics," *Fracture Mechanics*, Vol. 19, ASTM STP 969, American Society for Testing and Materials, Philadelphia, pp. 5-18 (1988).

14. C. W. Smith and A. S. Kobayashi, "Experimental Fracture Mechanics," Chap. 20, *Handbook on Experimental Mechanics*, 2nd edition, A. S. Kobayashi, editor, VCH Publishers, New York (1993).

15. V. A. Deason and M. B. Ward, "A Multipulsed Dynamic Diffraction Moiré Interferometer," *Laser Interferometry: Quantitative Analysis of Interferograms*, Vol. 1162, pp. 46-53, SPIE, Bellingham, Washington (1989).

16. V. A. Deason, J. S. Epstein and M. Abdallah, "Dynamic Diffraction Moiré: Theory and Applications," *Optics and Lasers in Engineering*, Vol. 12, No. 2, pp. 173-187 (1990).

12.6 Exercises

12.1 With the loading fixture of Fig. 12.3, large tensile stresses can be developed by a relatively low applied force P. List or discuss (a) the advantages and (b) the disadvantages of the system. (c) Propose an alternate design to develop large tensile stresses whereby the main disadvantage is circumvented.

12.2 Determine the crack opening displacement at the end of the deep notch in Fig. 12.8b.

12.3 In Fig. 12.8 the three fringe patterns exhibit distinctly difference qualities of the fringes. (a) Discuss the reasons, or probable reasons, for the differences. (b) Suggest changes of experimental procedure or apparatus that would improve each pattern.

12.4 Black holes and white holes appear near the crack in Fig. 12.9. (a) What changes of the apparatus could be made to minimize their extent? (b) Explain why they are not important regions for this analysis.

12.5 Referring to Fig. 12.10, describe (with diagrams and words) the force system that acts on the alumina layer that lies between cracks B and C.

12.6   Using the specimen of Fig. 12.11 as an example, explain (in paragraph form) the method of interior deformation measurements.

12.7   Although Fig. 12.12 is noisy, important information can be extracted. (a) Determine ε_x. (b) Determine (or estimate) γ_{xy} and plot γ_{xy} vs. y, assuming $\partial N_y / \partial x = 0$. (c) List any additional assumptions that were made.

13
Strain Standard for Calibration of Electrical Strain Gages

13.1 Introduction

Moiré interferometry can be used as a tool for precision measurements in numerous fields. In this chapter it will be applied to calibrate a test stand[1] which will be used for subsequent calibration of strain gages. Electrical resistance strain gages are in extensive use in two fields: experimental strain (and stress) analysis and transducer measurements. In the latter category, they are used in engineering and in commercial products to measure force, torque, pressure, and other parameters.

The strain gage is a thin filament of conducting material bonded to an insulating film. When it is adhesively bonded to a structural element that is deformed by external forces, local strains are transferred by the adhesive to the gage. The strain changes the length of the filament and the specific resistivity of the filament material, thereby changing the electrical resistance of the gage. The change of resistance is measured (directly or indirectly) and used to compute the strain in the underlying structural element. A *gage factor*, *GF*, defines the proportionality between strain and resistance by

$$GF = \frac{\Delta R / R_0}{\Delta \varepsilon} = \frac{\left(R_2 - R_1\right)/R_0}{\varepsilon_2 - \varepsilon_1} \qquad (13.1)$$

where
Δ means *change of*
R is electrical resistance of the gage
ε is longitudinal strain
and subscripts 0,1,2 prescribe these parameters when there is zero load and two different load magnitudes on the specimen, respectively.

A specimen that exhibits a uniaxial stress field is prescribed. In practice, the gage factor (GF) and initial resistance (R_0) are specified by the strain gage manufacturer and input into an instrument called a strain gage indicator. The electrical parameters are measured by the instrument and the strain reading is output.

Gage factors are determined for each gage design and checked by systematic sampling from manufactured batches. These routine measurements are made by installing test gages in the uniform strain field provided by the test stand. The active element of the test stand is illustrated in Fig. 13.1. A robust steel cantilever beam is used, where the width of the beam is tapered to produce a state of uniform strain along its top and bottom surfaces.

In order to calibrate the cantilever beam, a specimen grating was applied which responded to the strain when the beam was deflected. The strain in the grating was determined by moiré interferometry with high precision. However, it is not practical to employ moiré interferometry for the routine use of the test stand. Instead, a master strain gage was installed on the beam (Fig. 13.1) and its output is used to express the strain in the beam. The critical need, then, is to determine a calibration factor for the beam that relates the master strain gage readings to the strains measured by moiré. After calibration, the instrumented beam is called a *strain standard*, inasmuch as the strain on the surface of the beam is determined accurately by the readings of the master strain gage. Samples of production gages are subsequently bonded to the beam—the strain standard—to determine their gage factors.

A dual objective is posed:

- the calibration of the cantilever beam to 99.9% accuracy, i.e., ±0.1% error tolerance, and
- the traceability of accuracy to U.S. or international units of measurement.

In normal practice prescribed by ASTM standards, gage factors are determined at a strain level of 0.001, i.e., at 1000 μm/m. Hence, the strain standard is to be calibrated at that strain level. A tolerance of 0.1% means that strains must be determined accurately to 1 μm/m, a change of length of 1 part per million, to establish the calibration factor.

In the past, Class A extensometers had been employed to calibrate the beam. An example is the Tuckerman Extensometer, which utilizes a pivoting mirror and an optical lever to magnify the displacement between two knife edges.[2] However, such instruments are difficult to use. They are no longer marketed and existing instruments are in uncertain condition. Moiré interferometry was used instead to provide the precision strain measurement. Interest

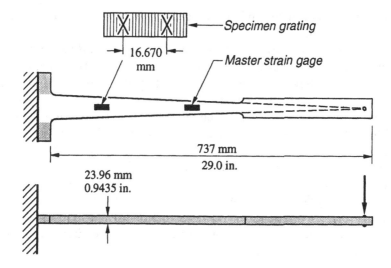

Fig. 13.1   Constant strain cantilever beam.

in the application goes back to 1981 when it was demonstrated that moiré interferometry had sufficient resolution.[3] The current work described here utilizes the same basic approach, but with refinements for practical application.

13.1.1 The Calibration Factor

It is the cantilever beam that is calibrated, not the master strain gage. The moiré system measures the strains and the master gage provides a convenient means of reporting the moiré measurements. The calibration operation determines the proportionality between the moiré measurements and the master gage readings. The constant of proportionality, which will be called the calibration factor (CF), is defined by

$$CF = \frac{\Delta \eta}{\Delta \varepsilon} = \frac{\eta_2 - \eta_1}{\varepsilon_2 - \varepsilon_1}$$ (13.2)

where η represents the master gage readings.

Special flexibility is afforded by the fact that the master gage is not being used to measure strains. Specifically,
- the gage factor of the master gage can remain unknown
- the alignment of the master gage can be arbitrary and undetermined
- the effective thickness of the master gage (or its distance from the neutral axis of the beam) can be arbitrary and undetermined

• the absolute accuracy of the strain gage readout equipment is inconsequential.

What is required is sufficient resolution, linearity and repeatability, and these requirements are satisfied admirably by strain gages and strain gage indicators. Regardless of gage factor, alignment, thickness and absolute accuracy, the output of the master gage is proportional to the strain in the beam. Once the calibration factor is established, the master gage output accurately defines the strain along the beam.

Equations 13.1 and 13.2 are similar, but they represent different stages of strain gage engineering. First the cantilever beam is calibrated using Eq. 13.2 to establish the strain standard. Then strain gages are calibrated on the cantilever beam to establish their gage factors. The moiré measurements are used in the first stage, and that is the main subject of this chapter.

13.2 The Specimen Grating

The specimen grating was replicated on the beam (Fig. 13.1) from a photoresist mold. In this case, a specific grating thickness was desired to match the effective thickness of typical strain gages that will be calibrated later. The desired thickness of 37 µm (0.0015 in.) was achieved by using shims under the mold in the replication process.

The specimen grating mold was made in a photoresist film on a glass substrate by the holographic process (Sec. 4.92). It was a linear grating (not a cross-grating) of 1200 lines/mm nominal frequency. Two x marks were scribed in the mold to establish the *gage length*, GL, as illustrated in Fig. 13.1. A gage length of 16.67 mm was chosen. The number of fringes, ΔN, in the gage length is calculated from the fringe frequency, F, where by Eq. 4.6,

$$F = f\varepsilon$$

(where f is twice the specimen grating frequency) and

$$\Delta N = F(GL)$$

These provide, for the nominal values of frequency and gage length and for a strain of 0.001 (i.e., 1000 µm/m),

$$\Delta N = f\varepsilon(GL) \approx 2(1200)(0.001)(16.67) \approx 40 \, fringes \qquad (13.3)$$

Conversely, the strain is calculated as

$$\varepsilon = \frac{\Delta N}{f(GL)} \approx \frac{\Delta N}{2(1200)(16.67)} \approx 25 \, \Delta N \; \mu m / m \qquad (13.4)$$

The sign for *approximately equal* is shown because nominal values are used instead of precise values.

13.3 Apparatus and Basic Procedure

13.3.1 The Interferometer

The moiré interferometer used in this work is illustrated in Fig. 13.2. It is a two-beam version of the system of Fig. 4.20. A collimated beam of coherent light strikes the 45° mirror from behind, to be redirected to the linear grating of 1200 lines/mm. The beams of +1 and –1 diffraction orders are deflected by the plane mirrors to intersect at the specimen grating. They create a virtual reference grating which is adjusted by fine-tuning the mirror alignment to produce a null field. Its nominal frequency is $f = 2400$ lines/mm.

Light that carries the moiré pattern emerges essentially normal to the specimen grating. It is redirected to the observer by the 45° mirror into the horizontal plane in front of the interferometer.

13.3.2 Experimental Procedure: Basic Description

Strains were induced in the cantilever beams by a deflection mechanism at the free end. The deflection was adjusted to produce an integral number of fringes, ΔN, in the gage length between the x marks (e.g., $\Delta N_1 = -40$). The reading η_1 from the master strain gage was recorded, with the strain gage indicator set at a gage factor of 2.000. The deflection was reversed and it was adjusted for an integral number of fringes (e.g., $\Delta N_2 = +40$), and the master gage reading η_2 was recorded.

The strain increment $\Delta\varepsilon$ determined by moiré was calculated by Eq. 13.4 for the total change of fringe order, whereby

$$\Delta\varepsilon = \varepsilon_2 - \varepsilon_1 = \frac{\Delta N_2 - \Delta N_1}{f(GL)} \qquad (13.5)$$

The corresponding master gage readings were totaled to give

$$\Delta\eta = \eta_2 - \eta_1$$

Then, the calibration factor was calculated by

$$CF = \frac{\Delta\eta}{\Delta\varepsilon} = f(GL)\frac{\eta_2 - \eta_1}{\Delta N_2 - \Delta N_1} \qquad (13.6)$$

Accordingly, the measurements were made for a total strain increment near 2000 μm/m. Any deviation from an initial null field at the zero strain level (zero beam deflection) was canceled by the subtraction in Eq. 13.5. Similarly, the presence of any initial strain gage reading was canceled by their subtraction.

An iterative process was used to adjust ΔN_1 (and ΔN_2) to an integral number. First, the beam deflection was adjusted to give a first approximation of the desired integral number of fringes between x marks. Then, the fringes were translated (i.e., they were shifted) to center a black fringe over the left x mark. The deflection was readjusted and the fringes were shifted again, until black fringes were centered over both the left and right x marks. The procedure was repeated several times to quantify random errors. The fringe shifting was accomplished by translating the upper grating in Fig. 13.2 by tiny amounts, using the mechanism described later in Sec. 13.5.2 and Fig. 13.6.

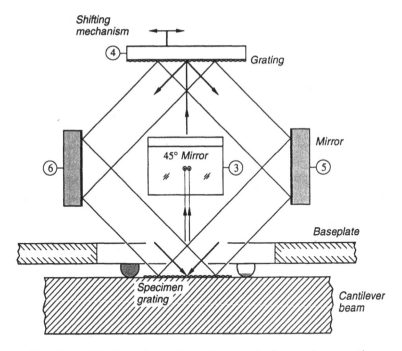

Fig. 13.2   Moiré interferometer to interrogate the specimen grating.

13.3.3 The Optical System

A top view of the complete optical system is illustrated in Fig. 13.3. The optical elements are mounted on an aluminum base plate, which was located above the specimen grating. The baseplate was suitably

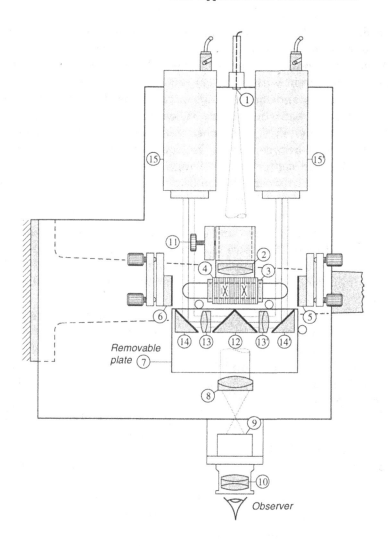

Fig. 13.3   Optical system mounted on the cantilever beam.

attached to the cantilever beam. The specimen grating is indicated in
the figure, although its view is actually obstructed from above by the
45° mirror (3) and the 1200 line/mm grating (4) shown in Fig. 13.2

Light from a 5 mW helium-neon laser diverged from optical fiber
(1) and it was collimated by lens (2). The collimated beam was
reflected upwards by the 45° mirror (3) to strike the upper grating (4).
There, beams of the +1 and −1 diffraction orders followed the paths
shown in Fig. 13.2, intercepting mirrors (5) and (6) and intersecting to
form a virtual reference grating at the specimen. Light that traveled

upwards from the specimen and carried the moiré pattern was directed forward by the 45° mirror toward the observer.

A secondary removable plate (7) carried lenses and mirrors. First consider the system with plate (7) removed. The specimen grating, with its x marks and moiré fringes, was focused by lens (8) onto a ground glass screen (9). There it was viewed with the aid of a focusing magnifier (10). The observer could see the x marks and count the fringes between them. He could shift the fringes by means of screw (11) and center a dark fringe above an x mark. He could count the fringes between the x marks, but as an aid, a transparency with 41 equally spaced lines was placed in contact with the ground glass. The size of the transparency was adjusted such that the 0th line coincided with the left x and the 40th line coincided with the right x. Superposition of the moiré fringes and the transparency created a supermoiré pattern. A supermoiré fringe that intercepted both x marks indicated a first approximation of 40 fringes in the gage length.

After the first approximation, plate (7) was inserted in the optical path. Its alignment was assured by contact with three pins. The field was split into two parts. The region surrounding the left x was focused by lens (13) and directed by mirrors (12) and (14) to video camera (15). The region surrounding the right x was focused onto the image plane of video camera (15').

Two video monitors were viewed by the operators. Figure 13.4 illustrates the images in one monitor before and after the fringe adjustment; Fig. 13.4b illustrates that perfect centering was not always achieved. The dark fringes were narrowed by overexposure and binarized by adjusting the monitors for high contrast. It was felt that these narrowed fringes could be centered more accurately than fringes with a cosinusoidal intensity distribution. In operation, one observer made fine adjustments of the cantilever beam deflection while another observer adjusted screw (11) until black fringes were judged to coincide with both x marks.

13.4 Tuning the Optical System

Tuning involved several steps which are described below.

13.4.1 Symmetry

The first step in tuning is to establish a first approximation of collimation. This is done by observing the plane of the fiber tip (1), shown in Fig. 13.3. Light that strikes the upper grating (4) reflects

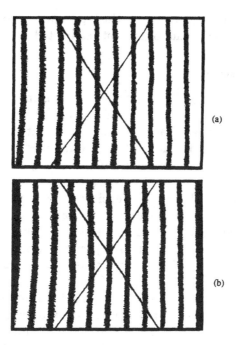

Fig. 13.4   Images on the monitor of a video camera (a) before alignment of a black fringe on a gage length mark and (b) after estimated alignment.

back in the zero order of the grating, and the back-reflected light is converged by the collimating lens (2) to an apex. The fiber tip (and the collet in which it is held) is moved back and forth until the apex lies in the same plane as the fiber tip. Then, the fiber tip is moved in its plane until it is coincident with the back-reflected apex. With these adjustments, the incident beam is essentially collimated and normal to the upper grating (4).

Next, mirror (6) is covered and the plane of the fiber tip is observed. Some of the light that reaches the specimen grating via mirror (5) is diffracted back in the second diffraction order and follows essentially the original path. Its apex is seen to move as mirror (5) is inclined back and forth. Mirror (5) is adjusted to direct the apex back into the source, i.e., into the fiber tip. Mirror (5) is covered and (6) uncovered, and the steps are repeated to direct the second-order diffraction from the specimen grating back into the source. These steps assure symmetry of the beams incident upon the specimen grating.

At this stage, moiré fringes are visible on the ground glass (9). Tiny adjustments of mirrors (5) and (6) are made to spread the moiré fringes out to a null field. By this procedure, the virtual reference grating is established at twice the specimen grating frequency. Thereafter, mirrors (5) and (6) are not readjusted.

13.4.2 Circulation

Referring again to Fig. 3.2, it is clear that some of the light that reaches the specimen grating from mirror (5) will diffract in the zeroth order and travel back to the upper grating via mirror (6). There, a portion will diffract in the zeroth order again and proceed along the original path. In their zeroth orders, the gratings act as mirrors and the light circulates repeatedly around the four-mirror circuit. Some of the light that reaches the specimen grating from mirror (6) circulates in the opposite direction.

However, upon each cycle of circulation, a fraction of the light is diffracted in the first order of the specimen grating and follows a path to the observer. Such extraneous beams create faint ghost patterns that reduce the visibility of the primary moiré pattern. A practical way to circumvent this problem is to move the fiber tip in its plane to a position above or below the apex of returning light. In relation to Fig. 13.3, the fiber tip is moved on a line perpendicular to the plane of the figure. This change does not affect the symmetry of beams approaching the specimen grating, but it introduces an out-of-plane inclination to the circulating beams sufficient to prevent them from overlapping and creating ghost fringes. An inclination of about 2° was used in this work.

13.4.3 Specimen Plane

An interferometer of this design tolerates spatial incoherence when it is properly tuned (Sect. 4.7.8). This means that it tolerates imperfect collimation. Perfect collimation could not be achieved because the collimating lens (2) was not fully corrected and because the 45° mirror (3) was not optically flat. Tuning to nullify the potential error associated with collimation was accomplished in two steps as described here and in the next section.

Referring again to Fig. 13.2, spatial incoherence is unimportant when the two diffracted rays that emerge from one point on the upper grating recombine, or cross, at a point on the specimen grating. The geometry of the figure shows that this condition requires that the upper grating must be raised to a specific, precise distance above the specimen grating.

The height of the upper grating was varied progressively by inserting thin shims under its supporting arm. For each step, the collimation was radically altered by moving the fiber tip forward by about 25 mm, whereupon the fringes were observed to see whether there was a change in the number of fringes between the x marks. No change signified the correct position of the upper grating. The position was optimized to within 40 μm (0.0016 in.) by this procedure.

Rotational alignment of the upper grating was upset in the course of removing and inserting shims. A thumbscrew arrangement was implemented to control the rotational alignment and the grating was realigned after each change of height.

13.4.4 Collimation

Optimal tuning of the collimation was performed in this step. The question of why we are so much concerned about collimation could be raised. The answer involves the out-of-plane displacement of the specimen and specimen grating that occurs when the cantilever beam is deflected. The mechanical arrangement is such that the out-of-plane displacement, W, relative to the upper grating is small, but it is finite. When $\varepsilon \approx 1000$ μm/m tensile strain, $W \approx 25$ μm at the x marks, and for compressive strain it is about -25 μm. If the light rays that pass through the x marks are not parallel, the virtual reference grating will change its frequency and the fringe count in the gage length will change as a result of the ± 25 μm (± 0.001 in.) displacement of the specimen grating.

As a final step in nullifying this effect, fine adjustments were made of the axial position of the fiber tip, to locate it at the optimum distance from the collimating lens. The axial position of the fiber tip was changed in small steps. For each step the upper grating was raised 2.5 mm (0.1 in.) and the change in the number of fringes between the x marks was observed. With the optimum position of the fiber tip, the change was reduced to 0.1 fringe for the 2.5 mm change of height. The corresponding influence of the ± 25 μm change of the specimen grating is only 0.001 fringes, which is certainly negligible.

13.5 Mechanical Systems

13.5.1 Method of Attachment

Figure 13.5 illustrates the attachment of the optical system to the cantilever beam. The baseplate that carries the optical system rested on the beam, with contact along two semicircular rails. Actually, there were three regions of line contact, each about one centimeter long, as indicated in Fig. 13.5b. The left side was clamped snugly to the beam, using a viscoelastic gasket that relaxed to reduce the clamping force. The clamping force on the right side was very small, merely enough to maintain contact.

The baseplate, with its attached components, was carefully balanced over the left rail so that all its weight would be reacted at that location. The 49N (11 lb.) weight of the system added a small bending moment to the left, but no bending moment to the right. The bending moments at the specimen grating location and master strain gage location were independent of the attachment. In addition, stresses and strains induced by the clamping forces were very small and locally self-equilibrated, and they were reasonably distant from the specimen grating.

With this method of attachment, the interferometer and optical system moved together with the beam. As the beam deflected, it bent into a circular arc of 12.0 m radius when $\varepsilon = 1000$ µm/m. Then, the maximum out-of-plane displacement of the grating, relative to the rails, was only 27 µm. At the location of the cross-marks, it was only 24 µm (0.001 in.).

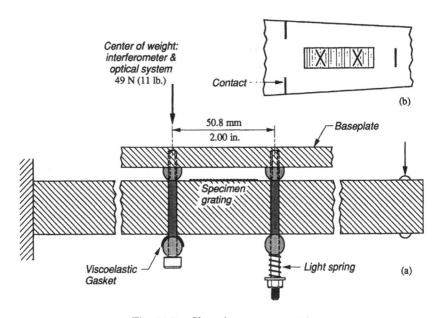

Fig. 13.5   Clamping arrangement.

13.5.2 Fringe Shifting Device

The arrangement used for fringe shifting is illustrated in Fig. 13.6. The upper grating (4) of the interferometer was mounted on a plate, which in turn was mounted on an elastic flexural fixture. The

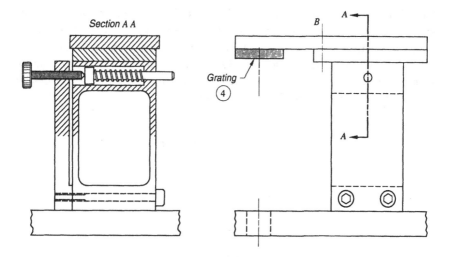

Section A A

Grating
(4)

B

A

A

Fig. 13.6   Flexural fixture for fringe shifting by tiny translation of grating (4).

fixture is basically a hollow tube that deflects to the right when the thumbscrew depresses the internal coil spring.

The displacement required to shift fringes by one full fringe order was only 1/2400 mm, or 0.417 μm. It is a tiny displacement, and only a tiny force on the flexural fixture was needed. A light coil spring was used so that a substantial rotation of the thumbscrew corresponds to a very small change of force on the spring.

The fixture was designed to provide in-plane rotational adjustment of the grating too. The plate to which the grating was mounted could pivot about centerline B; the rotation was controlled by a thumbscrew (not shown) at the opposite end of the plate.

13.5.3 Beam Deflection Device

Deflection of the cantilever beam was controlled by the mechanism illustrated in Fig. 13.7. Two hardened balls were fitted into the end of the beam to locate the position of load application. The mechanism was driven upwards or downwards by the screw-jack. Upwards motion induced compressive strains and downwards motion created tensile strains in the specimen grating. The end of the beam floated between two stiff coil springs as shown. Downwards motion of the assembly increased the force exerted by the upper spring and decreased that of the lower spring, and the differential force deflected the beam. Thus the screw-jack was used for coarse adjustments of the deflection. Fine adjustments were made by turning the hand-screw at the top, which engaged a coil spring of smaller stiffness.

The hand-screw increased or decreased the compression of the spring, thus altering the force exerted on the upper side of the beam by a small amount. Thus, the load or deflection of the beam could be adjusted to control the strain easily to 0.1 μm/m. The fine adjustment was made with the hand-screw, while viewing the video monitors, to center the black fringes on the x marks.

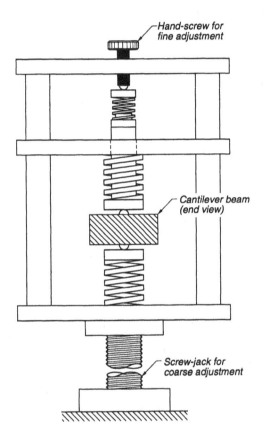

Fig. 13.7   Schematic illustration of beam deflection device.

13.6 Calibration of the Specimen Grating

The mold from which the specimen grating was replicated was a diffraction grating with *f*/2 lines (or furrows) per mm, and it had two x marks scribed on its surface to fix the gage length. Measurements of the grating frequency and gage length were performed on the mold prior to replication. The measurements were not made on the actual specimen, but instead, the dimensions of the mold and its replica were assumed to be equal. The assumption is justified because any uniform change of dimensions in the replication process would change the grating pitch and the gage length by the same factor and such a change would not influence the results. The rationale is explained in Sec. 13.8.1 under the heading *Temperature*. Accordingly, the measurements made on the mold served to calibrate the specimen grating.

The distance between the x marks in the mold was measured with a biaxial measuring machine manufactured by the David W. Mann Co. Its resolution is 1.0 μm per division. The average of six measurements of the gage length was 16.670 mm and the standard deviation was 1.23 μm. The standard deviation is 74 parts per million (ppm) of the gage length.

The frequency of the specimen grating mold was determined by interferometric comparison with a known reference grating. The reference was a nominally 1200 lines/mm diffraction grating on a fused silica substrate, manufactured by the Bausch and Lomb Optical Company (B&L). First, the accuracy of the reference grating must be addressed. Calibrations were performed by B&L to establish the linear accuracy of their ruling engines. The wavelength of the green line of the mercury 198 isotope lamp was the length standard for an earlier calibration,[4] and subsequent measurements were made with the red line of the helium-neon laser.[5] These wavelengths are internationally accepted standards for measurements of length. The frequencies of B&L gratings, produced on fused silica and specified at 20°C, were found to be 109 ±5 parts per million (ppm) greater than the nominal values. Consequently, the B&L reference grating has a traceable frequency of 1200.131 lines/mm with an uncertainty of ±0.006 lines/mm, or ±5 ppm.

The mold and the B&L grating were compared by moiré interferometry with the simple apparatus illustrated in Fig. 4.11. The B&L grating was inserted and the interferometer was adjusted to produce a null field. Thus, the virtual reference grating was 2400.26 lines/mm, or twice the B&L grating frequency. Then the mold was inserted; it was aligned to the same position by assuring that its second-order diffraction returned to the source aperture (A in Fig. 4.11). The interference pattern exhibited 7.0 fringes in the 16.670 mm gage length; the mold frequency was higher, inasmuch as angle α had to be increased to reduce the fringe density. The mismatch would be eliminated and a null field would be produced in the mold if the frequency of the virtual reference grating was increased by 7.0 lines in the gage length, i.e., to

$$f = 2400.26 + \frac{7.0}{16.670} = 2400.68 \text{ lines} / \text{mm} \qquad (13.7)$$

Thus, the frequency of the mold was $f/2 = 1200.34$ lines/mm. The uncertainty of 0.2 fringes in the 7.0 fringe observation leads to an uncertainty of the frequency of ±5 ppm. When added to the 5 ppm uncertainty of the reference grating, the total uncertainty is only ±10 ppm.

13.7 Strain Gage Instrumentation

The master strain gage response was measured with a Model 3800 Wide Range Strain Gage Indicator,[6] using a gage factor setting of 2.000. The instrument was calibrated with a Model 1550A Strain Indicator Calibrator.[7] The calibrator is traceable to National Institute of Standards and Technology (NIST) certification of absolute accuracy to 0.025%.

The strain gage indicator was corrected in its calibration operation to 0.1 μm/m at the 1000 μm/m strain level. The smallest unit in the digital output scale was 0.1 μm/m. After it was calibrated, the absolute accuracy of the strain gage indicator (at 1000 μm/m) was within 0.025%, or 250 ppm.

13.8 Systematic Errors, Uncertainties

A basic uncertainty of the strain gage indicator readings is half the least count. It influences the calibration factor as a random error within 0.05 μm/m in 1000 μm/m, or ±50 ppm.

The uncertainty of the absolute accuracy of the strain gage indicator does not influence the calibration factor of the strain standard. It does influence the subsequent determinations of gage factors of production gages, contributing a systematic error within ±250 ppm of the gage factor. Similarly, for subsequent determinations of gage factors of production gages, the uncertainty of the gage factor setting contributes another systematic error within half the least count (0.0005) in 2.000, or ±250 ppm. These uncertainties stem from different operations in the course of adjusting and reading the strain gage indicator and they are mutually independent.

Angular alignment of the specimen grating was done by the method of Sec. 4.9.3. In this case, an alignment bar was fabricated to match the taper of the cantilever beam. The alignment accuracy was estimated to be within 0.2°, which is consistent with predictions in

Ref. 8.   For any strain measurement, ε, the misalignment error, δ, is calculated from Mohr's circle of strain as

$$\frac{\delta}{\varepsilon} = -\frac{1+\nu}{2}(1-\cos 2\theta) \approx -(1+\nu)\theta^2$$

Using Poisson's ratio of $\nu = 0.289$ to define the biaxial strain field, and using a misalignment of $\theta = 0.2°$, the error in the measured longitudinal strain is within the range between 0 and -16 ppm.

There is an uncertainty of the thickness of the specimen grating of 3 mm, which influences the distance to the neutral axis of the beam by 0.025% or ± 250 ppm. This uncertainty produces a corresponding uncertainty of the calibration factor.

It should be noted that the strain standard is calibrated at the surface of the specimen grating. In subsequent use of the strain standard for determination of gage factors, no correction is needed when the thickness (t_g) of the installed strain gage is equal to the specimen grating thickness (t_m) of 37 μm. Otherwise, a systematic error occurs because of the unequal distances to the neutral axis of the beam; it is corrected by multiplying the calibration factor by

$$1 + \frac{(h/2) + t_g}{(h/2) + t_m}$$

where h is the thickness of the calibration beam.

13.8.1 Additional Considerations

Equation 13.5 can be updated with the measured values of grating frequency and gage length to

$$\Delta\varepsilon = \frac{\Delta N_2 - \Delta N_1}{2(1200.34)(16.670)} = 24.988 \times 10^{-6}\left(\Delta N_2 - \Delta N_1\right)$$
$$= 24.988\left(\Delta N_2 - \Delta N_1\right)\mu m / m \tag{13.8}$$

Other factors relating to the precision of measurements are considered carefully in this section, but they do not alter Eq. 13.8. Their influence is negligible.

Beam Curvature

Inasmuch as the tapered cantilever beam is designed for uniform strain, it bends into a circular arc. Figure 13.8a depicts the geometry, but the curvatures are greatly exaggerated. At the outer fiber, the change ΔL of any original length L_0 is

$$\Delta L = \varepsilon L_0$$

Then, by similar triangles

$$\frac{r}{L_0} = \frac{h/2}{\Delta L}$$

and therefore

$$r = \frac{L_0 \, h/2}{\Delta L} = \frac{h}{2 \, \varepsilon} \qquad (13.9)$$

For a strain level of 1000 μm/m, the radius of curvature is

$$r = \frac{23.96 \; mm}{2(0.001)} = 11,980 \; mm \approx 12 \, m$$

The virtual reference grating is fixed, while the specimen grating located between the x marks changes its length with the applied strain. Two factors influence the distance between x marks. One is the strain, but the other is an extraneous effect that stems from the curvature. When the beam flexes, the strain is defined in terms of the length along its surface, which is measured along the arc of a circle. On the other hand, the moiré corresponds to the projected distance between the x marks, i.e., to the chord or straight-line distance between the marks. At a strain level of 1000 μm/m, the chord differs from the arc by -1.36×10^{-6} mm. This difference is less than -0.1 ppm of the gage length. It is so tiny that no correction is required.

Beam Slopes

Referring again to Fig. 13.5, visualize the effect of beam curvature. Relative to the optical system, the slope of the beam remains zero at the center of the specimen grating and it changes monotonically with distance x from the center. The beam curvature and its slope Ψ are illustrated in Fig. 13.8b. In this case, the change of length that concerns us is measured along the surface of the beam. Thus, the slope introduces an extraneous fringe gradient given by Eq. 4.28, rewritten here as

$$F_e = \frac{dN_e}{dx} = -\frac{f}{2} \Psi^2 \qquad (13.10)$$

The extraneous fringe order N_e between x marks can be calculated by integrating Eq. 13.10. With the substitution

$$\Psi = \frac{x}{r}$$

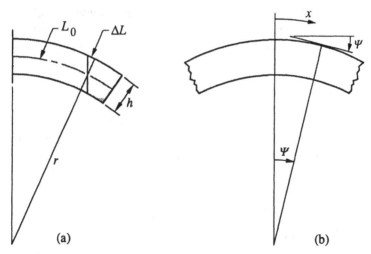

Fig. 13.8   (a) Relation between radius of curvature and strain $\Delta L/L_0$. (b) Slope of curved beam.

the integral is

$$N_e \;=\; -\frac{f}{2}\int \frac{x^2}{r^2}\,dx \tag{13.11}$$

At the x marks, i.e., where the integral is evaluated between $x = GL/2$ and $-GL/2$, we find

$$N_e \;=\; -\frac{f\,(GL)^3}{24\,r^2} \tag{13.12}$$

Equations 13.4, 13.9 and 13.12 can be combined to relate N_e to the fringe count ΔN. Thus,

$$N_e \;=\; -\frac{GL}{6\,f h^2}(\Delta N)^2 \;=\; -2.02 \times 10^{-6}\,(\Delta N)^2 \tag{13.13}$$

The extraneous fringes introduced by the slope of the specimen grating are always negative.

Because of these extraneous fringes, Eq. 13.8 should be rewritten as

$$\Delta \varepsilon \;=\; 24.988 \times 10^{-6}\Big[\big(\Delta N_2 - N_{e2}\big) - \big(\Delta N_1 - N_{e1}\big)\Big] \tag{13.14}$$

When $\Delta N_2 = -\Delta N_1 = 40$ fringes, the extraneous fringes N_{e2} and N_{e1} are equal in magnitude and sign and their effect is canceled in Eq. 13.14.

When $\Delta N_1 = -41$ and $\Delta N_2 = +39$ fringes, the extraneous fringes almost cancel; by combining Eqs. 13.13 and 13.14 and calculating $\Delta\varepsilon$, the influence of extraneous fringes is only 4 ppm. The correction will be disregarded.

Temperature

The mold used to replicate the specimen grating was made on a glass substrate. Its frequency was calibrated at a room temperature of 22.5°C against the traceable B&L grating, which was on a fused silica substrate. The gage length, i.e., the distance between x marks on the mold, was also measured at 22.5°C.

The frequency of the fused silica grating was established at 20°C and its length changed by only 1.05 ppm at the calibration temperature of 22.5°C. This tiny change was disregarded and the frequency at 22.5°C was established as $f/2$, where f is given by Eq. 13.7.

The mold was replicated on the steel cantilever beam at 23.4°C, and the moiré measurements were conducted at that temperature, 23.4°C. However, no additional corrections were required. The reason is that any subsequent expansion introduced changes of gage length and specimen grating pitch by identical factors. Equation 13.3 can be rewritten to express the strain sensitivity, i.e. strain per fringe, as

$$\frac{\Delta\varepsilon}{\Delta N} = \frac{1}{f(GL)} = \frac{g_s/2}{GL}$$

where g_s is the pitch of the specimen grating. Accordingly, temperature changes that expand the pitch and gage length equally have no influence on the strain sensitivity of the measurement. The critical requirement is merely that the grating frequency (or pitch) and the gage length must be established at a common temperature.

A separate requirement is that the moiré experiment must be conducted at a constant ambient temperature, for otherwise the frequency of the virtual reference grating might fluctuate during the experiment.

13.9 The Experiment: Data Sequence, Computations and Results

After the planning, assembly and tuning, the experiment—i.e., the data collection—took only one hour. This is the usual course, where the actual experiment seems like an anticlimax!

The cantilever beam was deflected in compression to give a fringe count of approximately −40 fringes in the gage length. Then it was adjusted more precisely for −40 fringes, using the video cameras and monitors, and the master gage reading was recorded. Next it was adjusted for $\Delta N_1 = -39$ fringes, using only the video system as a guide, and the master gage reading was recorded. Then data were taken for $\Delta N_1 = -40$ and −41 fringes.

The above procedure was repeated, with master gage readings taken for $\Delta N_2 = 40$, 39, 40, and 41 fringes. That concluded a full run of data collection. The data sequence is outlined in Table 13.1, where the master gage readings are represented by letters. The sequence was repeated twice more, to provide data for three complete runs.

TABLE 13.1 Schematic of Experimental Data

ΔN_2	40	39	40	41
η_2	A	B	C	D
ΔN_1	−40	−39	−40	−41
η_1	E	F	G	H

Note: letters represent the master strain
gage readings

Equations 13.6 and 13.8 were combined to give

$$CF = \frac{1}{24.988 \times 10^{-6}} \left(\frac{\eta_2 - \eta_1}{\Delta N_2 - \Delta N_1} \right) \qquad (13.15)$$

The measurements (Table 13.1) provided 16 independent combinations of master strain gage readings for fringe increments ranging from 78 to 82. The combinations are indicated in Table 13.2. By using these combinations in Eq. 13.15, sixteen values of calibration

TABLE 13.2 Sixteen Independent Strain
Increments for Each Run

$\Delta N_2 - \Delta N_1$	$\eta_2 - \eta_2$
78	B–F
79	B–E, B–G, A–F, C–F
80	A–E, A–G, C–E, C–G, B–H, D–F
81	A–H, C–H, D–E, D–G
82	D–H

factor were calculated for each run. The numerical values are given in Fig. 13.9 for the three runs, providing a distribution of 48 numerical values of gage factor. The average value,

$$CF = 1.03006 \qquad (13.16)$$

is shown in the figure by a solid line.

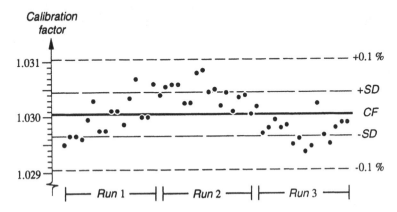

Fig. 13.9   Scatter of experimentally determined calibration factors. The lines show the average value (CF), the range of combined error for ±1 standard deviation (a = 1), and the range within the 0.1% tolerance.

The individual values of CF deviated from the average by D. The 48 values of D have an essentially Gaussian or normal distribution about the average value. The standard deviation, SD, was determined by

$$SD = \left(\frac{1}{47} \sum_{n=1}^{48} D_n^{\,2} \right)^{1/2} \qquad (13.17)$$

with the result

$$SD = 0.00039 \qquad (13.18)$$

which is 380 ppm of the calibration factor.

13.10 Accuracy

The result given by Eq. 3.16 provides a reasonable estimate of the calibration factor. The true value lies within an error band that surrounds the estimate. The purpose of this section is to estimate another parameter, namely the extent of the error band.

Table 13.3 lists the pertinent uncertainties (exceeding 10 ppm) and their origins. The influence on the calibration factor is given in ppm for each uncertainty. The first entry treats random errors involved in centering the fringes on the x marks and the standard deviation found above is listed in Column 3.

TABLE 13.3 Contributions to the Uncertainty of the Calibration Factor

Case of Uncertainty	Type	Range of Uncertainty (ppm)	Standard Deviation (ppm)	Symbol
Column	1	2	3	4
Centering fringes on x marks	Random		380	SD_1
Resolution of strain gage indicator	Random	$0^\dagger$	$0^\dagger$	SD_2
Gage length	Systematic		74	SD_3
Thickness of grating	Systematic	±250	250	SD_4
Grating alignment	Systematic	0 to–16	$16^\ddagger$	SD_5

† ±50 ppm is imbedded in SD_1
‡ approximated by a symmetrical distribution

Uncertainty SD_2 represents the error associated with the limited resolution of the strain gage indicator. Note that this random error is imbedded in the variations of the strain gage readings. Uncertainty SD_1 already contains SD_2. Therefore SD_2 is not considered independently, but its effect is automatically incorporated in SD_1.

The uncertainty of the gage length is represented by the standard deviation specified in Sec. 13.6, i.e. 74 ppm, as listed in column 3 of the table. The specified uncertainty of specimen grating thickness is 3 μm, which influences the calibration factor by ±250 ppm; it is listed in column 2. The range of error from angular misalignment of the specimen grating is taken from Sec. 13.8 and listed in column 2.

Statistical analysis can be applied to assess individual and overall errors. The analysis assumes normal distributions of measurement errors, which is justified by the nature of the error in each case. The range into which the error is expected to lie is expressed in terms of a confidence interval. The corresponding numerical values are listed in the first two columns of Table 13.4. For example, we can assume with 95.4% confidence that an error will fall within a range of two standard deviations ($a = 2$). Accordingly, the listing for SD_4 in

Table 13.3 assumes $a = 1$, i.e., a 68.3% confidence interval. Liberty is taken with SD_5 by assigning its value as if the uncertainty is symmetrical; however, the influence is extremely small and the contribution to the overall error is conservative.

TABLE 13.4   Coefficient of SD and Overall Error for Various Confidence Intervals

Confidence Interval	Error Band $\pm a\ (SD)$	Overall Error (ppm)
68.3%	$\pm 1.0\ SD$	267
86.6%	$\pm 1.5\ SD$	400
95.4%	$\pm 2.0\ SD$	534
99.0%	$\pm 2.58\ SD$	689
99.7%	$\pm 3.0\ SD$	801

The individual uncertainties must be combined to estimate the overall error. For the worst-case scenario, where every uncertainty attains its maximum magnitude, the overall error would be the scalar sum of the contributions. However, the worst-case scenario is extremely remote since the individual errors are expected to have intermediate magnitudes and various signs. Instead, the bounds for the combined error are predicted by [9]

$$e = \pm a \sqrt{\sum_{i=3}^{5} SD_i^2 + \frac{SD_1^2}{48}} \tag{13.19}$$

where a is the coefficient of SD listed in Table 13.4, i is the subscript in Table 13.3, and 48 is the number of readings from which SD_1 was determined. Note that the first term under the radical comprises the systematic errors and the second term comprises the random errors.

The assessment of overall error depends upon the confidence interval. The results are listed in Table 13.4 for overall error calculated by Eq. 13.19. Thus, for a confidence of 68.3%, where $a = 1$, the calibration factor is

$$CF = 1.03006 \pm 267 \text{ ppm}$$
$$= 1.03006 \pm 0.00027$$

and for 99.7% confidence ($a = 3$)

$$CF = 1.03006 \pm 800 \text{ ppm}$$
$$= 1.03006 \pm 0.00082$$

In a practical sense, the specification of confidence to three significant figures is artificial. It stems from the idealized rules of statistics. The experiment was designed to measure the calibration factor of the strain standard within 0.1% tolerance, and that accuracy was achieved with extremely high confidence. Rounded off to the 1% tolerance, the calibration factor is prescribed by Eq. 13.20. We note from Fig. 13.9 that no measurement exceeded this bound, which implies virtually 100% confidence.

$$CF = 1.030 \pm 0.001 \qquad (13.20)$$

13.11 Gage Factor Measurements

The strain standard, i.e., the calibrated cantilever beam, can be used to determine gage factors of production gages. To do so, the strain increment in Eq. 13.1 is determined by

$$\varepsilon_2 - \varepsilon_1 = CF\left(\eta_2 - \eta_1\right) \qquad (13.21)$$

where the conditions for the master strain gage readings are unchanged. This means that the gage factor setting for the master strain gage should remain 2.000.

To assess the overall accuracy of gage factors, the propagation of errors is predicted by Eq. 13.19, but additional contributions must be considered. They are $i = 6$ through 9, which account for potential systematic errors caused by uncertainties of the absolute accuracy of the strain gage indicator, the gage factor setting, the effective thickness of the strain gage being calibrated, and the alignment of the strain gage.

13.12 Future Refinements

Because of its importance to strain gage calibrations, the calibration factor of the strain standard should be determined on a regular schedule. Certain refinements of the experiment are appropriate for future measurements.

Fringe Shifting

With the present arrangement in which the operator turns the thumbscrew (11) shown in Fig. 13.6, he inadvertently exerts an uncontrolled axial force on the thumbscrew. This force introduces irregular movements of the fringes and optimal adjustment is hampered. An improvement would be remote control of the fringe shifting mechanism. Various electrical options are possible, including a motorized screw mechanism, a differential transformer actuator and a piezoelectric actuator. The last option might be the most practical. In addition to improved ease of operation, the refinement is expected to reduce the uncertainty associated with centering fringes on the x marks.

TV Observer

A substantial improvement in convenience would be achieved by viewing the full field of fringes with a TV camera instead of the focusing magnifier (10) shown in Fig. 13.3. The ground glass (9) would be removed and the image would be focused upon the receiving plane of a third TV camera. The fringes would be viewed on a remote monitor where they could be counted unambiguously and quickly.

Laser Diode Light Source

The current apparatus consisting of a helium-neon laser, a fiber coupler and an optical fiber is relatively clumsy. It could be replaced with a compact laser diode.

Grating Thickness

The relatively large uncertainty associated with grating thickness could be reduced by devising superior means to measure the grating thickness.

13.13 References

1. J. W. Dally and W. F. Riley, *Experimental Stress Analysis*, 2nd edition, McGraw-Hill, New York, p. 171 (1978).

2. R. B. Watson, "Calibration Techniques for Extensometry: Possible Standards of Strain Measurement," *J. Testing and Evaluation*, ASTM, Vol. 21, No. 6 (1993).

3. R. B. Watson and D. Post, "Precision Strain Standard by Moire Interferometry for Strain Gage Calibration," *Experimental Mechanics*, Vol. 22, No. 7, pp. 256-261 (1982).

4. D. Richardson and R. M. Stark, "Accurate Linear Scales Ruled on a Grating Engine," *J. Opt. Soc. Am.*, 47,1, pp. 1-5, (1957).

5. Dr. Erwin G. Loewen, Milton Roy Co., Rochester, N.Y. (successor to the Bausch & Lomb Optical Co. for grating operations), personal communication (1992).

6. "Model 3800 Wide Range Strain Indicator Instruction Manual," Instruments Division, Measurements Group, Inc., Raleigh, NC (1985).

7. "Model 1550A Strain Indicator Calibrator Instruction Manual," Instruments Division, Measurements Group, Inc., Raleigh, NC (1985).

8. D. Joh, "Optical-Precision Alignment of Diffraction Grating Mold in Moiré Interferometry," *Experimental Techniques*, Vol 16, No. 3, pp. 19-22 (1992).

9. C. Lipson and N. J. Seth, *Statistical Design and Analysis in Engineering Experiments*, McGraw-Hill, New York (1973).

13.14 Exercises

13.1 Explain (with words and sketches) why translation of grating (4) in Fig. 13.2 causes the virtual reference grating to shift with respect to the specimen grating.

13.2 Referring to Fig. 13.6, (a) design a practical remote control arrangement to replace the manual arrangement for translation of grating (4). (b) Discuss the advantages and disadvantages of your design.

13.3 List the advantages and disadvantages of utilizing a flexural member as the active element of the strain standard.

13.4 Propose a practical design that utilizes a robust tensile member instead of a flexural member for the strain standard. Avoid use of a testing machine to apply the strain.

Appendix A

Shadow Moiré with Enhanced Sensitivity; O/DFM for Geometric Moiré

A.1 Introduction

Shadow moiré persists as an important method of experimental analysis. In fact, its importance is growing. The method is used to study out-of-plane bending of structural plates, especially for buckling and prebuckling analyses. Its use in the electronics industry is intensifying for measurements of warpage of large circuit boards, both the initial warpage and temperature-induced warpage. While the methods of Sec. 3.3.1 have served nicely, there is a demand for higher sensitivity. The gap between the sensitivities of shadow moiré and holographic interferometry is more than two orders of magnitude, and a technique with intermediate sensitivity is needed for out-of-plane displacement measurements.

Although fringe shifting and image processing is popular for extracting fractional fringe orders, few methods are well adapted for the patterns of geometric moiré. The main reasons are (a) the intensity distribution across geometric moiré fringes is a complicated function of fringe order, rather than the simple harmonic function of classical interference fringes; and (b) in practice, the function is not fixed, but varies from region-to-region in the fringe pattern. However, the *optical/digital fringe multiplication* (O/DFM) method (Sec. 5.3) is very robust and circumvents the difficulties. It is compatible with diverse intensity distributions, which can vary from region-to-region in the field, and it copes with variations of bar-to-space ratios and optical noise. Consequently, the O/DFM method of image processing is well suited to geometric moiré. It will be employed to enhance the sensitivity of shadow moiré by an order of magnitude.

The approach meshes nicely with all applications of geometric moiré, not merely shadow moiré. It will be demonstrated for in-plane displacement analysis, too. We acknowledge, however, that this appears to be a less important application than shadow moiré. The reason is that moiré interferometry with a virtual reference grating can be utilized for intermediate sensitivity applications and that approach would usually be easier; Weissman[1] has extended the use of moiré interferometry to the relatively low sensitivity domain, too, with a virtual reference grating of 40 lines/mm (1000 lines/in.).

A.2 Shadow Moiré with O/DFM

Chapters 3 and 5 should be reviewed for in-depth treatments of geometric moiré and the optical/digital fringe multiplication method. Fringe multiplication by a factor β is employed, where β is an even integer.

The specimen and optical arrangement used for this demonstration are illustrated in Fig. A.1. They are the same as those in Fig. 3.26. Oblique illumination and normal viewing are used, with $L = D$. With this arrangement, Eq. 3.21 provides

$$W = gN_z = \frac{1}{f}N_z \tag{A.1}$$

where g and f are the pitch and frequency of the reference grating, respectively. With optical/digital fringe multiplication, the fringe contours are enumerated by N^*, where N^* is chosen with respect to an arbitrary zero-order datum (Sec. 5.3.3). The number of fringe contours in the field of view is increased by a factor of β. Then, the out-of-plane displacements become

$$W(x,y) = \frac{1}{\beta f}N_z^*(x,y) \tag{A.2}$$

Fringe shifting provides the data for the O/DFM method. It is accomplished by moving the reference grating in the z direction to increase W uniformly throughout the field. By Eq. A.1, N_z is increased by one fringe order when W is increased by g. Accordingly, the experimental data is comprised of β fringe patterns recorded after successive shifts of g/β of the reference grating. The corresponding phase shift is $2\pi/\beta$ for each successive pattern.

For this demonstration, the reference grating was mounted on a micrometer traverse as illustrated schematically in Fig. A.1b. The

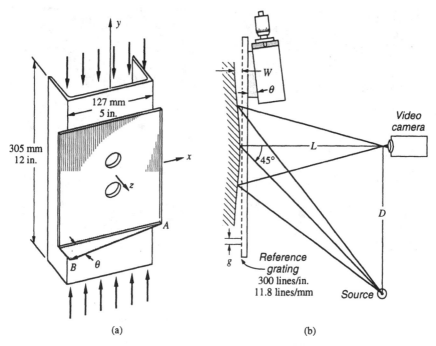

(a)   (b)

Fig. A.1   (a) Specimen with reference grating.   (b) Schematic diagram of
shadow moiré arrangement; the reference grating is attached to a micrometer
traverse.

traverse moved the grating in the direction *AB*, at a small angle to
the face of the specimen. Thus, relatively large displacements along
AB produced the tiny out-of-plane shifts g/β of the reference grating.
The motion was calibrated by determining the micrometer traverse
displacement, δ, required for an outwards (z direction) shift of g. In
the calibration procedure, the moiré pattern was recorded in the
video memory, the micrometer screw was advanced, and the new
pattern was recorded. The second digitized pattern was subtracted
from the first, and the steps were repeated iteratively until the
difference was zero. The displacement δ then corresponded to a shift
of g.

Figure A.2 demonstrates the method for $\beta = 2$. The initial moiré
pattern is shown in (a), and (b) is the corresponding pattern that is
fringe-shifted by π. They are complementary patterns, a name given
to patterns in which the phase at each point differs by π, and
consequently the maximum and minimum intensities are
interchanged.

A graphical representation of these two patterns is given in Fig.
A.3a, where the dashed curve indicates the intensity I^π of the

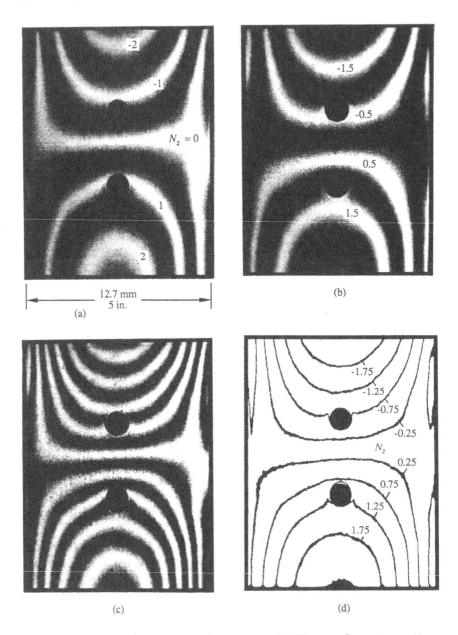

Fig. A.2 (a) The initial shadow moiré pattern. (b) The complementary pattern. (c) Pattern obtained by subtraction of (a) and (b), and inversion of negative values. (c) Sharpened contours corresponding to fringes in (c).

complementary pattern. The intensity distribution is shown as triangular with rounded corners. The triangular distribution is produced when the bars and spaces of the reference grating have equal widths and the rounding is caused by the finite resolution of the camera lens (Sec. 3.1.2). These are mathematically complicated intensity distributions, but the mathematical analysis for general periodic intensity distributions in Sec. 5.3.3 is applicable and the O/DFM method is fully effective.

The patterns are recorded by a video camera and the intensity at each pixel is converted to a digital value on a 256 gray scale. The O/DFM algorithm subtracts the intensities at each point, as illustrated in Fig. A.3b . Then it inverts the negative portions by

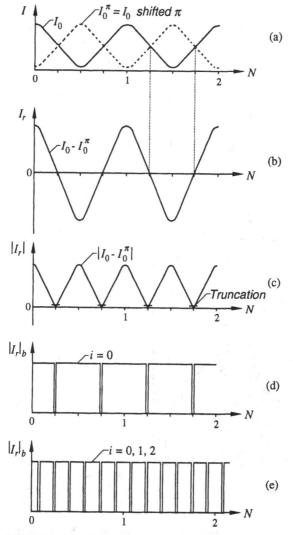

Fig. A.3   O/D fringe multiplication.   (a) Intensities of fringe-shifted complementary patterns. (b) Subtracted. (c) Absolute values. (d) Truncated and binarized. (e) Combined with other fringe-shifted patterns for $\beta = 6$.

taking absolute values, as illustrated in (c). The algorithm proceeds by truncating the data near $|I_r| = 0$ and binarizing by assigning intensities of zero and one to points below and above the truncation value, respectively. The result is graphed in (d). The whole-field patterns corresponding to graphs (c) and (d) are shown in Figs. A.2c and A.2d, respectively.

The result is a sharpened contour map that has twice as many contour lines as the number of fringes in the initial pattern. The sharpened contours occur at the crossing points of the complementary graphs, where the intensities of the complementary patterns are equal. Note that any noise or other factor that affects the two patterns equally has no influence on the locations of the crossing points. The crossing points and resulting fringe contours occur at odd multiples of $N/4$, as seen in Figs. A.2d and A.3a.

These steps complete the initial phase of the O/DFM method, whereby highly sharpened contours and multiplication by two are produced. The final step is illustrated in Fig. A.3e, which represents a case of $\beta = 6$. Data from pairs of complementary patterns that were shifted by $g/6$ and $4g/6$, and also by $2g/6$ and $5g/6$, were processed as

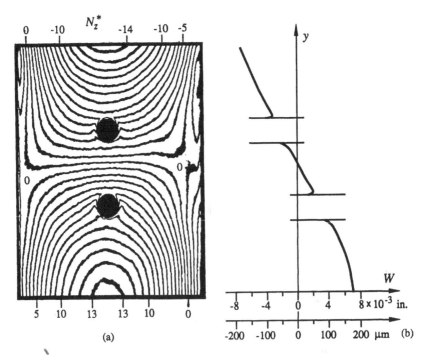

Fig. A.4   (a) O/DFM contours for multiplication by 6, corresponding to the shadow moiré pattern of Fig. A.2a. (b) Displacement W along the vertical centerline. The compressive load was 133 kN (30,000 pounds).

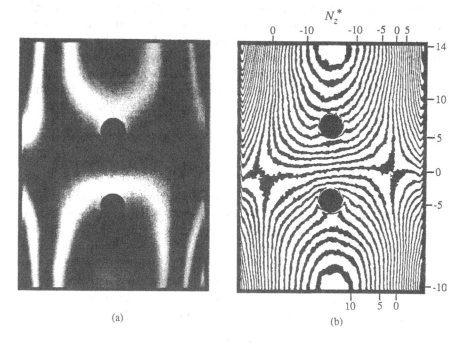

Fig. A.5   (a) Shadow moiré pattern. (b) Multiplication by 10. The reference grating was tilted to produce carrier fringes.

above and the three sets of sharpened fringe contours were combined to produce a composite pattern of fringe contours.

The whole-field contour map for $\beta = 6$ is shown in Fig. A.4. The left and right sides of the specimen were stiffened by the flanges of the channel and the out-of-plane distortions at the sides were very small. The web, or face of the channel experienced a second bending mode wherein the upper portion deflected inwards and the lower portion deflected outwards. In the immediate vicinity of the holes, the multiplied pattern shows a ridge on the left and right side of each hole; the ridge was in the positive z direction for both holes. At the top and bottom of each hole the surface curvature was inwards as indicated in Fig. A.4b, which is a graph of displacements along the vertical centerline. These features are not evident in the lower sensitivity pattern of Fig. A.2a. The contour interval for the multiplied pattern is g/β, i.e., 14.1 µm/fringe or 5.5 x 10^{-4} in./fringe order.

Figure A.5 shows the original and multiplied patterns for a different loading condition, but here the multiplication factor is $\beta = 10$. In this case, the reference grating was inclined to the face of the specimen, as revealed by the fringe gradients along the left and right sides. This inclination of the grating in the initial (no-load) alignment generates carrier fringes, which can be useful for elucidating the surface warpage. Again the mode 2 bending is evident. The optical arrangement of Fig. A.1 was employed and the pattern is interpreted by Eq. A.2. With $\beta = 10$ and $f = 11.8$ lines/mm (300 lines/in.), the contour interval is 8.5 μm/fringe order or 3.3×10^{-4} in./fringe order.

A.3 In-plane Geometric Moiré with O/DFM

The specimen and experimental arrangement for this demonstration are illustrated in Fig. A.6 The experiment is an extension of that shown in Fig. 3.20 and described in Sec. 3.2.2. Fringe shifting (for the V field) required tiny upwards motion of the reference grating in increments of $\dot{g} / \beta$. For this purpose, the reference grating was attached by a bracket to a micrometer traverse, as illustrated in Fig. A.6c, where θ is a small angle. Then relatively large adjustments of the traverse provided accurately controlled upward displacements of the reference grating. The calibration procedure was the same as that described in the previous section, whereby the factor of proportionality between the shift of the micrometer traverse and the shift of the grating was determined experimentally.

The experimental results are displayed in the following figures. Figure A.7a is the initial pattern for the V field and (b) is the complementary pattern obtained with a reference grating shift of $g / 2$.

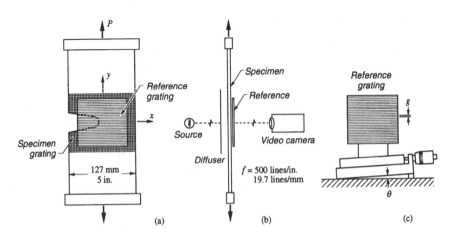

Fig. A.6   (a) Specimen with cross-line grating. (b) Optical arrangement. (c) Reference grating mounted on a micrometer traverse.

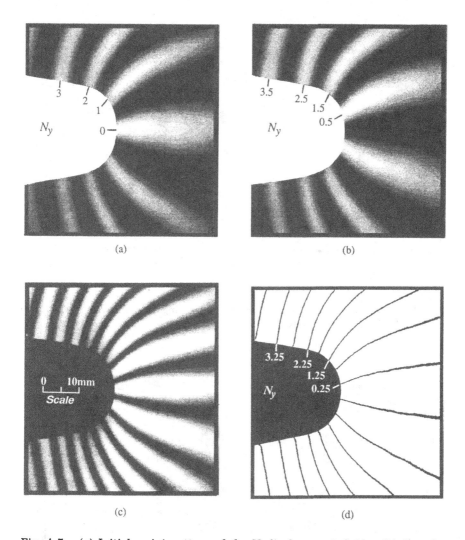

Fig. A.7   (a) Initial moiré pattern of the V displacement field.   (b) Complementary moiré pattern.   (c) Pattern obtained by subtraction and inversion of negative values.   (d) Sharpened contours corresponding to (c).

They correspond to the graphs in Fig. A.3a.   Pattern (c) is the image obtained after subtraction and inversion, corresponding to Fig. A.3c. Pattern (d) is the sharpened contour map after truncation and binarization, corresponding to Fig. A.3d.   The contours lie at the quarter points of the initial fringe pattern.

Additional data were taken with fringe shifts corresponding to $\beta = 6$ and $\beta = 10$ and the multiplied patterns are shown in Fig. A.8.   A wealth of displacement information is displayed in the multiplied

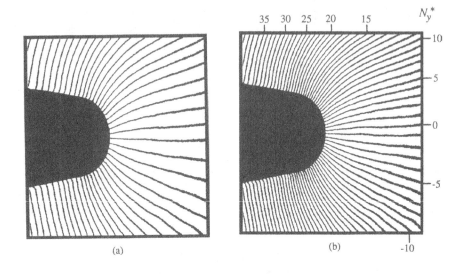

(a)   (b)

Fig. A.8   Multiplied fringe contours corresponding to Fig. A.7a.   (a) Multiplication by 6. (b) Multiplication by 10.

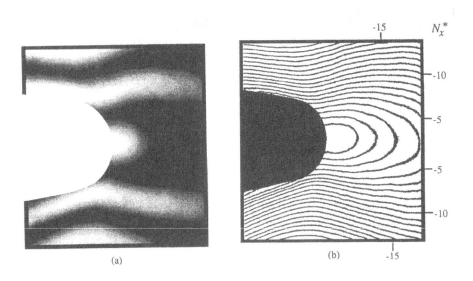

(a)   (b)

Fig. A.9   (a) Moiré pattern of the transverse, or U displacement field. (b) The corresponding O/DFM pattern for multiplication by 10.

patterns, compared to the sparsity of information in the initial pattern. Similarly in Fig. A.9, which reveals the U displacement field, the increase of displacement contours is dramatic. The displacements are determined for every point in the field by

$$U = \frac{1}{\beta f} N_x^* \qquad V = \frac{1}{\beta f} N_y^* \qquad \text{(A.3)}$$

The contour interval is $1/\beta f$, which is 5.08 μm (2×10^{-4} in.) per fringe order for the U and V fields with multiplication by 10.

The optical/digital fringe multiplication method is highly effective for processing fringe patterns of geometrical moiré. Enhanced sensitivity by an order of magnitude is readily achieved.

A.4 Reference

1. E. M. Weissman, "Experimental Tire Mechanics by Moiré," *Optics and Lasers in Engineering*, Vol. 13, No. 2, pp. 117-126 (1990).

Appendix B

Submaster Grating Molds

B.1 Introduction

In Sec. 4.9, we noted that fringe patterns of excellent clarity are achieved when specimen gratings are replicated from high quality molds. For most applications, the grating mold is not reused. Instead, it is considered to be a disposable tool. Master grating molds can be produced by the method of Sec. 4.9.2, or purchased from a commercial vendor. These are seldom considered disposable, but they can be used to produce submasters that are disposable.

Techniques for producing submasters are constantly evolving as new materials become available. One successful scheme will be described here. A virtually endless supply can be produced, while duplicating the high quality of the master grating. The scheme consists of three steps: first a silicone rubber submaster grating is replicated from the master grating; then, the grating mold is replicated in epoxy from the silicone rubber grating; finally, a reflective metal film is applied to the epoxy mold. The master grating can be reused to replicate several silicone rubber submasters, and these silicone rubber gratings can be reused to replicate numerous epoxy submasters. If desired, an epoxy submaster can be used to produce a new generation of silicone rubber submasters and subsequent epoxy submasters.

The master grating should not be replicated until the techniques described below are practiced and mastered. Instead, a clear glass plate of the same size as the master grating should be used for practice. The same replication steps should be followed, but in this case, the replica will have a flat silicone rubber surface instead of a grating surface. Subsequently, these flat silicone rubber replicas should be used to practice and perfect the replication process with epoxy (Sec. B.3) These epoxy plates can be used to gain experience with the deposition of the reflective coating (Sec. B.4).

B.2 Production of Silicone Rubber Submasters

The techniques described in this section apply to the replication of a silicone rubber submaster from a master photoresist grating, where the master was made by the methods of Sec. 4.9.2. Silicone rubber is used because it does not adhere to unprimed surfaces. Common float glass can be used for all submasters, since optically flat surfaces are not required.

The silicone rubber must be capable of faithfully replicating the fine features of the grating. It should not contain fillers, which might be coarser than the grating features and cause a secondary texture in the grating surface. Silicone rubber RTV 615[1] (manufactured by General Electric Co.) has been used successfully, but various other silicone rubber adhesives could be used instead. Since the silicone rubber bonds only to primed surfaces, the glass plate that is used for the submaster must be coated with the appropriate primer (Primer SS4120[2] by General Electric Co.). An effective method for applying the primer is illustrated in Fig. B.1. The primer is applied to the end of a lens tissue that is draped on the glass substrate. The tissue is dragged across the plate to distribute the primer in a uniform thin layer. Then, the primer is left to dry for at least 30 minutes. The plate should be dried in a vertical position to minimize dust contamination.

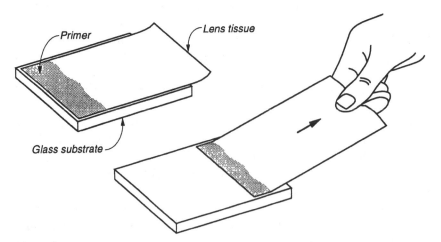

Fig. B.1   A simple method to apply a uniform, thin layer of silicone rubber primer to the glass substrate.

The silicone rubber is received as two liquids which subsequently crosslink (or cure) to form an elastomer. The two parts are mixed together in proper proportions and then the entrapped air bubbles are removed. Bubble removal can be accomplished with either a centrifuge or a vacuum. Even a small air bubble in the liquid will produce a defect in the submaster.

After mixing, the liquid is poured on the primed surface of the submaster plate to form a small pool (approximately 25 mm diameter) with the pool trailing off the plate on one side. The pool is illustrated in Fig. B.2a. The master grating, with its diffractive surface facing the pool, is lowered slowly, in such a way that air bubbles are not entrapped. The technique is illustrated in (b). The two plates are positioned for cure and held in place with transparent tape, as illustrated in (c). A disposable plastic sheet can be taped to the work table in advance, to protect it from the excess adhesive.

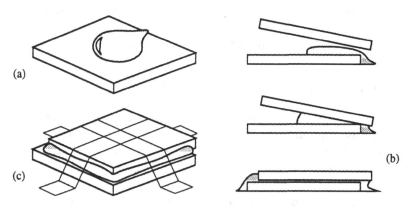

Fig. B.2   Steps for replicating subgratings. (a) Pour the liquid on the grating. (b) Lower the glass plate. (c) Restrain movement during cure. Note: The thickness of the silicone rubber (or epoxy) is not a critical parameter.

The submaster grating can be separated from the master grating after 24 hours, although full cure at room temperature requires about seven days. Alternatively, the two gratings can be stored in the sandwiched state for extended periods of time to prevent contamination and damage. Before separation, the excess silicone rubber that squeezes out is removed by cutting and scraping with a sharp razor. Then, the separation process should be assisted by using a special fixture, such as that illustrated in Fig. B.3. It is important to separate the gratings in a slow, controlled motion to prevent tearing the photoresist of the master grating.

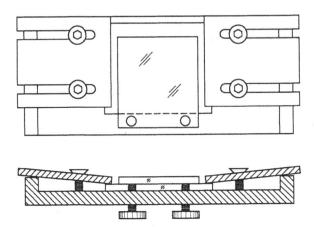

Fig. B.3   Fixture for grating separation.

Notice that the grating separation fixture requires glass plates of two different sizes, so that one can be clamped in the fixture and the other can overhang to contact the thumb screws.

B.3 Production of Epoxy Submasters

An epoxy grating mold can be replicated from the silicone rubber grating after the silicone rubber has fully cured. An epoxy that has worked well is marketed under the brand names Ultra Glo and Envirotex Lite.[3] It is essential that the steps for replicating the grating mold be followed closely. The epoxy is received as two liquid components, a resin and a hardener. Before mixing the two parts together, both must be degassed to remove volatiles. This is done by pouring equal portions of the resin and hardener into separate cups and placing them in a vacuum chamber that can reach an absolute pressure of 200 millitorr (0.2 mm of mercury), or less. Only small quantities are required. Both parts will begin to bubble. The vacuum should be held until most of the bubbles collapse, which usually requires about one hour. Then, the two parts are combined in a separate cup and mixed thoroughly for several minutes. The air bubbles that are entrained during mixing are removed by another cycle in the vacuum chamber, this one for about 15 minutes. The working time of the epoxy is less than one hour after the two components are combined, so replication should proceed without delay.

   A small pool of the liquid epoxy is poured on the grating surface of the silicone rubber submaster, which is lying face up on the work table. The pool should be about 25 mm in diameter, and it should trail off to one side of the grating, as illustrated in Fig. B.2a. A clean glass plate is then lowered slowly into the pool and taped into position, as illustrated in the figure. After one day, the excess epoxy should be removed from the plates with a sharp razor. The plates can be separated at this time, or they can be stored in the sandwiched state. Full cure for this epoxy system requires about four days.

It is convenient to produce a few silicone rubber submasters and use them to produce a few epoxy submasters in a single batch. Subsequently, the epoxy submasters can be used to replicate additional silicone rubber submasters which, together with the original set of silicone rubber submasters, can be used to produce another larger batch of epoxy submasters. Both types of submasters can be used repeatedly to produce additional submasters.

B.4 Application of the Reflective Film

To perform as a mold for replication of the specimen grating, the epoxy submaster is coated with a reflective metal film. The film serves two purposes. It acts as a parting agent for separation of the grating mold from the specimen grating; and by transferring to the specimen grating, the film provides the high reflectivity needed for good diffraction efficiency in the moiré interferometry system.

Typically, the film is aluminum, applied by a vacuum deposition system such as that described in Sec. 2.2.3. The process is similar to that used for applying metallic mirror coatings, but in the present case, weak adhesion is required instead of the strong adhesion desired for mirror coatings. To achieve this, the grating is shielded from the radiant heat generated by melting the aluminum. Once the aluminum melts and evaporation commences, the shield (or baffle) is moved away and the aluminum deposits on the grating surface. Another reason to shield against the radiant heat is that the epoxy grating will soften when it is heated, and the epoxy will degrade if it becomes hot enough.

Two layers of aluminum are used to ensure transfer of the reflecting film to the specimen. After the first layer is applied, air is admitted into the vacuum chamber to oxidize the aluminum. In the most conservative procedure, a parting agent is applied between the two layers as follows. The submaster grating is removed from the vacuum chamber after depositing the first layer of aluminum. It is dipped into a dilute detergent (e.g., Kodak Photo-flo 200[4] diluted 1:300 in distilled water) and allowed to dry in a vertical position. The second layer of aluminum is applied after 24 hours.

B.5 References

1.   RTV 615 Silicon Rubber, General Electric Company, Silicon Products Division, RTV Products Dept., Waterford, NY 14650.

2.   SS4120 Primer, General Electric Company, Silicon Products Division, RTV Products Dept., Waterford, NY 14650.

3.   Ultra Glo and Envirotex Lite, ETI Environmental Technology Inc., Field Landing, CA 95537 (can be purchased at hobby and craft stores).

4.   Photo-flo 200, Eastman Kodak Company, Rochester, NY 14650 (can be purchased at photographic supply stores).

Appendix C

Adhesives for Replication of Specimen Gratings

C.1 Adhesives

This appendix lists adhesives that have been used successfully for specimen grating replication. Equivalent and alternative materials undoubtedly exist, and new materials are constantly becoming available. The experimentalist is encouraged to try different materials and techniques, and to report successful systems.

Room Temperature Cure
    Photoelastic PC-10C[1]
    Photoelastic PC-6C[1] (when $\varepsilon > 3\%$)

Elevated Temperature Cure
    Tra-Bond F211[2]

Low Temperature Cure
    Norland Optical Adhesive 61[3] (UV cure acrylic)

C.2 Addresses

1. Photoelastic Div., Measurements Group Inc., P.O. Box 27777, Raleigh, NC 27611, U.S.A.

2. Tra-con Inc., 55 North Street, Medford, MA 02155, U.S.A.

3. Norland Products Inc., P.O. Box 7145, North Brunswick, NJ 08092, U.S.A.

Subject Index

Mechanical Engineering Series *(continued)*

Laminar Viscous Flow
V.N. Constantinescu

Thermal Contact Conductance
C.V. Madhusudana

Transport Phenomena with Drops and Bubbles
S.S. Sadhal, P.S. Ayyaswamy, and J.N. Chung

Fundamentals of Robotic Mechanical Systems:
Theory, Methods, and Algorithms
J. Angeles

Electromagnetics and Calculations of Fields
J. Ida and J.P.A. Bastos

Mechanics and Control of Robots
K.C. Gupta

Wave Propagation in Structures:
Spectral Analysis Using Fast Fourier Transforms
J.F. Doyle